ION IMPLANTATION: BASICS TO DEVICE FABRICATION

by

Emanuele Rimini

Catania University

D1529598

KLUWER ACADEMIC PUBLISHERS
Boston / Dordrecht / London

Distributors for North America:
Kluwer Academic Publishers
101 Philip Drive
Assinippi Park
Norwell, Massachusetts 02061 USA

Distributors for all other countries:
Kluwer Academic Publishers Group
Distribution Centre
Post Office Box 322
3300 AH Dordrecht, THE NETHERLANDS

Library of Congress Cataloging-in-Publication Data

A C.I.P. Catalogue record for this book is available
from the Library of Congress.

Printed on acid-free paper.

Printed in the United States of America

to Anna, Rebecca, Roberto and Thea

CONTENTS

Chapter 3

RANGE DISTRIBUTION

Chapter 4

RADIATION DAMAGE

Chapter 5

ANNEALING and SECONDARY DEFECTS

Chapter 6

ANALYTICAL TECHNIQUES

PREFACE

Ion implantation offers one of the best examples of a topic that starting from the basic research level has reached the high technology level within the framework of microelectronics. As the major or the unique procedure to selectively dope semiconductor materials for device fabrication, ion implantation takes advantage of the tremendous development of microelectronics and it evolves in a multidisciplinary frame. Physicists, chemists, materials scientists, processing, device production, device design and ion beam engineers are all involved in this subject.

The present monography deals with several aspects of ion implantation. The first chapter covers basic information on the physics of devices together with a brief description of the main trends in the field. The second chapter is devoted to ion implanters, including also high energy apparatus and a description of wafer charging and contaminants. Yield is a quite relevant issue in the industrial surrounding and must be also discussed in the academic ambient. The slowing down of ions is treated in the third chapter both analytically and by numerical simulation methods. Channeling implants are described in some details in view of their relevance at the zero degree implants and of the available industrial parallel beam systems. Damage and its annealing are the key processes in ion implantation. Chapter four and five are dedicated to this extremely important subject. The complexity of the involved phenomena is stressed by considering the influence on the secondary damage of several parameters, such as wafer temperature, ion beam energy, ion flux density and dopants. The interaction of defects among themselves and with impurities can be used advantageously and is at the basis of the defect engineering field. Some examples illustrate this quite interesting and promising field. A large number of analytical techniques is used to characterize the implanted layers. Some of them are described in chapter 6, as SIMS, spreading resistance profilometry, RBS and channeling effect, transmission electron microscopy. Particular attention has been paid to those techniques that provide two-dimensional profiles of damage and of dopants. Any application of ion implantation deals with masked regions, so that two dimensional information is needed. Chapter 7 is devoted to silicon based devices, and several topics are described, the threshold voltage control, shallow junctions, minority carrier lifetime control by metallic ion implants, high energy implants etc. The last chapter deals with compound semiconductors, such as GaAs and InP, and the use of ion implantation not only for doping but also for isolation. The formation of buried dielectric and metal layers by high dose ion implantation is also described together with device fabrication.

The book is mainly a collection of material from a number of sources, ranging from textbooks to research journals. The presentation relies also on the Handbook of Ion Implantation Technology. This text is based on a few days course that preceeds the biennal Ion Implantation Technology Conference whose tenth edition has been held in Catania in June 1994. Almost all the proceedings of these meetings are published in Nuclear Instruments and Methods in Physics Research- B.

I am very indebted to Anna Balsamo for the patience and skillful in typing the camera-ready manuscript. Natale Marino helped me a lot for figures and for drawings. I like to thank my young co-workers and students who provided me results of their published and unpublished work: Corrado Spinella, Mario Saggio, Vittorio Privitera, Giusy Galvagno, Vito Raineri, Francesco Priolo, Salvo Coffa, Salvo Lombardo, Anna Battaglia and Francesco La Via. The long and fruitful collaboration with Giuseppe Ferla from SGS - Thomson Microelectronics, and with Nuccio Foti and Ugo Campisano has had a tremendous impact on my work in the ion implantation field and then, ... I hope, on the book. I would like also to thank those who have allowed me to use their figures, tables and micro graphs.

Catania, July 1994

Emanuele Rimini

LIST OF TABLES

CHAPTER 1

SEMICONDUCTOR DEVICES

1.1 Introduction

Industry of semiconductor devices drives the overall electronics industry whose market is currently estimated to be in the range of 900 B$ much larger than the automotive or steel. Solid state devices are the building blocks of any electronic system. In the 1993 the device market was 77 B$ and it is estimated to be 112 B$ in 1998 at an average growth rate of 10.0% per year [1.1].

The principal motivation for the long-continued and rapid development of solid state devices is their use in several fields such as information, computing, comunication and commodity. Solid state devices are pervasive, they are used in the memory of a computer where, to store a bit of information, require femto watts but also in the train motor to handle megawatts. It is amusing that in all so quite different applications (computers, washing machines, cars, trains etc.) the heart and the intelligent part is a small piece of a semiconductor material mainly silicon.

The content of semiconductor devices in the electronic equipments is raising with time at a faster rate than the electronic industry production as shown in Fig. 1.1, where the two trends are reported as a function of the year. The dashed area represents quantitatively the meaning of silicon pervasiveness. The market share between the different industrial segments is shown in Fig. 1.2a. Computer covers 49% of the market and it represents the driving force for the development of semiconductor devices. In the next years it is predicted a rise still of the computer segment with a fall of the military market. The semiconductor market share by product family is shown in Fig. 1.2b. For simplicity all

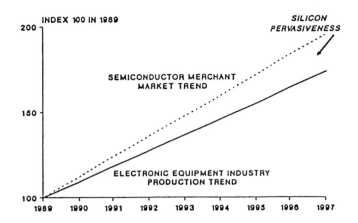

Figure 1.1 *Silicon pervasiveness: the growth rate of the semiconductor merchant market is higher than that of the electronic equipment industry. Source: Dataquest Dec. 1993*

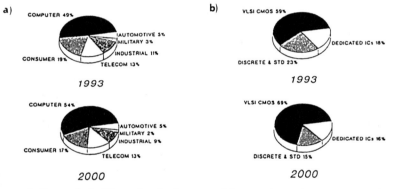

Figure 1.2 *(a) Trend of the world semiconductor market by segment: (b) trend of the market share by product family. Source: Dataquest Dec. 1993*

the products have been aggregated in only three classes. VLSI CMOS (*V*ery *L*arge *S*cale *I*ntegration *C*omplementary *M*etal *O*xide *S*emiconductor) includes all the different kinds of memories: *D*inamic *R*andom *A*ccess *M*emory *D*RAM, *E*lectrically *P*rogrammable *R*ead *O*nly *M*emory EPROM, *S*tatic *R*andom *A*ccess *M*emory SRAM, *R*ead *O*nly *M*emory ROM, and microprocessors. The dedicated part of integrated circuits refers to specific applications in telecomunication and in other sectors. The discrete and standard class includes a variety of devices, such as power bipolar transistor, silicon controlled rectifier, diodes, digital logic circuits, voltage regulators, operational amplifiers and optoelectronic devices [1.2].

The latter ones are based on III-V semiconductor compounds (e.g. *GaAs*) instead of silicon and include light emitting diodes (LED), solid state lasers used also for compact disks, photodetectors, microwave devices, solar cells, etc. [1.3]. The world market for *GaAs* based devices was 276M$ in 1992 to be compared with 55B$ for *Si* based devices. In spite of the relatively low amount of money involved, devices based on compound semiconductors offer quite interesting properties specially in the high frequency regime and in the photonic field where electrons are coupled to photons. Some silicon-based devices are shown in Fig. 1.3.

The silicon controlled rectifier (SCR) or thyristor of Fig. 1.3a is a single device of 75 mm in diameter. It is used in power switching and in control circuits [1.4]. A current up to 2600 A can flow through it in the conducting state ("on" state) and a voltage of 4500 V can be sustained in the blocking state ("off" state). The large amount of handled power requires a careful design of the package for the thermal dissipation of the heat produced during the working. The silicon pellet is fixed to a molybdenum disc whose thermal expansion coefficient is very close to that of silicon, and it is encapsulated between the copper of the anode and the silver washers of the cathode. The ceramic case is shown in Fig. 1.3b. To contrast the SCR a 16 Mbit EPROM already mounted to the frame is shown in Fig. 1.3c. The silicon chip, a rectangle of $10mm \times 20mm$ is divided into four smaller rectangles each of them of 4 Mbit capacity. Inside the chip about 10^7 elementary devices or cells are located. Each of them can store, as a charge, the information (1 bit). The minimum feature size of the cell is 0.5 μm and the adopted technology is named therefore $0.5\mu m$. The transmission electron micrograph of Fig.1.3d represents the cross section of 1 Mbit "Flash-Eprom". The device is formed by the silicon substrate and by several layers of different materials deposited on it, such as silicon dioxide, polycrystalline silicon, silicide, aluminum, etc. The memory cell (see also Fig. 6.32 for an enlarged view) is formed by a thin (~ 10 nm) oxide layer (gate oxide), a polycrystalline silicon layer (floating gate) a dielectric layer, another polycrystalline layer (control gate) and a silicide as a metal contact. The single cell can be electrically programmed by the injection of electrons in the floating gate where they remain unless are rejected back to the silicon substrate by an electric field. These memories are more complex than the popular DRAM but a similar technology is used in the fabrication of both devices. The trend is that of decreasing the size of the device, i.e. the lenght

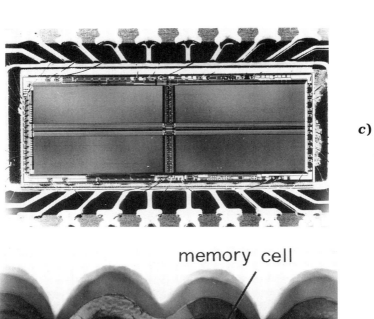

c)

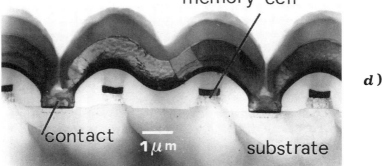

memory cell

d)

contact   1 μm   substrate

Figure 1.3   *(a) Silicon controlled rectifier (SCR) 75 mm in diameter, (b) ceramic case of the SCR, (Courtesy of Ansaldo) (c) 16Mbit EPROM memory chip of 2 × 1cm, (Courtesy of SGS Thomson Microelectronics) (d) cross section transmission electron micrograph of a few transistors in the 1Mbit flash-EPROM.*

of the channel to allocate more and more elementary cells on the chip and to increase the integration scale.

Power and logic devices can be integrated in the same chip giving rise to what has been named smart or intelligent power technology. The layout of an intelligent device used for automatic electronic ignition is shown in Fig. 1.4a. The lower part of the layout refers to the logic circuit, the upper part to the power component. Both the logic and the power parts are based on bipolar transistor. The power transistor can switch 8A at 400 V, with an energy handling capability up to 1 Joule. The control part features logic protection circuit and digital feedback signal. The schematic diagram of the working circuit in the electronic ignition is shown in Fig. 1.4b. The device includes a logic interface that allows it to be controlled by a standard CMOS logic signal from the engine control system. The flow of the current is vertical across the wafer thickness as in the SCR, in the memories it is horizonthal along the wafer surface. The technology for fabricating intelligent power devices is quite complex because several requirements must be satisfied on the same piece of silicon.

These are just a few examples of the enormous variety and complexity of semiconductor devices. Semiconductor technology has made dramatic advances over the last decade as devices have been scaled to micron dimension. The miniaturization of the single device has allowed to put several millions of them on a single chip of area $\sim 1cm^2$. Progress is often limited by lack of basic understanding of materials, processes and their interactions. These conditions include drastic concentration gradients, large electric fields, materials compositions that must remain far from thermodynamic equilibrium during subsequent hot processing, and extreme sensitivity to defects.

This book is devoted to just a single issue, among all the different materials aspects, i.e. to the doping of semiconductors, mainly silicon, by ion implantation which is the process of introducing energetic ions into a substrate. The last chapter of the book is devoted to the applications of ion implantation to other semiconductors, such as III-V compounds, where it is used not only for doping but also for the electrical isolation of complex structure. Electrons in compound semiconducting materials such as $GaAs$ move with much higher velocities under applied electric fields than do electrons in silicon. This and other properties of semiconducting compounds are opening the way for new devices, including logic circuits that operate at higher speeds and circuits that interface easily with optical fibers. To appreciate the relevance of the ion implantation technology in the semicon-

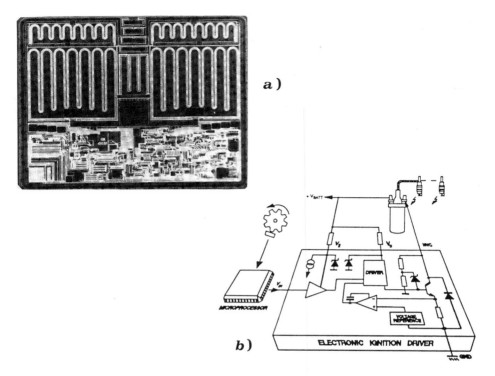

Figure 1.4 *(a) Layout of vertical intelligent power device able to switch more than 8A at 400V and 25mm² of area, (b) schematic diagram of the working circuit in the automatic electron ignition (courtesy of SGS Thomson Microelectronics).*

ductor industry it is necessary to know at least the basics of the most common device structures, i.e. the bipolar transistor and the metal- oxide silicon field effect transistor, being them the building blocks of almost all the devices. In the next sections these two elementary structures will be briefly described.

1.2 Semiconductor Physics

Basic to all semiconductors devices is the possibility to control, by the addition of suitable impurities, the resistivity and the kind of electrical carrier. As shown in Fig. 1.5 it is possible to change the resistivity of silicon from $10^{-3}\Omega \cdot cm$ to $10^3\Omega \cdot cm$, i.e. by six orders of magnitude by varying the dopant concentration from $10^{20}/cm^3$ to $10^{13}/cm^3$, i.e. by seven orders of magnitude [1.5]. The two curves of Fig. 1.5 refer respectively to the doping

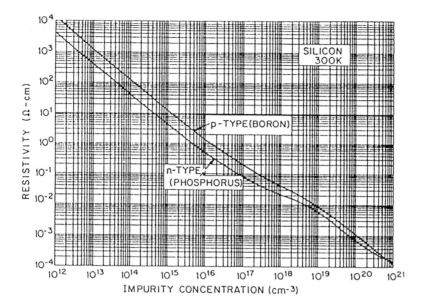

Figure 1.5 *Relationship between carrier concentration and resistivity for n-type (phosphorus doped) and p-type (boron doped) silicon. The difference between the two curves is due to the different values of electron and hole mobility respectively (from ref. 1.5).*

by the introduction of phosphorus atoms, n-type material or of boron atoms, p-type material in silicon. In both cases phosphorus and boron atoms replace silicon atoms in the crystalline lattice.

Williams Shockley in his article "Transistor Electronics: Imperfections, unipolar and analog transistors" published in Proc. IRE 40 (1952) p.1289 pointed out clearly that... **transistor electronics exists because of the controlled presence of imperfections in otherwise nearly perfect crystals** [1.6]. In a perfect silicon crystal all the electrons forming the tethrahedral covalent bonds between nearest neighbours atoms fill the available energy states in the so called valence band. To free an electron a bond must be broken and an energy of about 1 eV must be given. The free-electron is able to change easily its state and to move then under an applied electric filed. It belongs to the conduction band. The empty state, left by the missing electron, in the valence band is called hole. If an electric field is applied, the collective motion of the electrons in the band is equivalent to the motion of the empty state in the direction of decreasing crystalline momentum.

Electrons carry a positive mass and a negative charge while holes carry a positive charge with a positive mass. The energy difference between the minimum energy of electrons in the conduction band and the maximum in the valence band is the energy gap, E_g, of the semiconductor material. At room temperature E_g is 1.14 eV for Si, 0.9 eV for Ge and 1.4 eV for $GaAs$ respectively. Moreover in $GaAs$ electrons at the maximum of the valence band and at the minimum of the conduction band have the same crystalline momentum or wave vector, k=0, while in Si and Ge the conduction band minimum energy is at a k value different from zero that of the electrons at the maximum energy of the valence band. Si and Ge are said to have an indirect energy gap, while $GaAs$ is said to have a direct energy gap. The difference playes a relevant role in all the processes of generation and recombination of hole-electron pairs. In perfect single crystal semiconductors no allowed electronic states exist within the gap. At temperatures different from 0K some bonds are broken, for each broken bond the electron, thermally excited into the conduction band, creates a hole in the valence band. In this case the number of electrons equals that of holes.

Intrinsic (perfect and undoped) silicon is at room temperature an insulator of $10^6 \Omega \cdot cm$ resistivity. The concentration of thermally excited electrons and holes is $\sim 10^{10}/cm^3$ to be compared with the atomic silicon density of 5×10^{22} at/cm^3. The conductance of a semiconductor is given by $\sigma = qn\mu_e + qp\mu_p$ where q is the absolute electronic charge, n and p the electron and hole density, and μ_e and μ_p the electron and hole mobility respectively. The mobility depends on the effective mass of the carriers on thermal vibrations, on impurities and lattice defects.

If an element from the V group of the periodic table (e.g. P, As, Bi) replaces a silicon atom in the diamond lattice, it forms four tethraedral covalent bonds with the nearest neighbouring silicon atoms. The fifth bond is easily ionized at room temperature having a binding energy of 0.05 eV, so that an electron is introduced in the conduction band. A similar situation happens when a trivalent element like B, Al, Ga replaces a silicon atom in the lattice. This time only three covalent bonds can be formed, the fourth one requires an additional electron from the neighbouring covalent bonds. So that an empty electron state is found in the valence band, i.e. an hole. In n-type material the electron density is higher than the hole density, electrons are named majority carriers and holes minority carriers respectively, the viceversa is true for a p-type material. At a given temperature the product np is a constant, $np = n_i^2(T)$, for Si at 300 K it amounts to $2 \times 10^{20}/cm^6$

and for $GaAs$ to $3.2 \times 10^{12}/cm^6$. The difference is mainly due to the higher energy gap of $GaAs$ with respect that of Si.

At room temperature in a n-type material the electron concentration is given by $N_d - N_a$, being N_d and N_a the donor and the acceptor concentration $(N_d > N_a)$, while the hole concentration is given by $n_i^2/(N_d - N_a)$. Analoguos formulas are valid for p-type material. In any case semiconductors are characterized by dramatic change of carrier concentration. The mobility of electrons in silicon is nearly three times higher than that of hole. This explain the difference between the resistivity curves of p and n-type material reported in Fig. 1.5. In $GaAs$ the electron mobility is a factor five higher than in silicon and for this reason $GaAs$ is used in devices operating at high frequency (see Fig. 8.7).

In the band scheme the probability that an electronic state of energy E is occupied at a given temperature is given by the Fermi - Dirac distribution function f(E) where

$$f(E) = \{1 + exp[(E - E_F)/kT]\}^{-1} \qquad 1.1$$

E_F is the Fermi level, and k the Boltzmann's constant. The quantity E_F is the chemical potential, i.e. the change in the free energy of the system when one electron is added or removed from the material. The Fermi level in an intrinsic semiconductor (n=p) is located with a good approximation at the center of the energy gap (see Fig. 1.6a).

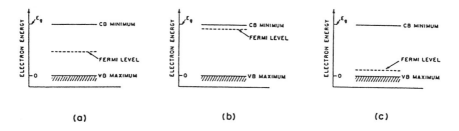

Figure 1.6 *Position of the Fermi level, E_F, in an intrinsic (a), n-type(b), p-type(c) semiconductor of energy gap E_g.*

In extrinsic semiconductors (n$\#$ p) the location of the Fermi level depends on the amount and on the type of doping. In n-type material E_F is in the upper half (Fig. 1.6b), and in p-type material in the lower half (Fig. 1.6c) of the energy gap. With increasing the dopant concentration the Fermi level moves toward the conduction band in n-type material and toward the valence band in p-type material respectively. At high concentration $(N_a, N_d > 5 \times 10^{19}/cm^3)$ the dopant states are no longer discrete, with a well-defined ionization energy; an impurity band is formed wide enough to overlap the conduction or the valence band. The Fermi level is above (below) the conduction (valence) band and the semiconductor is degenere. The arguments described in this section are found in more details in ref. [1.7, 1.8, 1.9].

1.3 p-n Junction and Diode

A fundamental characteristic of semiconductors is the formation, by suitable doping of the substrate, of two contiguous regions with different kind of conductivity, i.e. a p-n junction. The semiconductor p-n junction is not only of interest tecnologically but also for a wide range of interesting phenomena in applied solid state physics. For simplicity let us consider a semiconductor in which there is a change from n-type to p-type over a small distance as shown in Fig. 1.7(a) schematically, i.e. an abrupt p-n junction. The n-type region has an excess of electrons relative to the p-type side and similarly the p-type side has an excess of holes relative to the n-type side. The concentration gradient gives rise to a flux of electrons from n to p-type regions and viceversa for holes. Electrons in p-type regions are then in higher concentrations and for thermodynamic reasons tend to recombine with holes.

Recombination of electron with an hole is the opposite process of generation. As a result of the recombination free holes are removed from the p-type side and this region has an excess of negatively charged ionized acceptor centers. Since these ionized acceptors are at fixed positions in the semiconductor crystal lattice a region of negative space charge (with a concentration of N_a negative charges per unit volume) is created in p-type region near the junction. In a similar way a region of positive space charge is created in the n-type region. The space charge creates an electric field and a potential barrier that prevents diffusion of holes (electrons) from the n(p) type region to the p(n) type region. To a good approximation the space charge within the transition region can be considered as due only to the uncompensated donor and acceptor ions and the few carriers can be neglected. The width of the depleted region, W, depends on the dopant concentration profiles, and it extends more on the less doped region. For an abrupt junction made up of material with N_a acceptors/cm^3 on the p side and of N_d donors $/cm^3$ on the n-side the width W is given by

$$W = \sqrt{\frac{2\epsilon}{q} \left(\frac{1}{N_a} + \frac{1}{N_d} \right) V_0}$$ 1.2

while the potential difference, V_0, between the two regions is given by

$$V_0 = \frac{kT}{q} \ln \frac{N_a N_d}{n_i^2}$$ 1.3

q is the electronic charge and ϵ the dielectric constant of the material.

V_0 is called contact potential or built-in potential barrier, it is an equilibrium quantity and no net current can result from it. At equilibrium the Fermi levels of the two regions are at the same position (see Fig. 1.7b) and the current density of electrons from n to p due to diffusion equals the current density of electrons from p to n due to the build-in electric field in the depleted region. The junction is also adopted to electrically isolate one device from a contiguous one. The width of the space charge region in Si ranges between 0.1 and $1\mu m$ for the usual doping concentrations. The potential barrier is in general lower than 1 V. The electric field in the depleted region is in the order of $10^4 \div 10^5 V/cm$. In the case shown in Fig. 1.7b the contact potential amounts to 0.709 Volt and the donor and acceptor concentrations are $4.4 \times 10^{17}/cm^3$ and $3.2 \times 10^{14}/cm^3$ respectively.

So far we have discussed qualitatively the properties of a p-n junction at equilibrium. Application of an external potential to a p-n junction can increase or decrease the potential V_0 across the transition region by aiding or opposing the equilibrium electric field, and the width of the transition region as well. A feature of the p-n junction is that current flows quite freely from the p to the n side when the p region is positive biased by the external voltage relative to the n region (forward bias and forward current), whereas virtually no current flows when p is made negative relative to n (reverse bias and reverse current). This asymmetry of the current flow makes the p-n junction diode very useful as a rectifier.

The current-voltage characteristic of a diode whose cross section is shown in Fig.1.7(a) is reported in Fig. 1.7(c). The current increases exponentially with the forward voltage in the range 0.4 - 0.7 V. At higher voltage, i.e. high current new phenomena take place in the p-n junction, the voltage drop in the quasi neutral

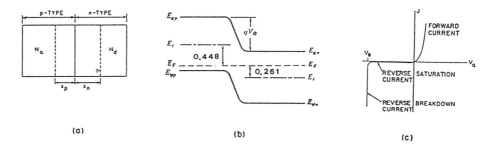

Figure 1.7 *(a) Structure of a p-n diode, the hatched area represents the width of the depleted region at equilibrium, (b) Fermi level and band skeme in a p-n junction, (c) current -voltage characteristic of a diode for forward and for reverse bias.*

regions of the diode cannot be neglected and the minority carrier concentration becomes at large level of injection comparable with the majority carrier concentration. In the reverse regime the current depends slightly on the applied voltage until a critical value is reached where the magnitude of the reverse current increases sharply, as shown in Fig. 1.7(c). This effect is called reverse breakdown of the p-n junction. In several devices the mechanism responsible of the breakdown is the avalanche multiplication of electron-hole pairs by the free carriers. These last ones are accelerated by the electric field in the junction, thereby gaining kinetic energy. This kinetic energy may be large enough ($> E_g$) to produce free carriers by impact ionization. These newly generated free carriers can also be accelerated by the electric field and produce by impact others electron -hole pairs. The resulting production of free carriers is called an avalanche. The breakdown voltage according to the device can range from few to several kilovolts. See for more details ref. [1.10].

In the case of forward bias a large number of electrons (holes) is injected in the p(n) neutral regions where diffusion takes place. The system reacts by the mechanism of recombination to reduce the excess of electrons or of holes. In steady state a flow of current results. Recombination can occur also in the depleted region although its contribution to the forward current is usually small at voltage above 0.3 V for silicon. In the case of reverse bias instead the equilibrium concentrations of free-carriers are reduced and the system reacts by generation of electron-hole pairs. The main contribution to the reverse current is the generation within the depleted zone in silicon diodes. The increase with the applied voltage of the depletion region width is responsible of the increase of the reverse current. The reverse and the direct current depend on the recombination - generation mechanisms by which electron-hole pairs disappear or appear. If a carrier is excited into the conduction band, it will eventually recombine and disappear after a time whose characteristic average is called lifetime. The lifetime of a carrier in a given semiconductor will depend on the kinetics of its mode of recombination, either by direct recombination, or by recombination via an intermediate state or deep level.

In semiconductor with direct band gap (e.g. $GaAs, InSb$) direct recombination of an electron and a hole, both with the same crystalline momentum, to produce a photon can occur easily and dominates the carrier lifetime. However, in a semiconductor with an indirect band gap (e.g. Si and Ge) direct recombination of electron-hole pairs requires a lattice vibration, called phonon, to fulfill the momentum conservation law. This is then a three-body process (electron, hole and a lattice vibration) of low probability. For this reason the dominant recombination process in indirect band-gap semiconductors usually proceeds through intermediate states whose efficiency is maximum if they are located at the middle of the gap. These levels are associated to metallic impurities (such as Pt, Au, Fe) to point defects and to extended defects. These levels can act also as generation centers and are responsible of the reverse current [1.11]. In a well designed and built diode the reverse - density current is of the order of $10^{-9} A/cm^2$.

1.4 Unipolar and Bipolar Transistors

The most advanced device, from the technological point of view, is the dynamic random access memory (DRAM). It is based on the Insulated Gate Field Effect Transistors (IGFET) whose serious technological activity began in 1963-64. These devices are particularly well suited to control switching between a conducting state and a nonconducting state, and are useful in digital circuits. The specific initial implementation was termed MOSFET, standing for Metal Oxide Semiconductor Field Effect Transistor [1.12, 1.13]. Today the term is technically incorrect being the gate electrode made not of metal but of polycrystalline silicon (polysilicon), with on top a metal to reduce the resistivity. In 1970 the 1K bit was produced, today the 16 Mbit is on selling. A similar memory was shown in Fig. 1.3d. Every three years a new generation of memories is ready for the market [1.14].
The basic MOS transistor prior to its final metallization, is illus-

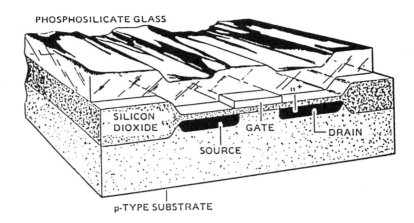

Figure 1.8   *View of an n-channel MOSFET (from ref. 1.15).*

trated in Fig.1.8 for the case of an n-type channel formed on a
p-type substrate [1.15]. The top layer is a phosphorus-doped silicon dioxide (P-glass) which is used as an insulator between the
polysilicon gate and the gate metallization. The two heavily doped
n regions are termed source and drain respectively. No current
flows between these two regions without a conducting n channel
between them, since depleted layers exist between drain-substrate-
source (n-p-n regions). When a positive voltage is applied to the
gate electrode relative to the substrate, positive charges are deposited on the gate material. As a consequence negative charges
are induced in the underlying Si, by the formation of a depletion region and a thin surface region containing mobile electrons.
These electrons are referred as channel and establish a conductive
path between source and drain.

A positive voltage applied to the drain causes then a current
flow. An important parameter in MOS transistors is the threshold
voltage V_T, which is the minimum gate voltage required to induce
at the surface a concentration of electrons equal to that of holes
in the p-substrate. This value, as we will see in Chap. 7, where
a more detailed description of the MOS behaviour will be presented, depends on the work function difference between the gate
material and the substrate doping, on the presence of charges at
the silicon-oxide interface and inside the oxide, in addition to the
other parameters as the oxide thickness and the substrate doping.

Some n-channel devices may have a channel already at zero
gate voltage, and a negative gate voltage is required to turn the
device off. Such a "normally-on" device is called a depletion-

mode transistor, since gate voltage is used to deplete a channel that exists at equilibrium. The more common MOS transistor is "normally-off" with zero gate voltage, and operates in the enhancement mode by applying a suitable positive gate voltage to induce an inverted layer at the surface. MOS is the basic element in digital electronic, it is switched from the "off" state (no conducting channel) to the "on" state. The control of the drain current is obtained at the gate electrode which is insulated from the source and drain by the oxide silicon layer. The impedance in d-c of an MOS circuit can be very large. From what said it is necessary an accurate control and adjustment of the threshold voltage value V_T because it determines the on-off operation.

MOSFET is the dominant device used in VLSI circuits because it can be scaled to smaller dimensions than other types of devices do. The channel lenght of a commercial device is now $0.5\mu m$ (see Fig. 1.3e) and it will reach 0.25 for 64 Mbit DRAM or EPROM memories. Figure 1.9 shows the sequence of the processing steps for the fabrication of an n-channel MOS [1.15]. The starting material is a p-type material lightly doped. At first the isolation region between one device and the contiguous is formed.

A thin oxide layer ($\sim 50nm$) thick is thermally grown followed by a silicon nitride ($\sim 100nm$) deposition (1.9a). The active device area is defined by a photoresist mask and boron dopant is introduced by ion implantation through the composite nitride-oxide layer (1.9b). The composite layer is then removed in the uncovered region and an oxide is grown thermally on these regions. Oxide cannot grow in the presence of a nitride layer since the nitride has a very low oxidation rate. During the oxidation boron atoms are pushed underneath the oxide to form a p^+ layer. This is called p^+ channel stop, because the high concentration of p-type semiconductor will prevent surface inversion. The boron implant (1.9c) is used to control the threshold voltage to a predetermined value. The next step is the gate formation on the thin oxide layer by deposition of polycrystalline silicon heavily doped by phosphorus or arsenic atoms (1.9d). Source and drain regions are then formed by *As* implantation, (1.9e) the gate structure is used as a mask and a self-aligned structure is obtained. A phosphorus-doped oxide (P-glass) is deposited (1.9f) over the entire wafer and is flowed by heating the wafer to give a smooth surface tophography. Contact windows are defined and etched in P-glass. A metal layer (usually Al) is then deposited and patterned (1.9g). The top view of the MOSFET is shown in Fig. 1.9h.

The concept of a field-effect transistor is to establish a current-carrying channel between two contacts, source and drain,

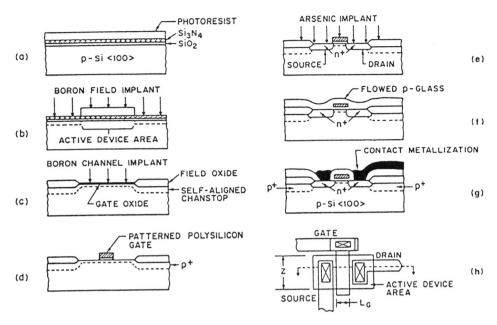

Figure 1.9 *Cross sectional views of NMOS fabrication sequence: (a) formation of SiO_2, Si_3N_4 and photoresist layer, (b) boron implant, (c) field oxide, (d) gate, (e) source and drain. (f) P-glass deposition. (g) metallization, (h) top view of the MOSFET (from ref. 1.15).*

and to have a third electrode to modulate the current flow. For high speed operation, high carrier mobilities and velocities are required to reduce the transit time of the carriers along the channel. A good opportunity is given by $GaAs$ to take advantage of its high electron mobility [1.16]. The lack of a reliable oxide and the problems in realizing a p-n junction in the compound $GaAs$ semiconductor have been overcomed by using a different skeme for the FET. The metal-oxide layers are replaced by a metallic layer in direct contact with the lightly - doped n-type thin layer epitaxially grown on semiinsulating substrate. Source and drain are heavily doped n-type material. (see Fig. 1.10). A Schottky barrier is created between the metal and the semiconductor with the formation of a depletion layer that extends only in the semiconductor. By reverse biasing the Schottky gate, the channel can be depleted up to the semiinsulating substrate and the current-voltage characteristics are similar to those of the MOSFET. The fabrication sequence is simpler than that of MOS, no diffusions are involved and gate lenghts smaller than $1\mu m$ are common in these devices. Doping and isolation of a device from the contiguous one can be realized by ion implantation and the processes are described in Chapter 8.

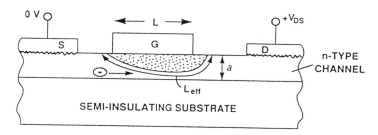

Figure 1.10 *Metal - semiconductor field effect transistor (MES-FET) realized with GaAs (from ref. 1.16).*

The other kind of device commonly adopted is the bipolar transistor. It is a three-terminal current device in which the current through two terminals can be controlled by small changes in the current or voltage at the third terminal. The bipolar transistor is formed by two contiguous junctions, so that we can have n-p-n or p-n-p transistors. The three regions are called emitter, base and collector respectively, and the two junctions are formed between the emitter and the base and between the base and the collector. The view of a bipolar n-p-n transistor is shown in Fig. 1.10a, figure 1.11b represents schematically a cross-section of a planar n-p-n transistor and the dopant profiles of the three regions are shown in Fig. 1.10c [1.17].

Several possibilities arise in the biasing of the junctions. In the forward-active mode the emitter base is forward biased while the base-collector is reverse biased. Under these conditions electrons are injected into the center p region, or base.

The base is so thin that practically all of these electrons diffuse across it to the junction with the collector, where the electric field accelerates them into the collector. Since the base is doped with acceptors there are holes into the base that may recombine with electrons and the forward bias on the emitter junction also causes a current of holes to flow into the emitter and to recombine with electrons in the emitter. Recombination can occur also in the emitter-base depletion region. The holes that disappear should be replaced by others that must be supplied by current through the base contact. This base current is detrimental and should be kept as low as possible so that the current gain, i.e. the ratio

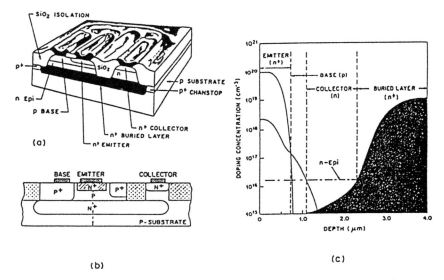

Figure 1.11 *(a) View of an oxide-isolated bipolar transistor, (b) schematic cross section, (c) doping profiles along the dashed line in (b) (from ref. 1.17 - (c)1980 IEEE).*

of the collector current to the base current has to be elevated. Typical steady state values of gain are 50-100.

If an alternating current source is added to the base-emitter input a large variation is obtained at the collector output and amplification results. Bipolar transistors are also used as switch from an on-state to an off-state and back. In the on-state both junctions are forward biased while in the off-state are reverse biased.

Another parameter of relevance for the bipolar transistor is the switching-speed, which is determined by the total time taken to transfer charge backwards and forwards across the device and it is associated to several time constants of various components of the structure and to the transit time for electrons to cross the base region. Bipolar transistors control current at high densities and allow to operate at fast rate. The speed and power differences between bipolar (BJT) and MOSFET are not fundamental but rather are measures of different technological capabilities. At $1\mu m$ technology level bipolar base widths, as defined by diffusion distances of the n-type and p-type dopants, are typically $0.2\mu m$ or one-fifth of the MOS channel length of $1\mu m$ which is lithographycally defined, the speed advantage of bipolar technology is reduced if litography is improved. Conversely improvements in shallow junction diffusion technology would gave bipolar a greater advantage.

The majority of bipolar are of n-p-n type because of the higher mobility of minority carriers (electrons) in the base region. In integrated circuit each device is electrically isolated from the nearest one by p-n junction reverse biased or by oxide layers. Both of them are used in the transistor whose fabrication sequence is shown in Fig. 1.12 [1.18].

The starting material is a p-type lightly doped $< 100 >$ or $< 111 >$ oriented silicon wafer. The first step is the fabrication of the n-buried layer by As^+ implant in the window region opened in a thick SiO_2 layer (1.12a). The layer is required to reduce the series resistance of the collector. After high temperature annealing a thin epitaxial silicon layer is grown by chemical vapor deposition of silane in a reactor. (1.12b) The thickness and the doping of the epilayer depends on the use of the device. Analog circuits require thick layers ($\sim 10\mu m$) and low doping ($5 \times 10^{15}/cm^3$), power transistors even thicker epitaxial layers (up to $100\mu m$) and lower doping ($\sim 10^{13}/cm^3$) while digital circuits require thin layers ($\sim 3\mu m$) and high doping ($2 \times 10^{16}/cm^3$). The next step is the formation of the lateral oxide isolation region. A thin thermal oxide and nitride layer is formed on the wafer, a photoresist mask allows the selective etching of the oxide-nitride layer and of part of the epitaxial layer (1.12c).

In the uncovered silicon areas boron ions are implanted (1.12d). After removal of the photoresist the wafer is placed in an oxidation furnace. Oxide grows only in the areas not protected by the nitride layer and during the growth boron ions are pushed underneath the oxide to form a p^+ layer, a channel stop (1.12c). Another mask and photoresist selects the areas where the base must be formed (1.12g) by boron implant. The other figures 1.12g, 1.12h, refer to the emitter formation.

1.5 Ion Implantation and Semiconductor Devices

Several steps of the fabrication sequence of MOSFET and bipolar transistors require the controlled doping of selected area of the wafer. Ion implantation is used to this aim. In 1954 W.Shockley [1.19] submitted a patent describing the **"Forming of Semiconductor Devices by Ionic Bombardment"**. This original patent describes the field as we practice it today and detailed the use of implantation for chemical doping and electrical activation. Several concepts were elucidated, such as range of the implanted ion, damage, annealing, dopant activation and electrical activation. To day a typical modern process for a memory chip

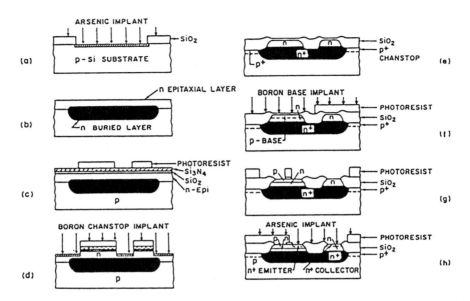

Figure 1.12   *Sequence of n-p-n bipolar transistor fabrication steps:(a) buried-layer implant, (b) epitaxial layer, (c) photoresist mask, (d) channel stop implant, (e) oxide isolation, (f) base implant, (g) removal of thin oxide, (h) emitter implant (from ref. 1.18).*

consists of about 350 steps. Among them 10-15 are different ion implantation steps.

Ion implantation [1.20, 1.21] is the introduction of energetic charged particles into targets with enough energy to penetrate beyond the surface region. A beam of dopant ions of fixed energy, typically between 10 and 100 keV, is rastered across the surface of the semiconductor. The penetration depth is determined by the energy of the incident ions. They enter the crystal lattice, collide with silicon atoms, and gradually lose energy, finally coming to rest at some depth within the lattice. The total number of implanted ion/cm^2, dose, $N_\Box$ or fluence is given by the product of the flux F of incident ion/$cm^2 \cdot s$ and the implantation time, t, during which the ion beam is impinging on the sample: $N_\Box = F \cdot t$ (for a costant flux). The flux F of positive ions of charge q represents a current that can be measured directly by a meter connected between the sample and the ground.

A schematic of an ion implantation system target chamber is

ION IMPLANTATION:
BASICS TO DEVICE FABRICATION

The Kluwer International Series in Engineering and Computer Science

ELECTRONIC MATERIALS: SCIENCE AND TECHNOLOGY

Series Editor
Harry L. Tuller
Massachusetts Institute of Technology

Other books in the series:

Sol-Gel Optics: Processing and Applications, L.C. Klein, editor
Solid State Batteries: Materials Design and Optimization, C. Julien and G. Nazri, authors

---

The Series **ELECTRONIC MATERIALS: Science and Technology** will address the following goals

* Bridge the gap between theory and application.

* Foster and facilitate communication among the materials scientists, electrical engineers, physicists and chemists.

* Provide publication with an interdisciplinary approach in the following topic areas:

* Sensors and Actuators	* Optoelectronic Materials
* Electrically Active Ceramics and Polymers	* Composite Materials
	* Defect Engineering
* Structure-Property-Processing -Performance Correlations in Electronic Materials	* Solid State Ionics
	* Electronic Materials in Energy Conversion-Solar Cells, High Energy Density Microbatteries, Solid State Fuel Cells, etc.
* Electrically Active Interfaces	
* High Tc Superconducting Materials	

With the dynamic growth evident in this field and the need to communicate findings between disciplines, this book series will provide a forum for information targeted toward

● Materials Scientists ● Electrical Engineers ● Physicists ● Chemists

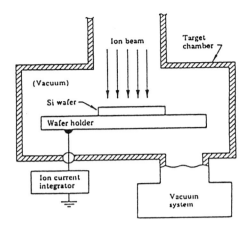

Figure 1.13 *Schematic of an ion implanter target chamber (from ref. 1.16).*

shown in Fig. 1.13 [1.16]. In the case of MOS the adjustment of V_T the threshold voltage is done by ion implantation, as already shown in Fig. 1.9c. The first large scale application of ion implantation was infact related to the precise threshold control of MOS device [1.22].

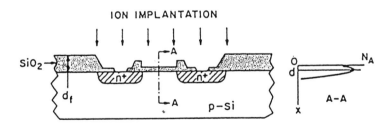

Figure 1.14 *Boron implant for the control of the threshold voltage V_T.*

Boron ions are implanted through the gate oxide (Fig. 1.14) of the device such that the peak occurs just below the Si surface. The implant introduces a negative charge of boron acceptors that in a first approximations varies V_T of the quantity $qN_\square/C$ being q the electronic charge and C the capacity of the metal-oxide- silicon substrate capacitor. For istance a variation of 1V, with an oxide layer $0.1\mu m$ thick, requires a fluence of 2×10^{11} boron ions$/cm^2$.

Decreasing the thickness of the oxide to $0.02 \mu m$ the fluence reaches the $10^{12}/cm^2$ value. This is an extremely small amount of dopant, it corresponds to 10^{-3} of the number of atoms$/cm^2$ forming a monolayer of material.

Before the advent of ion implantation, dopant was diffused into the wafer from a surface source such as a doped glass, or from gas ambient. To achieve reproducibility the dopant concentration at the surface is maintained at solid solubility limit during the predeposition stage [1.23]. In addition the surface must be clean and the gas source of dopants must provide a uniform flux of dopants impinging on the semiconductor surface. Localized contaminated regions or native oxide layers might impede the transport of dopants across the gas- semiconductor surface. The accurate control of gas flow and of surfaces was quite difficult, so ion implantation became the new dopant technique.

The current can be easily measured in the range above 10^{-9}A. At a boron current of $1 \mu A$, a typical value of the implanters available twenty years ago a 4" wafer, i.e. a wafer of $10cm$ in diameter required an implantation time of $\sim 12.5s$ for a $10^{12}/cm^2$ fluence. Today the industrial implanters run at a current of about 10 mA so that several hundred wafers of 8" (20 cm) diameter can be processed per hour. The number of wafers that are processed (e.g. implanted) per unit time is called throughput. Measurement of the accelerating voltages determines the ion energy and integration of the beam current over the implant time provides the total delivered charge and hence the implantation dose. By control of the ion beam current and implantation time, values of Φ between $10^{10}/cm^2$ and $10^{16}/cm^2$ can be obtained according to the different applications. If the implanted dopants are distributed uniformly over $1 \mu m$ the concentration can be controlled between $10^{15}/cm^3$ and $10^{21}/cm^3$, over six orders of magnitude i.e. the range spanned in Fig.1.5.

The energy of the beam determines the depth distribution inside the target of the dopants, surface contamination or native oxide layers, have a negligible affect: in several applications moderate beam energies are adopted and the implanted dopants are diffused by a subsequent thermal treatment. The implant is used as a predeposition step. Recently, as we will see later, profiles close as possible to the as-implanted ones are required for new applications in device fabrication and the beam energy has been shifted either toward high values in MeV regions or low values in the few keV regime.

As an example of the wide range of thicknesses at which the

implanted ions can be located by changing the beam energy, the profiles of P implanted at 80 keV, 250 keV, 50, 80 and 120 MeV are shown in Fig. 1.15. The last three very high energy implants were performed in the same sample.

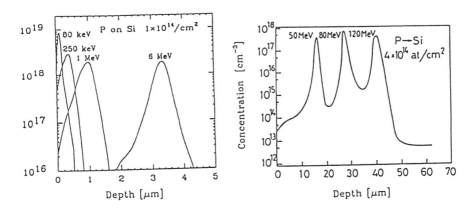

Figure 1.15 *Profiles of P implanted in Si at several energies. The left hand implants were performed in different samples, while these at the right hand in the same sample.*

The ion implantation technology allows selected areas to be implanted by masks which leave well-defined areas of the semiconductor exposed to the beam and other areas masked from the beam (Fig.1.16). The thickness of the mask must be greater than the penetration depth of the ions. At low energy, i.e. for depth distribution of dopants in .1μm thick layer, mask thickness of 200 nm are enough to block the ions. At high energy the thickness of the mask can reach several microns and this is a serious problem for the large scale applications of high energy implants. The masks can be made by silicon oxide layers obtained by thermal growth or by chemical deposition, by organic layers or by metal films.

The source and drain regions of MOSFET are formed by dopant implantation, of n type species (P or As) at doses in the $10^{15}/cm^2$ range (see Fig. 1.9a). The mask layers are patterned by photolithographic techniques, as used in other steps of integrated circuit processes, and the mask material is removed in areas where the semiconductor wafer is to be exposed to the ion beam. After implantation all the mask material is removed so that the wafer can receive further processing steps. Autoregistration, in which the same feature is used successively to define several selected

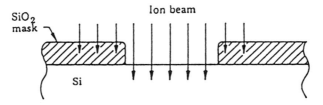

Figure 1.16   *Mask layer of suitable thickness stops the ions.*

doped regions, allows much denser packing of components, and a reduction of parasitic capacitance.

The use of ion implantation in the fabrication of bipolar transistor has been delayed with respect MOSFET. The first decade of planar bipolar integrated circuits development (1965-1975) proceeded almost entirely without the use of ion implantation, in spite of the 1954 Shockley patent on the formation of the bipolar base by ion implantation.

The first application of ion implantation was the emitter formation by As^+ implant. High dose was required in the range $5 \times 10^{15}/cm^2$ to $2 \times 10^{16} cm^2$, so that the industrial application was possible where stable high current ion source was available [1.24]. This contrasts the first low dose application in MOS technology.

To day the emitter is still arsenic-implanted, but the implant is performed into a polycrystalline silicon overlayer from which out-diffusion of dopant during thermal annealing occurs in a free-damage region (see also sect. 7.8).

The implanted base allows even narrower base widths to be achieved reproducibly. One device must be isolated from the other, the depletion layer at the p-n junction has been replaced by a recessed oxide layer that reduces capacitances and increases packing density (see Fig. 1.11a). This has been achieved by the fabrication of a precisely doped channel stop region under the oxide which prevented the occurrence of n-type surface channels in the silicon substrate material adjacent to the oxide as in a MOSFET. The buried heavily doped collector needed to decrease the collector resistance is also formed by implantation of n-type species before the growth of the epitaxial layer. In some cases Sb is used as n-type dopant instead of As, for its low diffusion coefficient. The spreading of the buried layer during the subsequent thermal processes is then minimized.

1.6 Damage and Yield

The major problem with ion implantation is the elimination of the damage created during the slowing down of the projectiles in the target. The violent collision of the ions with the target atoms displace them from their equilibrium lattice sites. The minimum energy that should be transferred to a silicon atom to be displaced is about 15 eV. If the knock-atom has enough energy it can create other displacements, giving rise to a collision cascade process. For each implanted ion 10^3 to 10^4 silicon atoms are displaced from the lattice sites. The primary damage after implantation is a quite complex function of several parameters: ion mass and energy, dose rate and dose, target temperature, substrate orientation, dopants and impurities present in the substrate.

A thermal process is necessary to activate electrically the dopant and to eliminate the damage. It is possible to activate electrically, after a suitable annealing process, all the implanted atoms. They are now located in substitutional lattice sites where replace silicon atoms and give electrons or holes. Quite often residual damage in the form of extended defects like dislocation line, dislocation loops, stacking-fault etc., is left in the annealed wafer. The residual or secondary defects depend on the annealing process, on the primary damage, and on processes like oxidation, nitridation, silicidation performed during or after the annealing. These secondary defects if present in particular regions of the device can be detrimental. The device yield can be impaired, so that the main trend in this field is the understanding of these complex phenomena to reduce the density of defect and to increase the yield. Defects are the major factors of yield losses and reliability problems in integrated devices.

Decreasing the device dimensions and increasing process complexity the susceptibility of device technology to the formation of process-induced detrimental effects is enhanced. Fig. 1.17 illustrates the relationship between yield and chip area as a function of defect density according to a statistical relationship called Murphy's law [1.25]. It states that the yield (Y) of good devices on a wafer depends on the product of the average defect density (D_0) and on the area (A) of the chip by

$$Y = [(1 - exp-AD_0)/D_0A]^2 \qquad 1.4$$

The normalized defect distribution is supposed of triangular shape with a maximum at D_0 For yield levels above 50%, in the case of a 16 Mbit DRAM defect densities less than 0.5 defect $/cm^2$ have to be achieved. Thus strategies and remedies for reduction defect

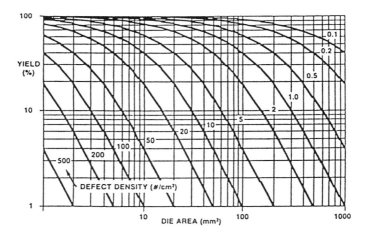

Figure 1.17 *Device yield as a function of the chip area for several defect densities according to eq. 1.4. The chip size increases with integration and it is shown for several DRAM (from ref. 1.25 (c) 1964 IEEE).*

densities expecially in the ion-implantation related processes are a prominent and main task in the mass production of VLSI (Very Large Scale Integration) and ULSI (Ultra Large Scale Integration) devices. Improved defect density control is essential since the size of a defect that causes a chip loss scales right along with technology scaling.

The defects affect device properties both by interacting with dopants during thermal processing and thus changing e.g. the final profile of the implanted species and also directly by their own electrical effects on the device operation. To the first category belongs the transient diffusion of dopants that takes place during the thermal treatment [1.26]. Mobile point defects can enhance the diffusivity of dopant atoms thus broadening the profile with a change in the junction depth.

As an example of the second category consider a bipolar transistor. In the fabrication of collector region because the implanted dose must be sufficiently high to obtain a low collector resistance, high damage levels are produced. If residual damage in the form of a dislocation line intersect a p-n junction, an increased leakage current can result due to mid-gap electronic levels associated with defects. Furthermore, when both the collector /base and the emitter/base junctions are connected via a dislocation, collector -emitter shorts may arise by enhanced diffusion of the emitter dopants along the dislocation. n^+ bridges are formed between emitter and collector responsible of high leakage ohmic current paths in the off state, as shown in Fig. 1.18 [1.27]. In the MOS device the presence of dislocations in the depletion layers causes an increase in the parasitic leakage current between drain and substrate.

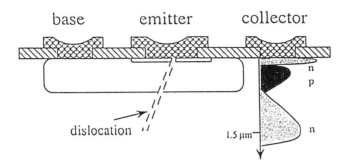

Figure 1.18 *Schematic illustration of a defect, dislocation, that shorts the emitter to the collector in a n-p-n bipolar transistor.*

At very small channel widths, the dislocation can connect the source to the drain with the formation of an n^+ lateral region. In both devices it is the "off" state which is most strongly affected by the defects.

The aim of the annealing is the production of defect-free silicon with the minimum amount of thermal budget. The residual defects, if any, should be located in non-active areas of the devices. In some cases a compromise is established between the defect density and the change in the dopant profile caused by the annealing. Ion implanters introduce, as side effect, some metallic contamination during the path lenght traversed by the ions. This contamination is of relevance in the VLSI technology and must be maintained at very low level expecially in industrial implanters.

1.7 Future Trends

Reduction or scaling of the device dimensions has been the overriding trend in the semiconductor technology for the last two decades. The Moore law [1.28] which states that the number of MOSFET per device double each year is still followed and it will be up to the first decade of the next century. Several criteria have been adopted for the scaling of the device dimensions. In the constant field scaling the electric fields in the scaled device are identical to those in the unscaled device [1.29]. All device dimensions are reduced by a scaling factor, α, this requires an increase of α of the doping and a reduction of the same factor, for the voltages. In the constant voltage scaling the power supply is maintained constant while device dimensions are scaled by α. The doping in the semiconductor is increased not of the scale factor but of a suitable value to obtain the desired threshold voltage [1.30]. The electric field in the channel is then not constant and it increases with the scaling, hot electrons are produced with detrimental effect on the device performance. These hot electrons can overcome by tunnel effect the energy barrier and can be trapped in the thin gate oxide where they change the threshold voltage and the I-V characteristics of the device. New technological processes are devised to alleviate the hot electrons effect and drain and source doping is modified by adding other implants even at large angle of incidence. The aim is to reduce the field between the drain and the channel regions.

Shrinkage of the device relies also on the possibility to form junction depth as shallow as $0.1\mu m$. Shallow junction formation is another field where ion implantation playes a relevant role. By decreasing the beam energy the penetration depth is reduced so that in principle shallow junction can be formed. For light ions as B the energy range should be in the keV regime, but at this energy beam current is faible so that industrial processes cannot be performed unless ion sources and implanters operating at this low energy become available. The other effect which limits the depth is the single crystal nature of the target. Particles which enter the target at a small angle respect a low index axis or plane can be deviated by elastic collisions with the matrix atoms along an axial or planar channel (feeding-in). They continue their trajectories as channeled projectiles. The energy losses are considerable reduced for channeled particles and then the depth profile is characterized by penetrating tails. As an example of the dramatic changes in the dopant profiles caused by the ordered structure of the crystalline substrate, several carrier concentration distributions are reported in Fig. 1.19 for $2 \times 10^{13}/cm^2 - 1MeV\,P^+$ implants at different angle of incidence [1.31]. The beam energy is quite high just to evidence more clearly the effect. The analysis reports the carrier (electron) profiles that are assumed coincident, under some experimental conditions, with the dopant distributions. The range changes of a factor three between the implants performed in an amorphous structure and along the $< 100 >$ axis of the Si single crystal substrate. Even at 7^0 tilt angle of incidence with respect the $< 100 >$ axis a substantial change in the profile occurs. The phenomenon depends quantitatively on several parameters that cannot be controlled easily, so that the aim so far has been that to avoid channeling in industrial applications.

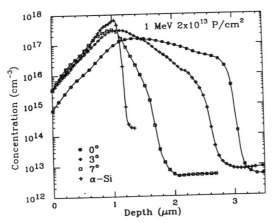

Figure 1.19 *Carrier profiles in Si implanted with $2 \times 10^{13}/cm^2 -$ $1 MeV P$ ions impinging in amorphous layer, along the [100] axis at 3^0 and at 7^0 tilt angle from the axis. All the samples have been annealed at $1000^0 C$ for 10s (from ref. 1.31).*

The trend might change in the future if parallel beam scanning systems become available together with careful selected oriented wafers. In the case of heavy ions the large amount of damage destroyes even at low dose ($< 10^{14}/cm^2$) the long range order and directional effects disappear. For light ions several procedures have been considered as the preamorphization by an inert species like Ge or Si. The boron ions, that are implanted later see the target as an amorphous structure and the profile is in this case reproducible and quite well described theoretically [1.32].

To maintain shallow junction depth the subsequent thermal processes should be performed at the lowest budget to avoid any substantial broadening of the as-implanted profile. New thermal rapid processes have been designed and used in these last years. It is now possible to heat a silicon wafer near its melting point ($\sim 1400^0 C$) for a duration of a few seconds. In the most common apparatus the sample is heated by optical radiation emitted by very intense lamps [1.33]. Several problems, as we will see in details later, have been solved but others are under investigations before its large scale application for device fabrication.

RTP or RTA (Rapid Thermal Annealing) is the fall-out of other advanced methods based on the deposition, in the near surface region, of a large amount of energy in a short time. The most

popular was fifteen years ago a laser pulse of about $1\,J/cm^2$ and of
several nanosecond duration [1.34]. At this energy the near surface Si layer was melted and during cooling, resolidification takes
place by a process similar to liquid phase epitaxy. The process
was termed "laser annealing" because the first experiments were
performed on implanted silicon samples with a surface amorphous
layer. The fusion and solidification induced the recrystallization of
the damaged layer. From an industrial point of view the method
is quite far from any short time applications but it has stimulated
several interesting studies and has opened the field of directed
energy processes to whom RTP belongs.

The combination ion implantation - RTP is one of the fabrication process responsible of the answer "How small can a device be?". The horizontal dimensions are determined by litography, the vertical dimensions are controlled through the means by
which dopants are introduced in the wafer surface regime [1.35].
The complete two dimensional profile of the implanted ions is required. In addition to the vertical straggling in the ion range, ions
are spread out laterally and they are stopped under the mask
layer. With decreasing the channel width the lateral straggling of
the implanted ions for source and drain may become comparable
to it, and then the performance of the MOSFET will be impaired.
This is another aspect of the implant process that deserves attention. Ion implantation can be modelled relatively easily by
computers, it can be well described by a few parameters. The
use of simulation reduces process development and implementation time, of course a detailed knowledge of side-effects such as
feeding-in, transient diffusion is required to obtain a reliable simulation.

Recently the availability of very-high-current implanters ($\sim$
$100\,mA$) allows implanting concentrations up to three orders of
magnitude higher than those required for doping and enough to
change the chemical composition and to form compounds.

It is now possible to synthesized by high-dose oxygen implantation
($10^{17} \div 10^{18}/cm^2$) stoichiometric SiO_2 films buried under a thin
film of single crystal, as illustrated in Fig. 1.20 [1.36]. Such Si-
on-insulators (SOI) structures provide complete dielectric isolation used in conventional integrated circuits. The devices become
radiation hard and have high voltage capabilities. The implantation method from the founder is known as SIMOX acronym of
$\underline{S}$eparation by $\underline{Im}$planted $\underline{Ox}$ygen [1.37]. The beam energy exceeds usually 150 keV and a continuous oxide layer buried under
$0.1 \cdot 0.3\mu m$ of crystalline Si is formed. The devices are built in
this top layer. To form 100 nm of $SiO_2, 4.4 \times 10^{17}$ 0 at $/cm^2$ are

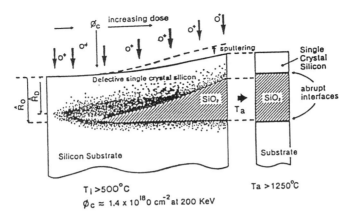

Figure 1.20 *Sequence illustrating the formation of buried* SiO_2 *layer by oxygen implantation (from ref. 1.36).*

required, assuming a stoichiometric composition for the oxide.

The formation of buried compounds has not limited to silicon dioxide, but has been extended to other insulators as Si_3N_4 [1.38] and to metallic silicide as $CoSi_2$ [1.39]. This silicide has the CaF_2 structure with a small lattice mismatch with Si ($\sim 1.2\%$). The formation process is quite complex for the $CoSi_2$ and it will be considered in some details in Chapt. 8, together with other applications to III-V compound semiconductors.

High dose implants in Si are not limited to buried dielectric and metallic layers fabrication. Semiconducting Si_xGe_{1-x} alloys are formed also by Ge high dose implants ($10^{17} \div 10^{18}/cm^2$) [1.40]. These alloys form heterostructures with Si substrates and can be used as base for the fabrication of bipolar transistors working at very high frequency [1.41]. The band gap of Si_xGe_{1-x} alloy decreases with the Ge content and heterojunctions are created along the line of band gap engineering. There is a 4.2% of lattice mismatch between Si and Ge lattice constant. This means that heterojunctions formed between Si and Si_xGe_{1-x} alloys are either strained below a certain critical thickness, or contain misfit dislocations if this limit is exceeded. The critical thickness is a function of the alloy composition x [1.42].

As any crystallographic defects are generally to be avoided, it is strained layer epitaxy that is of technological interest. The strain has the additional effect of modifying the band gap of $SiGe$ alloys. At 10% of Ge the band gap becomes 1.05 eV. Ion implantation offers an alternative procedure to the other epitaxial thin-film growth techniques, such as the ultra high vacuum chemical vapor deposition carried out at a temperature of $550^0 C$ in an ultrapure environment [1.43]. Ion implantation forms graded composition alloys and places less severe requirement on the lattice strains.

Another field in which ion implantation might be play a unique role is that of optoelectronic and integrated photonic devices based on silicon. One may take advantage of the mature silicon technology to combine optical and electrical components on a single chip. However as already mentioned the application of silicon in optoelectronic requires real break throughs in the engineering of the optical properties of this material. Recently Er doping of silicon or of semiinsulating polycrystalline silicon layers (SIPOS), by ion implantation has been recognized as a promising approach to obtain light emission from silicon [1.44]. Er is a rare-earth element that, when incorporated in a solid in its trivalent charge state, exhibits a sharp atomic-like transition at $1.54\mu m$, due to a transition in the internal 4 f shell. When Er is introduced in the Si host matrix this transition can be excited both optically and electrically. Using non-equilibrium solid phase epitaxy of co-implanted oxygen and erbium layers it is possible to incorporate Er at concentration as high as $1^{20}/cm^3$ without the formation of precipitates [1.45]. Similar effects are also formed if the Er implant is performed in SIPOS layer containing an atomic concentration of oxygen in excess of 10% [1.46]. In both cases light emitting diodes are fabricated and operate at room temperature. Of course this is a field just beginning and it is in the stage of basic research.

Another application of ion implantation which instead is entering the market maturity is the surface treatment of material used for orthopaedic prosthesis [1.47]. Titanium alloys containing Al and V ($Ti - 6Al - 4V$) have excellent biocompatibility, good strength and corrosion resistance. Unfortunately they lack hardness and wear performance in articulating contact with the polyethylene socket used in the total joint replacements. Nitrogen implantation at doses in excess of $10^{17}/cm^2$ improves the wear performance of these devices by a factor 10^3, thus allowing the use of titanium-based alloy as hip-joints for a much longer time. The reduced wear is associated to the formation of a surface oxide layer that seems to act as a lubricant. Several other surface procedures have been developed recently, such as the deposition of layer with a concurrent ion bombardment [1.48], but these methods will not be considered any further in this book.

CHAPTER 2

ION IMPLANTERS

2.1 Introduction

An ion implanter (see Fig. 2.1) consists of the following major components: an ion source, an extracting and ion analyzing mechanism, an accelerating column, a scanning system and an end station [2.1]. The ion source contains the species to be implanted either as solids, or as liquids or as gases and an ionizing system to ionize the species. The source produces an ion beam with very small energy spread enabling high mass resolution. Ions are extracted from the source by a small accelerating voltage and then injected into the analyzer magnet.

A spatial separation of ions subjected to the Lorenz force occurs due to the differences in the mass and charge. Only the selected ions are injected into the accelerating column, the others are stopped by the presence of suitable screens. In this case the system operates in the pre-analysis configuration. If the ions are accelerated to their full energy before mass separation the system works in the post- analysis configuration. The extraction voltage ranges usually between 15 and 40 kV. The selected ions are accelerated by a static electric field, are focussed and shaped in the column and ready to be implanted with an energy up to 80-400 keV. The ions are distribute uniformly over a target by electrical fields varying in the x and y directions with a sawtooth or by a mechanical shift of the target. The measurement of the ion-current requires the use of Faraday cup, a secondary electron suppressor, and in some cases, the wafer holder can be oriented by rotation and tilt. This description illustrates schematically the main features of the implanters.

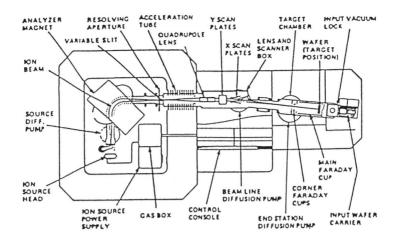

Figure 2.1 *Schematic illustration of a medium current implanter with a maximum potential of 200 kV and a current of ~ 1mA (from ref. 2.2).*

The industrial apparatus are quite sophisticated machines, computer controlled and capable of a throughput of several hundred implanted 8" wafers per hour with a cost of ~ 2M$. In low and intermediate energy implanter, the ion source produces positive ions that are accelerated after extraction.

Nearly all of the available commercial ion implanters are built with the source and the magnet at high voltage. The implanters are classified according to the beam intensity: (i) low current ~ 0.1mA, (ii) medium current ~ 1.0mA, (iii) high current ~ 10mA and (iv) very high current ~ 100mA. The last one is still at the prototype level and is devoted mainly to oxygen and nitrogen implants. The beam energies range between 10-200 keV or 25-400 keV, except at high current where the energy is usually less than 80 keV. The vacuum is maintained usually by three pumping systems, one for the ion sources, another for the scanner system and the third one for the wafer chamber.

In a low current machine, the current at the target is limited by the output of the ion beam. The beam is scanned electrostatically over the area of the target. In the medium current implanters the beam current at the target is limited by space charge problems. The beam is scanned over the target in at least one direction electrostatically. In the high and very high current machines the beam is space-charge neutralized. Thermal electrons can be maintained within the beam in the region between the extraction electrode and the resolving slit. The target is scanned mechanically through a stationary beam.

In high current or high energy implanters the target can reach elevated temperature unless a suitable wafer cooling system is adopted. Hot implants alter the structure of primary and secondary damage, and must be avoided for reproducibility of the process. Recently high energy beams, ~ 1 MeV, are used for several applications in CMOS and bipolar technology. High energy accelerators with voltage in excess of MV are based usually on two different structures: tandem and linear accelerators. In the tandem structure the ion source produces negative ions that are accelerated toward the high voltage part of the machine located in the middle of the tank (see Fig. 2.6). The ions pass through a gaseous or thin foil targets to be stripped of electrons and thus they become positive charged and are repelled from the positive voltage toward the terminal of the accelerator tube at the earth potential. If V is the terminal voltage, the final energy is given by $(1 + n)qV$, being n the charge state of stripped positive ions. Particular attention requires the production of negative ions. The linear accelerators are made by a series of cavities that are biased by a radio-frequency voltage (see Fig. 2.8). The positive ions are accelerated by the voltage at the gap between one cavity and the next up to reach the final energy. These two high energy accelerators are described in more details in sect. 2.3.

2.2 Ion Sources

Ion source is the most important component in industrial implanters and requires the most frequent preventive maintenance. The lifetime of an ion source varies between several tens of hours to a few hundred of hours, and it is very dependent on the operating conditions. All the sources [2.3 - 2.4] produce ions by means of a confined electrical discharge which is sustained by the gas or vapor of the material to be ionized. Differences arise in the ionization method and in the formation of the gas phase of the required species. The most common sources in ion implanters are positive ion sources. In the case of tandem accelerators for high energy implants negative ion sources are needed and they will be briefly described in the next section on high energy implanters.

The species to be implanted must be present in the beam as charged particles. The ionization process is then quite relevant in determining the amount of beam current that can be extracted from the source. Each chemical species is characterized by a well defined ionization potential, i.e. the energy required to remove that electron from the atom or from the ion. Metals of valence one, as sodium, lithium, cesium, have low ionization potentials whereas closed shell atom as argon, neon have large electron binding energies. It is possible, of course, to remove more than one electron from a species so to create multiply charged ions, but the ionization energy increases rapidly. The first, second and third ionization potential of oxygen, (i.e. of 0, 0^+, 0^{++} are 13.6, 35.1 and 54.9 eV respectively.

Positive ions are primarily created by anelastic collisions with energetic electrons produced by hot filaments or cold cathodes, in some cases the energy is provided by electromagnetic radiation at radio frequency.

One of the most adopted source, the Freeman ion source, [2.5] is shown in Fig. 2.2. The arc chamber with the extraction system is reported in Fig. 2.3a. The cathode is made by a tungsten bar, it is normally heated by Joule effect at $\sim 2500K$, so that a flux of electrons of 0.3 A/cm^2 is emitted by thermoionic effect.

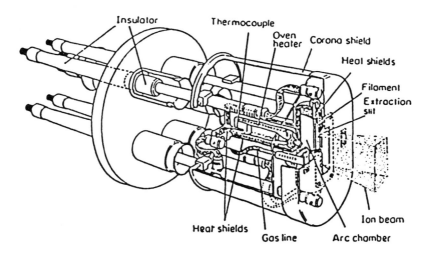

Figure 2.2   *Schematic view of the high current Freeman ion source (from ref. 2.5).*

These electrons are spiralized by the presence of magnetic fields, one external of 100 G and another internal created by the current through the tungsten wire. The overall length of the electron trajectory inside the chamber is then considerable increased so that the collision probability of the electron with the gas species present increases. As a result a fraction

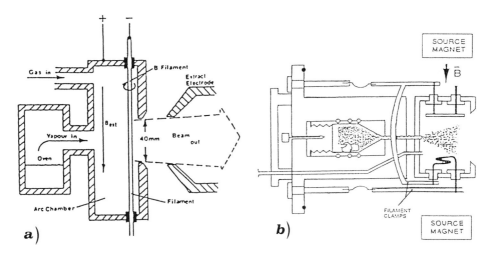

Figure 2.3 *Arc chamber and extraction system of Freeman (a), and of Bernas ion source (b) (from ref. 2.6).*

of the gas is ionized. Positive ions and electrons form a gas plasma with no space charge. Such a gas contains therefore an equal number of positive and negative charges, it is field free and can be treated as a conductor whose conductivity depends on the electron density.

During the operation the current between cathode and anode and the gas inlet should be adjusted to maintain a stable gas discharge or an arc in the chamber, so that positive ions can be extracted by a properly shaped and biased (negative potential for positive ions) electrode extraction. The cathode in the Freeman source is parallel and extremely close to the extraction slit in order to achieve a strong discharge in the most favorable position. This source is capable to deliver currents up to 20 mA. A voltage of 50-100 V is applied between the cathode and the anode and a gas or vapor is introduced in the arc chamber at a pressure in the $10^{-2} - 10^{-4}$ Torr to initiate the arc discharge. This source has been modified by Bernas, see Fig. 2.3b and an enhanced yield is obtained [2.6]. Both sources are commonly adopted in the industrial implanters. The dopant species is provided by a cylinder of gas or liquid containing the feed material and connected to a pressure regulator and a fine leake valve to control the introduced amount of material. Solid materials are heated in a small oven at temperatures up to $1000^0 C$, and the vapor is transported to the arc chamber. The solid material should have a vapor pressure of 10^{-3} Torr at temperatures not above $1000^0 C$. The optimum operating pressure for high beam currents with good quality depends on the desired species and on the charge state. To enhance the amount of high charged state ions (e.g. B^{++}, P^{++}, P^{+++}) the pressure must be kept low enough to reduce charge exchange reactions and to increase the electron mean free path. Higher operating pressure is required to initiate the arc. Solid with low vapor pressure in the elemental form cannot be used, boron, for istance, where temperature above $2000^0 C$ would be necessary to reach a reasonable vapor pressure. In these cases compounds are used, such as BF_3. Different ionized species are then present in the arc chamber, B^+, BF^+, BF_2^+ in addition to that of the residual ambient gas. Phosphorus element may be used as a feed material for P^+ production, molecular ions such P_2^+ or P_3^+ are present in a considerable amount. The P^+ component amounts to 70% of the total output.

Some other sources are used in research accelerators, they are more versatile but provide low current. In a cold cathode the electrons are not emitted by thermoionic effect but as secondary electrons from the cathode as a result of positive ion bombardment. These sources require voltages in the 1-2 kV range to maintain a stable discharge and magnetic fields up to 2KG to obtain efficient electron collimation.

Cold cathode is used in the penning ion source, [2.7] shown in Fig. 2.4.

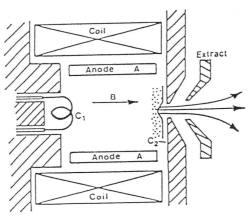

Figure 2.4   *Penning ion source (from ref. 2.7).*

The magnetic field is directed along the main axis. The electrons are spiralized and are trapped by the field lines in a central column. Extraction of positive ions is either axial, from the centre of anticathode as shown in figure, or lateral through a slit cut in the anode. This source finds applications in low and medium current ($10\mu A$ to several $100\mu A$) accelerators.

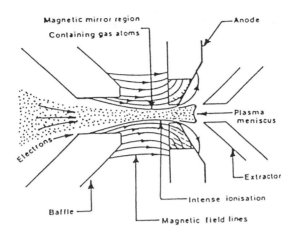

Figure 2.5 *Duoplasmatron ion source (from ref. 2.8).*

The duoplasmatron source is characterized by a high density plasma , $10^{14}/cm^3$, produced near to the anode aperture by using an intermediate electrode and a strong magnetic lens [2.8] (see Fig. 2.5). The source is efficient mainly for gases, positive or negative ions are extracted along the axis through the anode. The source consists of two plasma regions. The lower density cathode plasma between cathod and buffle (or intermediate electrode) and the high density plasma between intermediate electrode and anode.

2.3 High Energy Implanters

Tandem accelerators are widely used in research laboratories for high energy implantation. In Fig. 2.6 a schematic drawing of a Tandem accelerator is shown [2.9].

In tandem accelerators negatively charged ions are produced in the source, pre-accelerated to $\sim 20 - 100 keV$, separated by a magnet and injected into the acceleration tubes inside the tank. This region presents the two extremes at ground voltage while the central part is at positive voltage (maximum 1.7 MV in the tandem shown here). The negatively charged ions are then accelerated towards the central terminal where they are stripped of some of the electrons in a stripping canal filled by a gas (typically nitrogen or argon) and then accelerated again (being now positive) towards the other end at ground. The total final energy will therefore be $E = qV_0 + qV_T(1 + n)$ being V_0 the pre-acceleration potential, V_T the terminal voltage and n the charge state of the ions produced in the stripper. Charge states from +1 to +3 are easily produced in the stripper so that energies from 300 keV to 6.9 MeV can be obtained with considerable current ($\sim 20 - 100\mu A$). Higher charge states, and hence higher energies, can be obtained with smaller current depending on the ion species. In the case of Cu, for example, it is possible to obtain a charge state +10 (corresponding to an energy of 18.8 MeV for 1.7 MV maximum voltage) with a current of $0.1\mu A$.

Once accelerated the ions are focussed by a quadrupole, selected by a second magnet and injected in the implantation beamline. Scanning, also at these high energies, is usually obtained electrostatically.

Table 2.1, just to have a general idea, reports for the case of Si the final currents obtained on target at the different charge states and the corresponding energies. The ion flux is obtained dividing the current by the charge state.

Two major features of tandem accelerators should be discussed. First of all the high voltage generator. Typically two kinds of tandem systems exist: pelletrons and tandetrons. In pelletrons the high voltage on terminal is obtained through the charging of several moving chains while in tandetrons acceleration is produced by a solid state Walton- Cockroft system (this last system, shown on Fig. 2.6, is clearly visible on the low middle-section of the tank).

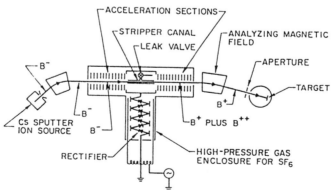

Figure 2.6 *Schematic drawing of a tandem ion implanter (from ref. 2.9).*

In both cases the tank containing the acceleration tubes (and the chains or the Walton-Cockroft system) is filled by the insulating gas SF_6 to a pressure of $\sim 8\ bar$.

Table 2.1 *Beam currents at the target and energies for the different charge states of Si ions.*

Si charge state	Current	Energy
	μA	MeV
1	60	3.4
2	100	5.1
3	50	6.8
4	25	8.5
5	6	10.2
6	2	11.9

A second important feature of tandems is the ion source that, in this case, should in part produce negative ions [2.10]. Negative ions are formed by the attachment of an electron to an atom or to a molecule. The process is exothermic and energy is given out by the system, viceversa the positive ion formation is an endothermic reaction and energy must be supplied to the system. The binding energy of a negative ion is called electron affinity. A stable negative ion exists if the electron affinity is a positive quantity, i.e. the negative ion has a lower potential energy than the neutral atom. Several processes lead to the formation of negative ions. Electron impact in a dense gas, charge exchange with neutral atoms or molecules, charge exchange with surface atoms during absorption and desorption are normally used to form negative ions. If an ion beam passes through a vapor of atoms of group I or II metals which have low ionization potentials, there is a good chance that a substantial fraction of the incident ions will capture two electrons to form negative ions. In the case of H and He a duoplasmatron source is used. High currents of H^- or of He^+ are produced. The positive He^+ ions are then injected into a charge exchange canal. This is a region containing Li vapors obtained by heating a Li filled reservoir to $\sim 500^0 C$.

The following reactions take place [2.11]:

$$He^+ + Li \rightarrow He^{\circ\star} + Li^+$$

$$He^{\circ\star} + Li \rightarrow He^- + Li^+$$

$He^{\circ\star}$ is an exited state of He and the reaction is important for negative ion formation of elements with a negative electron affinity. A suitable electrostatic field is then used to get read of the remaining He^+ beam while the He^- ions are pre-accelerated and injected in the tandem. With this source He currents of $\sim 1\mu A$ are easily obtained on target.

In other ion sources the species to be ionized is obtained by sputtering a solid target cointaining the desired material with energetic ions present in the chamber [2.12]. The sputter source is used for just about all the elements in the periodic table with the exclusion of noble gases (which are not produced in negative states). A detailed drawing of this source is shown in Fig. 2.7.

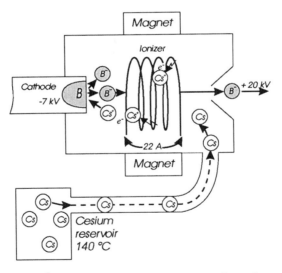

Figure 2.7 *Cs^+ Sputtering ion source for the production of negative ions (adapted from ref. 2.12).*

The sputter ion is Cs^+ which strongly increases the yield of sputtered negative ions. Cesium is contained into a reservoir which is heated up to $\sim 170^0 C$ to produce Cs vapors which are introduced into the source through a small tube. An ionizer, heatep up to $\sim 1200^0 C$ through a $\sim 22A$ current flow, is used to produce Cs^+ ions which are accelerated by $\sim 7kV$ voltage towards the cathode and hence produce the sputtering of the cathode material. Negative ions sputtered from the cathode are extracted by a 20kV voltage and later on further accelerated by a 0 - 80 kV potential of a pre-acceleration stage prior to injection into the acceleration tube. Cathodes are usually made by small copper cylinders with a central hole which is filled with the material to be sputtered (and hence implanted).

A crucial issue is the choice of the right cathode material to maximize the negative ion yield of the element to be implanted. For some species, which are copiously produced into negative states, (such as, for example, Si or Au) the elemental material can be introduced in the cathodes. In the case of other species, whose negative ion yield is very low, some tricks should be instead used. For instance in the case of N, cathode can be made by a mixture of graphite and boron nitride which will produce a good yield of CN^- molecular ions; these ions can be extracted and injected in the tandem where they will be broken in the stripper region producing a good N^+ current on target. The sputter source just described is able to produce good negative ion current, with final current in the 10-100μA range. Moreover the switching from one cathode to another takes $\sim$ 5 minutes. This means that in such a short time the machine can be switched from the implantation of an element to that of a different one. This characteristics is particularly attractive for research purposes. In the case of machines dedicated to industrial purposes (for which high currents are needed), however, sputtering sources are not used due to their medium currents. In these machines, usually, high current positive sources are used. The positive ions are then converted into negative ions by a charge exchange canal (of Li or Mg) and then injected into the tandem [2.13]. With these techniques current as high as $\sim 1mA$ can be obtained on the sample.

Since high energy implantation is now becoming more and more important for industrial application also different technical solutions have been adopted. Among these it should be mentioned the production of linear accelerators (LINAC) [2.14] for this aim. These accelerators are made by a series of cavities that in some cases are polarized by a radio-frequency voltage. The linear accelerator is a resonance device in which the electric component of a time varying rf electromagnetic field acts on charged particles to produce acceleration. In Fig. 2.8 a schematic view of the acceleration steps is shown. The positively charged particles enter from the left hand side a cavity. The accelerating electric field is applied to the central electrode across the gap. It is generated by an rf voltage of a few tens of kilovolts. As soon as a positively charged ion arrives at the first gap its voltage is driven to a negative value in such a way that the ion is accelerated. While the particle is

moving within the drift tube (which is equipotential) the polarity
is reversed in such a way that when the ion reaches the second
gap the polarity is now positive and the ion is accelerated again
towards the rest tube (which is at ground). If the phase and the
rf frequency are set properly the ion is continuously accelerated
passing through the different gaps. Of course only the ions arriving at the first gap with a negative voltage can be accelerated. All
the other particles are instead lost. Therefore acceleration occurs
in bunches of particles. These bunches typically contain 30% of
the ions injected into the machine.

Moreover acceleration depends on the particle which is in-
jiected. This concept is clearly illustrated in Fig. 2.8 where the
voltage oscillation as a function of time in the resonator cavity C

Figure 2.8   *Steps in the*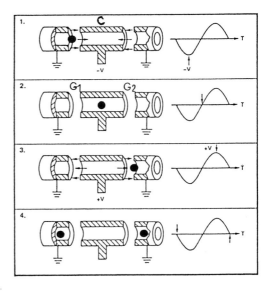
acceleration of a positive
ion. 1) The ion is at-
tracted toward the negative
central electrode across the
first gap, gaining energy.
2) The ion moves in a field
free region while the phase
of the applied voltage ad-
vances. 3) The ion leaves
the central electrode when
this is now positive and it
is pushed across the gap,
gaining again energy. 4)
The process is repeated in
the next cavity (from ref.
2.14).

is shown. If the injected particle is a light one (for instance ^{11}B)
the time required to travel through C from point G_1 to point
G_2 will be short, on the other hand if the particle is an heavy
one (for instance ^{121}Sb) this time will be longer. As a result the
phase of the oscillating voltage should be properly adjusted for the
particle to be accelerated and the energy gain will also depend on
the particle. Typically frequencies of $\sim 5MH_z$ are used to drive
the electrodes. Clearly as the particles are accelerated and their
velocity increases the transit time through the single electrode

decreases. To maintain acceleration possible, as the particle is travelling through the machine, the dimension of the electrodes is therefore varied and the frequency in the last tubes is doubled.

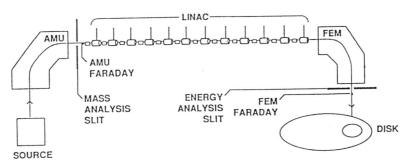

Figure 2.9 *Schematic of an high energy implanter based on the linear acceleration system. The linac has twelve cylindrical electrodes separated by grounded electrodes. Two magnets are indicated, that at the entrance (AMU) and that at the end, final energy analysis magnet (FEM) (from ref. 2.14).*

A schematic picture of the whole machine is reported in Fig. 2.9 [2.14]. It consists of a standard ion source for positive ions, a magnet for mass analysis, a preacceleration stage, the cavities of the linear accelerator, a second magnet which is now acting as an energy analizer and the end station. There are several advantages for this approach to high energy implantation. First of all the maximum voltage to deal with is $80kV$ while implantation can be performed at 1 or 2 MeV. Secondly there is no charge exchange and the ions are always in the positive charge state. This is a major advantage since the final current at the end of the machine is higher than for tandem-type machines, in which charge exchange is needed. Beam current of $\sim 1mA$ are typical in commercial linear accelerators.

2.4 Magnetic Analyzer and Beam Transport

The ions extracted from a source will contain as many different ion species as there are atomic and molecular species in the discharge chamber. Magnetic mass analyzers are used to separate the desired ionic species from all the other charged species due to contaminants from residual air, hydrocarbons from the vacuum pumps, impurities from the solid components of the source and carrier gas. The ions extracted from the source enter a region where a magnetic field normal to their paths is present and are then deflected. The force $\vec{F}_i$ on the ions of charge q_i and velocity v_i is given by the Lorentz formula:

$$\vec{F}_i = q_i(\vec{v}_i \wedge \vec{B})$$ 2.1

where $\vec{B}$ is the magnetic flux density. In a homogeneous magnetic field the charged particles move in a circular path of radius r_i as indicated in Fig. 2.10 for a field directed out of the page. The particles are subjected to a centrifugal force

$$F_{p,i} = \frac{M_i v_i^2}{r_i}$$ 2.2

combining eqs 2.1 and 2.2 we obtain

$$r_i = \frac{M_i v_i}{q_i B} = \frac{(2 E_i M_i)^{1/2}}{q_i B}$$ 2.3

For extraction voltage V, the kinetic energy $E_i = \frac{1}{2} M_i v_i^2 = q_i V$, so that

$$r_i = \frac{1}{B} \left(\frac{2 M_i V}{q_i} \right)^{1/2}$$ 2.4

Figure 2.10   *Schematic of an analyzing magnet deflecting an ion beam.*

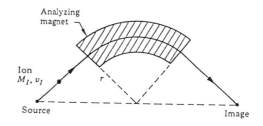

The product $B r_i = (2 M_i V / q_i)^{1/2}$ is called the magnetic rigidity of the particle. Particles with the same charge will have the same energy so that the radius of curvature of trajectories is proportional to the square root of the mass, or for a given acceleration potential V and a radius of curvature r, the magnetic flux density, B, is proportional to the square root of the mass-to-charge ratio. For singly ionized P^+ ions accelerated through 100kV, in a magnetic field of $0.5 W b/m^2$ the radius is

$$r = \frac{1}{0.5} \left(\frac{2 \times 31 \times 1.66 \times 10^{-27} kg \times 10^{5}}{1.6 \times 10^{-19}} \right)^{1/2} = 0.51m$$

The magnet for the analysis is then quite compact if the ion energies or ion masses are not too great. A typical spectrum of the output of an ion source is shown in Fig. 2.11, for solid elemental phosphorus.

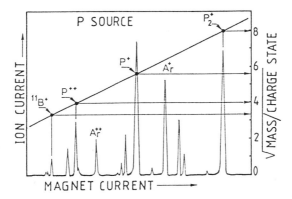

Figure 2.11 *Measured ion current versus the analyzing magnet current for a solid phosphorus source.*

The current intensity of the ions composing the extracted ion beam is plotted versus the magnet current or the magnet field. Several masses are present, single, doubly and triple charged phosphorus ions, molecular phosphorus ions, Argon ions etc. The magnetic spectrum provides useful information on the presence of unexpected peaks and on the ratio between different charge state ions.

It can also be used to determine the mass to charge ratio of the observed peaks. In the spectrum of Fig. 2.11 a vertical scale has been added to the right in units $(M_i/q_i)^{1/2}$. According to eq. 2.4, this ratio is proportional to the magnet field or magnet current. The radius of curvature is now fixed by the analysing magnet as the extraction acceleration potential, so that all the $(M_i/q_i)^{1/2}$ ratios must follow a straight line vs the magnet field. The values of four peaks are plotted at the appropriate location on the horizontal and vertical scales. The straight line through these four points can be used to quantify the mass resolving power of the implanter, i.e. the ability to separate ion masses differing by one or more unit atomic masses. The mass resolution is a fundamental property of the ion source/extraction design and analyzing magnet optic.

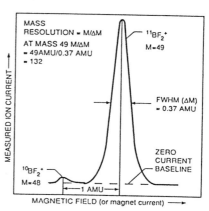

Figure 2.12 *Mass resolution measurement with a magnetic spectrum (from ref. 2.15).*

It can be determined experimentally following the method illustrated in Fig. 2.12 for a BF_3 source gas [2.15]. A small portion of the magnetic spectrum has been expanded near mass 49 ($^{11}BF_2^+$), and the mass resolution is given by the ratio of mass of an ion beam in atomic mass unit (AMU) to the width at half-maximum. In the case of figure 2.12 the mass resolution $M/\Delta M$ is 132. The peak of mass 48 ($^{10}BF_2^+$) is clearly resolved so that only ^{11}B is implanted if mass 49 is selected.

The ion beam in its path from the source to the target interacts with electric and magnetic fields. In addition to increase the energy and to deflect the ions, these fields are used to focus the beam. Focussing is required because the envelope of the ions leaving the source is naturally divergent. The equilibrium shape of the plasma from which the ions are extracted is not a plane so that even at the starting the beam envelope is not cylindrical. The ions extracted from the source have random velocities in direction and in value up to energies of several tens of electron volts. After acceleration the individual ions possess this random velocity component superimposed to the directional velocity, as a result the ion beam has an angular and an energy spread of a few degrees and of a few tens of eV respectively.

The space charge of the beam causes repulsion so that even a initially parallel beam of particles becomes divergent. In all the implanters focussing elements are present along the beam transport system to constrain the particles and to transmit them efficiently to the the end station. The field of ion optic is quite well developped, the shape of the electric and of the magnetic field is determined by computer and several procedures to follow the ion trajectories are available.

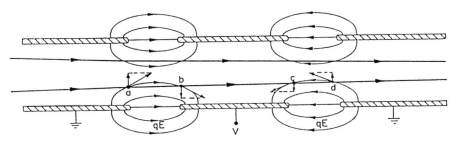

Figure 2.13 *Einzel electrostatic lens. The field lines shown are "lines of force", that is of q_E (from ref. 2.16).*

As an example the operation of the electrostatic einzel lens, used for focussing low energy beams is described. The lens consists of three adjacent electrodes as shown in Fig. 2.13. The outside electrodes are usually connected to the ground potential and the central electrode to a high voltage supply which provides a bias that accelerates or decelerates the incident beam. The beam enters and lives the lens with the same energy. Its operation can be understood by considering what happens to a parallel beam that enters from the left. When the ions arrive in the region **a** they experience a force with a lateral component and get a certain momentum that bends them toward the axis. In the region **b** the ions experience an opposite effect but they have gained energy on going from **a** to **b** so that they spend less time in **b**. The forces are the same but the times are shorter, so the momentum is less. In leaving the high voltage region, the particles get another shift toward the axis. The force is outward in region **c** and inward in region **d**, but the particles stay longer in the latter region, so there is again a net momentum toward the axis. For distances of the charged particles not to far from the axis the total momentum gained by the ions is proportional to the distance from the axis so that focussing takes place.

The focussing operation occurs if the potential of the middle electrode is either positive or negative with respect the other two. The working principle is then to expose the particles for a shorter time to the defocussing field than to the focussing field. In the defocussing region the particles must have higher energies than in the focussing one to stay more in the latter.

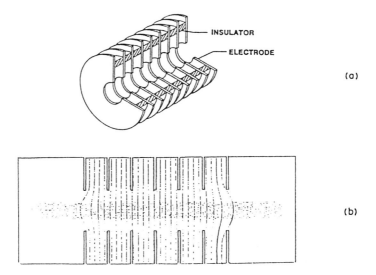

Figure 2.14   *(a) Typical acceleration tube, (b) calculated equipotential surfaces, ion trajectories are represented in the middle (from ref. 2.17).*

The same principle works in a typical acceleration tube (see Fig. 2.14a) made by a series of electrodes separated by insulators. The overall acceleration voltage is distributed over the electrodes. The calculated equipotential surfaces are shown in Fig. 2.14b [2.17]. The accelerating electrostatic fields are at right angles to the equipotential surfaces. At the entrance the surfaces of constant voltage emerge from the end of the tube and the electrical field has a converging radial component with a focussing action. The shape of the field is due to the fact that electromagnetic fields cannot be terminated abruptly but in a well defined and smooth manner giving rise to fringes. In the center of the tube the equipotential surfaces are planes parallel to the electrodes. The particles are subjected to a uniform field and no focussing occurs in it. The exit region is again characterized by fringing fields that give an outward net momentum to the ions. But this defocussing action takes place at high particle energies so that the overall action of the acceleration tube is again a focussing of the beam together with its acceleration. It must be noted that similar focussing action can be obtained by magnetic fields, and the optics of an ion implanter is quite sophysticated and provides beam of suitable shape in both horizontal and vertical directions.

The extraction voltage is of the order of 10-20kV, while the acceleration voltage ranges between 60 and 200 kV according to the implanter.

2.5 Energy Contamination

As shown in Fig. 2.11 multiply charged ions are present in the source and if selected by the magnet can give rise to high energy beam. With a 200 kV implanter the energy of a doubly charged ion beam is 400 keV and of a triply charged ion 600 keV respectively. High energy implants are used for the formation of retrograde well structures and for isolation.

Several problems are associated to the use of multiply charged ions. The yield of high charged ions decreases considerably with the charge state due to the higher ionization potential and to the required number of anelastic collisions with the electrons. The other problem is the energy contamination for the presence in the beam of ions of the same species but of different energies. Energy contamination can result in a degradation of the implant uniformity and in junction depth variations. If the contaminant ions are not scanned over the wafer in the same way that the primary ions are a dose nonuniformity over the implanted area of the wafer will result. Energy contamination in excess of 1-3% causes implant profile distortion with deleterious effects on the device electrical characteristics.

A type of energy contamination occurs when double charged ions as P^{++} are implanted [2.18]. Molecular ions such as P_2^+, extracted from the ion source, can break before to reach the mass analyzer magnet. The process results in a positive ion P^+ and in atom P with an energy of one quarter of the energy gained by the double charged ions P^{++}. The curvature radius of P^{++} is given by $r_{P++} = \frac{1}{B} \left(\frac{2MV}{2q} \right)^{1/2} = \frac{1}{B} \left(\frac{MV}{q} \right)$ and that of P^+ by $r_{P+} = \frac{1}{B} \left(\frac{2MV}{2 \cdot q} \right)^{1/2}$ with V the extraction potential and q the electronic charge. The two radii or the two magnetic rigidities are equal so that the dissociated P^+ ions undergo the same deviation as the P^{++} beam. The spurious beam can be eliminated by the use of an electrostatic filter, like the einzel lens or an electrostatic mirror, i.e. a lens biased at a suitable positive potential to repel the P^+ ions coming from the dissociation of the molecular P_2^+ in the path between the ion source and the magnet. For istance if the extraction voltage is 40kV it is enough to set the bias of the electrostatic mirror at a positive voltage higher than 20 kV to eliminate the P^+ component from the analyzed beam.

The large intensity of the molecular ions P_2^+, P_3^+ and P_4^+ with respect to the desired atomic ion P^+ is responsible of contamination even in the implant of P^+. If P_4^+ molecular ions breakup in the beam line between the source and the entrance to the analyzing magnet in P_2^+ and P_2 the P_2^+ molecules have half the energy (one quarter the energy per atom), and the same magnetic rigidity as the main P^+ beam. They therefore pass through the analyzer slit and can be implanted into the wafer.

Similar energy contamination has been found also in arsenic implantation [2.19]. The molecular ions As_4^+, As_3^+ and As_2^+ are identified and the presence of low energy As_2^+ break-up product from As_4^+ was detected. The amount of contaminants depends on the working conditions of the ion source. The beam filter is located just after the exit slit of the magnet.

The filter is ineffective for other single charged ions generated elsewhere in the beam through charge exchange with a residual gas molecule, A. As an example the following charged phosphorus beam reaction $P^{++} + A \rightarrow P^+ + A^+$ can occur for a double charged phosphorus beam. If the capture of the electron by the double charged ions occurs after the magnet analyzer but before receiving their full acceleration voltage, they have lower energy than that of the main beam. A similar effect occurs with single positive ions that become neutrals after charge - exchange. The charge -exchange process increases with the residual gas pressure, with the beam path length and depends on the collision cross-section. In formula the yield of the process is given by $Y = 3.3 \times 10^{16} pL\sigma$ where p is the pressure in Torr, L the lenght traversed by the beam, in cm and σ the collision cross-section of the order of $10^{-16} - 10^{-15} cm^2$. For a residual pressure of 5×10^{-6} Torr, a path length of 2m and $\sigma \sim 10^{-15} cm^2$, [2.20] Y amounts to 0.033. The amount of contamination is then 3.3%.

A contamination at high energies is present also in boron implants using BF_3 as a feed material. This gas presents several advantages, high beam currents are easily achieved, the boron energy can be lowered by implanting the molecular BF_2^+ species so that shallow depth distribution can be obtained, the damage is quite consistent, amorphous regions are formed and the residual extended defects are reduced. Dissociation of the BF_2^+ molecule after magnetic analyzer but prior to final acceleration produces boron ions with energy higher than that of the boron reaching the target as BF_2^+ [2.21]. A deeper tail on the primary profile is introduced. When the BF_2^+ molecule hits the target surface it immediately breaks up into its component and the energy, E, is divided among the different atoms, according to the relative mass. $E_B = \frac{M_B}{M_{BF_2}} E = \frac{11}{49} E$. Energy contamination is also present when the beam suffers a deceleration instead of an acceleration stage. Modern and industrial implanters have now developed a quite sophysticated system of energy filter just before the wafer to avoid or to reduce considerably the energy contamination. In these implanters a single wafer at time is implanted and the contamination is eliminated or reduced by electrostatic field normal to the beam direction.

2.6 Scan System and Current Measurement

The main advantage of the ion implantation technology over the chemical introduction of dopants is that it relies on an electrical approach to monitoring both the total dose and the dose rate. Electrical current can be easily measured to an accuracy of a tenth of a percent or less and can be measured over a very large dynamical range, from nanoamperes to amperes. Let's consider how this is accomplished in the implanters. After the acceleration the beam is ready to hit the wafer. The beam size is of the order of a few square centimeters and it must be scanned over the wafer $\sim 20cm$ in diameter to distribute uniformly the dopant.

The scan is performed usually by electrostatic fields or by mechanical movement of the target, or by a combination of both. In the first case (see Fig. 2.15a) the ion beam traverses a region in between two systems of plakets where a sawtooth voltage is applied. The resulting electric fields in the x and y directions sweep the beam over the wafer. One of the two scan speeds is faster than the other of a factor between 10 and 100. The ratio between the two frequencies should be such to avoid Lissajous patterns.

A good beam overlap on successive passes must be obtained. In the pure mechanical scan the wafers are placed on a rapidly spinning disk (see Fig. 2.15b). The slow scan is achieved by translating the disk assembly. This system will be described later in details. The hybrid system is shown in Fig. 2.15c, the beam is swept electrostatically in the x direction and mechanically in the y direction. The first and the third system are used in a serial implanter where a single wafer at time is implanted. The second system is used in batch machines where several wafers (~ 24) are implanted simultaneously at high current in the 10mA range. The

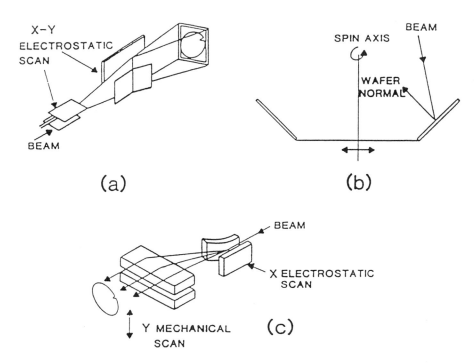

Figure 2.15   *(a) Electrostatic scanning system in both X and Y directions, (b) Mechanical scanning system, (c) Hybrid system: electrostatically along X and mechanically along Y (Varian Manual).*

serial approach is possible for low and medium current implanter.

In some implanters the beam before to traverse the scanning region is deflected electrostatically of a few degrees to eliminate the neutral component formed by charge-exchange mechanism in the beam line. These neutrals not only invalidate the correlation between integrated current and total dose, but for those systems which use electrostatic scanning they can introduce dose non-uniformity.

The measurement of the dose is made usually by placing a current integrator between the target and ground and counting the total collected current from the incident ions. The dose is given by

$$dose = \frac{Q_{tot}}{q_{ion}} = \frac{1}{q_{ion}} \int_0^T I_B dt \qquad 2.5$$

or

$$\frac{dose}{cm^2} = \frac{1}{A(cm^2)} \int_0^T \frac{I_B}{q_{ion}} dt \qquad 2.6$$

One assumes also that $I_B \equiv \frac{dN_{dop}}{dt}$, i.e. the measured beam current $I_B(A)$ applied for time T(seconds) reflects the dopant-material arrival rate, of surface density N_{dop}.

The aforementioned simple relationship does not hold when:

 (i) the incoming beam consists of a mixture of ion species,
(ii) the ions interact with the target giving rise to several secondary processes,
(iii) the Faraday cup is not properly designed.

When the ion beam impinges upon the target, secondary electrons, sputtered and ionized atoms, photons, and adsorbed gas molecules are emitted. The yield of secondary electrons ranges between 2 and 20 per incident ion. The meter measures then two currents, the impinging ion beam and the escaping secondary electrons. As a consequence the measured current is significantly larger than the ionic current.

Two techniques have been developed to prevent the escape of the secondary electrons. They are based on the presence of an electric or magnetic field between the target and the beam line to suppress the transmission of unwanted particles. The electric field suppression is made by a suitable design of a Faraday system. In the internal system the wafer and its holder are mounted within the Faraday cup itself. In the esternal Faraday system the dose measurement is performed outside and away from the wafer holder [2.22].

The structure of a Faraday cup with an internal wafer is shown in Fig. 2.16a for a scanned beam.

The electron component of the beam is eliminated at the entrance by a short electrode biased at $-500V$. A grounded aperture in front of it defines the target area scanned by the beam. A second larger electrode, biased at -100V forms the cup walls. Both electrodes are connected to the wafer target. The suppression works if the field provided by the bias voltage penetrates the center of the bias electrode. At high beam current the potential of the beam itself may shield the suppression field and a computer analysis is required to design properly the Faraday system. In the external Faraday system, four Faraday cups are located in the overscan region of the beam, ahead of the wafer (Fig. 2.16b).

The suppression of secondary electrons can be also achieved by a magnetic field as shown schematically in Fig. 2.17.

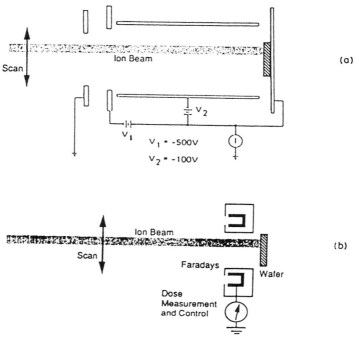

Figure 2.16 *(a) Faraday cup for an internal wafer, (b) Wafer external to Faraday cup (from ref. 2.22).*

Figure 2.17 *Suppression of secondary electrons by magnetic field (from Eaton Manual).*

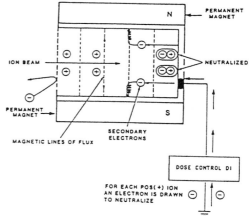

The field prevents either the escape of secondary electrons or the entrance of externally generated particles, as electrons that neutralize in part the beam. Even in this case the electric field $\mathcal{E}$ creates by a high current beam can combine with the orthogonal magnetic field, B, to produce a null force on the electrons of velocity $\mathcal{E} \cdot B$. Several problems arise in high current machines. The mechanical scan velocity is about 10^3 cm/sec, i.e., 1/10 of the electrostatic scanning.

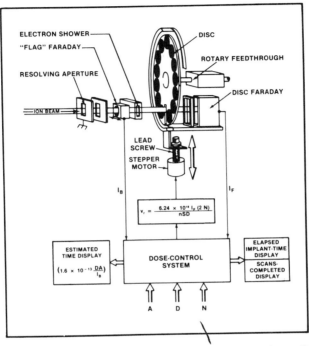

Figure 2.18 *Schematic representation of complete dose control disc system (from ref. 2.23).*

A scan system for a high current machine is shown in Fig. 2.18 [2.23]. The measurement of current is made through a precision slot of constant width in the disc. The transmitted current signal will vary inversely with the speed at which the slot passes through the beam (α 1/WR) and directly with the instantaneous current intensity (α I). The scan speeds are not uniform in either axis. The tangential velocity (fast can) varies directly with R and the radial velocity (slow scan) must therefore be varied inversely with R to achieve a uniform rate of areal doping. The beam signal is not measured on the wafer surface and the dose error due to variation in surface composition and secondary effects is eliminated. The slot technique provides on-line control of doping uniformity.

The entire beam path in front of the wafers is free of electrostatic fields (Fig. 2.17). The suppression in the Faraday cup is realized by a magnetic field. The wafers can also be flooded with low energy electrons from "an electron shower" to inhibit surface charging. The implant time, T, for each batch is given by the following equation $T = 1.6 \times 10^{-13} DI^{-1} A$; D the dose in ions/cm^2, I the beam current on target in μA and A the total implant area in cm^2. The throughput can reach 300-400 wafers per hour. In a recent design [2.24] the wheel can be tilted and twisted at varying angles thus allowing a better control of the wafer orientation with respect the beam incidence (see Fig. 2.19).

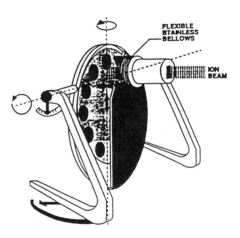

TWO AXIS TILT

Figure 2.19   *Disc scanner system with two-axis tilt (from ref. 2.24).*

Implants can be then performed at well defined angles and these new achievements are already used in the fabrication of advanced devices. Systems are also available that allow the implants to be performed with a parallel beam. A better uniformity is reached when wafer tilt-angle implants or channeling implants are performed.

As a result of several studies and designs it is now possible to achieve routinely in industrial implanters dose uniformities of $0.5\% - 1\%$. The absolute value of the dose is instead subject to a much higher uncertaines, it may reach $10\% - 15\%$ among the different implanters. This is due to the influence of the pressure on the Faraday systems and of charge-exchange processes after magnetic deflection.

2.7 Wafer Cooling

The energy of the beam that strucks the target is dissipated into heat. The wafer temperature rises during the implant and a cooling system is necessary to maintain the temperature at a suitable low value. The dopant is introduced in selected areas through the opening of windows in the photoresist layer. Photoresist is damaged during implantation and if the temperature is excessive it will blister and flake off the wafer. Chemical reactions can occur at temperatures above $100^0 C$ and make very difficult the removal of photoresist after the implant.

The disorder created by the implant depends crucially on the substrate temperature. For a given ion energy and fluence an increase in the target temperature from liquid-nitrogen temperature to a few hundrends of degrees above room temperature may lead, after the bombardment, to a variety of defect structures spanning from amorphous zones to a highly defective crystal. The mobility of defects play a key role in these processes. If the temperature rises during the implants the formation of amorphous might be prevented at high dose with the occurrence of dynamical annealing. Channeling can take place and the resulting ion profile will depend, for a given dose, on the substrate temperature.

Consider now briefly the physics of the wafer heating and cooling processes. As the beam sweeps across the implanted wafer, the wafer surface immediately beneath the beam is heated since the beam energy is converted into heat essentially instantaneously. Heat is in part removed by black body radiation, by diffusion through the silicon wafer thickness and if a cooling system is provided, such as gas or elastomeric material, by conduction to the cooled metal target holder. The remaining part increases the wafer temperature. Consider a serial implanter, where a wafer at time is implanted and suppose that the period of the scan system is small with respect the thermal diffusion time through the wafer of thickness L. The energy is deposited in a layer of thickness much smaller than that of the wafer so that one may assume a surface heating at an average power flux.

The heat flow equation is given by [2.25]

$$LC_p\rho\frac{dT_w}{dt} = \frac{P_B(t)}{A_s} - h(T_w - T_{wH}) - 2\sigma\epsilon_w(T_w^4 - T_s^4) \qquad 2.7$$

where C_p is the specific heat, ρ the density, P_B the beam power, A_s the area scanned by the beam, h the cooling coefficient and ϵ_w the wafer emissivity, T_{wH} the temperature of the wafer holder and T_s the temperature of the surrounding ambient in which the implant is performed, σ the Stefan-Boltzmann constant, $= 5.67 \times 10^{-12} W/cm^2/K^4$. The factor 2 takes into account the front and the back surface of the wafer.

For Si $C_p = 0.6$ J/g, $h \simeq 20 mW/cm^2 \cdot ^0 C$, $\epsilon_w \simeq 0.35\text{-}0.5$.

The term in the left hand side of eq. 2.7 is the amount of heat to rise the temperature of ΔT in the time Δt per unit area of the wafer. This power per unit area is equal to the incident beam power per unit area minus the heat conducted away by radiation and by conduction. In a vacuum conductive cooling is not easily achieved. The thermal contact between the back surface of the wafer and the holder is in any cases limited to a few points so that the main mechanism of loss occurs by black-body radiation.

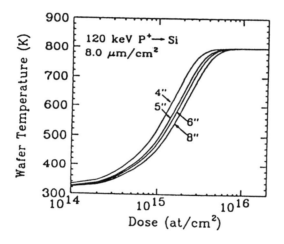

Figure 2.20 *Temperature rise vs dose assuming only black-body radiation as loss mechanism for a serial implanter and for different wafer sizes and thicknesses.*

The solution of eq. 2.7 can be performed numerically and some examples are given in Fig. 2.20 where the temperature rise is plotted versus dose for different wafer sizes. Each size has a particular thickness. For istance the curves of Fig. 2.20 were calculated assuming the following relationships: 4"(10 cm in diameter), L=500μm; 5"(12.5 cm), L=600μm; 6"(15 cm), L=650μm; 8"(20 cm), L=730μm. At high dose above $10^{15}/cm^2$ a steady state is reached and the impinging power is dissipated only by black body radiation i.e. $2\sigma\epsilon_W(T^4 - T_s^4) = P_B/A_s$. Temperatures above $300^0 C$ can easily be achieved.

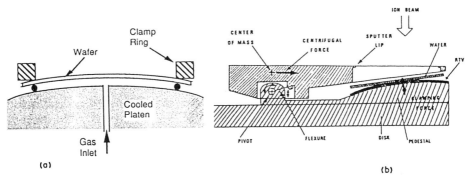

Figure 2.21   *(a) Gas cooling platen design (from ref. 59); (b) elastomery (RTV silicone compound) as a conductive cooling in a full-ring centrifugal clamp (from ref. 2.22).*

Two cooling vacuum systems have been developped: gas cooling [2.26] and contact cooling. The first is used mainly in serial implanters. The wafer is sealed by an O-ring against a platen (see Fig. 2.21a) and a gas is introduced behind the wafer. The gas provides a thermally conducting path between the wafer and the cooled platen. The cooling coefficient h is about $10 mW/cm^2 \cdot^0 C$ for nitrogen at a pressure of 5-10 Torr and it ranges from 10 to 30 $mW/cm^2 \cdot^0 C$ for helium in the pressure range 3-10 Torr.

In the elastomeric cooling the cooled metal platen is covered with a thin layer of elastomer. This elastomer under pressure follows the roughness of the back surface and allows a much larger contact area [2.27]. In the batch implanters elastomeric cooling is adopted, and the contact pressure during the implant is provided by the centrifugal force of the spinning disk, (Fig. 2.21b). For both cooling systems the heat dissipation by radiation is negligible so that the heat equation becomes

$$\rho L C_p \frac{dT_w}{dt} = \frac{P_B}{A_s} - h(T_w - T_s) \qquad\qquad 2.8$$

again P_B/A_s represents the average power flux on the wafer and the overall scan period is short compared to the cooling decay time $\tau = \rho L C_p/h$, for conduction from the back of the wafer to the cooled platen [2.22].

The solution of eq. 2.8 is given by

$$T(t) - T_s = \frac{1}{h}\frac{P_B}{A_s}\left(1 - exp -\left(\frac{t}{\tau}\right)\right) \qquad\qquad 2.9$$

At t=0 $T(t)=T_S$, and if the total implant time is long compared to the cooling decay time a steady state is reached and the maximum temperature rise $(T_{max} - T_s)$ is given by $\frac{1}{h}\frac{P_B}{As}$.

In high current implanters with a beam current of 25 mA and energies up to 200 kV a total beam power of 5000 W is delivered to the spinning disk. The scan areas are generally large, as high as $7000cm^2$ for 200 mm wafers. The average power flux is then $0.7W/cm^2$ and with a cooling coefficient of $10mW/cm^2/^0C$, the steady state temperature rise is 70^0. Even at this high power the temperature is below 100^0C and the photoresist is not considerable altered. The damage, as we will see in chapter 4 is instead influenced by this relatively slight increase in temperature.

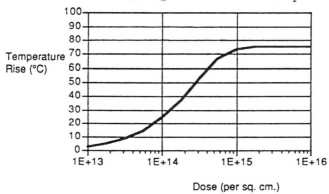

Figure 2.22   *Temperature rise vs dose for a serial implanter: $I = 1.5mA, V = 200kV, A_s = 160cm^2$ and $h = 25mW/cm^2/^0C$ (from ref. 2.22).*

The conditions of eq. 2.9 are also verified by serial implants, although a higher average beam flux is reached. For istance considering a beam current of 1.5mA and a voltage of 200kV with a scan area of $160cm^2$ the average beam flux is $1.88W/cm^2$. Eq. 2.9 is plotted in Fig. 2.22 for this case assuming $h = 25mW/cm^2/^0C$. The abscissa has been changed from time to implant dose. The graph clearly show the saturation in temperature rise for doses in excess of $10^{15}/cm^2$. In the low dose region $(t << \tau)$ $T(t) - T_s = \Delta T(t) \simeq \frac{P_B t}{A_s \cdot L C_p}\rho$ the temperature rise is independent of the cooling coefficient and is proportional to the total energy deposited per unit area divided the total specific heat per unit area.

The wafer temperature is measured by infrared radiometry or by temperature stickers. The first method is subject to the wide variation of the emissivity with the surface material covering the wafer and it needs a calibration. The second is based on the change of colour of wax following its melt. The melting points differ for the different waxes in the range $50-150^0 C$. The problem with the stickers is their location. They cannot be heated by the beam and if positioned on the back front they cannot be placed between the wafer and the platen.

2.8 Wafer Charging

Implants are usually performed on structured wafers with insulating layers such as oxide, nitride and photoresists. The building-up of charges transported by the beam at these layers can be detrimental because of damage and of electrical break-down. Wafer charging can also affect the beam propagation and may cause a blow up of it with a deterioration of the uniformity [2.28]. To illustrate the phenomenon consider for simplicity the voltage reached by an insulating layer of area A and thickness L in a single pass of the beam (Fig. 2.23a).

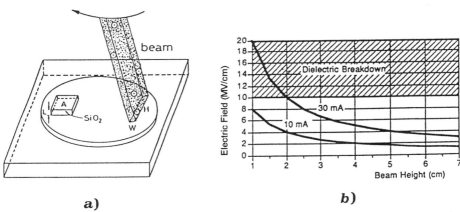

a)   **b)**

Figure 2.23 *Schematic of the geometry adopted for calculation (a), electric field reached by an insulated device as a function of beam height H for a spinning disk system of high current implanters (b) (from ref. 2.22).*

If the emission of secondary electrons is neglected the charge Q transferred from the beam of height H and width W to the device is given by $\int i(t)dt$ where $i(t)$ is the net current, i.e. the positive ion beam and the negative electron beam: it is then given by $i = I_B(1 - \beta)\frac{A}{H \cdot W}$, being β the neutralization factor, and A

the area of the device smaller than that of the beam cross section $H \cdot W$.

The charge Q produces in the device a voltage $V = Q/C$, where C is the capacitance of the device, $C = \frac{\epsilon_r \epsilon_0 A}{L}$ with ϵ_r the relative dielectric constant of the insulator. If the wafer is placed in a spinning disk end station the exposure time T to the beam is $T = \frac{W}{2\pi R\nu}$ where R is the radius of the implant site on the disk and ν the frequency. The electric field, called in this case dielectric stress, is given by

$$\mathcal{E} = \frac{V}{L} = \frac{I_B(1 - \beta)}{2\pi \epsilon_r \epsilon_0 H R\nu} \qquad\qquad 2.10$$

The field in this simplified model is independent of the device size and thickness. The breakdown field of a good thermal silicon dioxide is $\sim 10MV/cm$, and according to the graph of Fig. 2.23 it is reached after 5 passes at H=4cm and $I_B = 10mA$ or 2 passes at $I_B = 30mA$ (the neutralization fraction was assumed zero). Although the previous relation has been obtained in an oversimplified model and several relevant surface phenomena were neglected, it provides a useful guide line for the dependence of the wafer charging on the main parameters.

The increase in the beam height of the low scan direction, in the disc radius and in the rotational frequences will reduce charging effect as the increase in the neutralization fraction β. This is the most important parameter because can be varied of a factor ~ 50 between a partially neutralized beam and a nearly completely neutralized beam.

Beam neutralization occurs through ionization of gas molecules present in the beam line, but very little electron current is available at normal implant pressures. Electrons can be with drawn from the beam by the wafer to maintain its neutrality. In implanters where the Faraday apparatus is external to the wafer, the beam can be unneutralized for a long distance. The positive ions of the beam are not any more screened by the electrons and they repel each other.

The beam diameter increases by this kind of dynamic expansion. Soon as it will hit the metal holder the large number of secondary electrons will neutralize it and a beam contraction occurs. As a result of the dynamic expansion and contraction the dopants is distributed non-uniformely over the wafer with a lower dose in the center. In some cases the sheet resistivity may vary of 20% or more. In the case of internal Faraday geometry where the wafer is inside the Faraday cup, electrons are prevented to entry the cup and then the path length of the unneutralized beam must be quite short to reduce the dynamic expansion effect. In this case the intrinsic charging of the wafer is quite relevant.

For what said it is necessary to supply electrons to the wafer during implant to neutralize the positive beam charge. This can be done by flooding electrons directly to the wafer or to the beam. The electrons used for wafer neutralization are usually secondary electrons of low energy to reduce the damage. This is just one of the possible techniques for beam neutralization [2.29].

Secondary electrons showers operate in the manner depicted in Fig. 2.24. Energetic electrons are emitted by a thermoionic filament and strike metallic surface to produce a shower of secondary electrons. If the wafer is implanted within the Faraday apparatus, the shower must be mounted inside and electrons cannot escape from it. In the apparatus of Fig. 2.24a adopted in the Varian E1000 [2.30] the escape of electrons from the Faraday cup is prevented by the electric field at the entrance and by a short range magnetic field between the disk and the Faraday block. A different approach, illustrated in Fig. 2.24b is used for the Eaton NV20A [2.31]. The electrons are emitted by a thermoionic filament biased at 300V. The electrons striking the grid create secondary electrons that cannot reach the external housing due to the presence of a -12V potential between grid and housing. The primary electrons are then reflected back and forth within the housing until they hit the grid or collide with gas molecules creating secondary electrons. The neutralizing electrons are carried out from the beam to the wafer. In a serial machine the wafer charging is not a serious problem because of lower current and higher scan speed. In some cases (Fig. 2.24c) the defining aperture of the scanning system is used as a target for the emission of the secondary electrons that being very near to the wafer can ensure a complete neutralization.

The influence of the wafer charge on the device yield is more pronounced for thin oxide layer, $\sim 10nm$, as adopted in VLSI and ULSI. The structure to determine the effect of the charge is usually a MOS capacitor. After implantation two different tests are performed: breakdown voltage and J-T testing. In the first the voltage across the capacitor is increased in time until the current through the device exceeds a fixed value, the failure point, usually $1\mu A$. The voltage at which the current reaches this value is the breakdown voltage (in a good silicon oxide it is of the order of 10MV/cm). In J-T [2.32] the current through the capacitor increases with time and failure is indicated by a sudden drop in the voltage. This technique measures the charge, as the time integral of the current density, to breakdown, Q_{bd}. The changes of

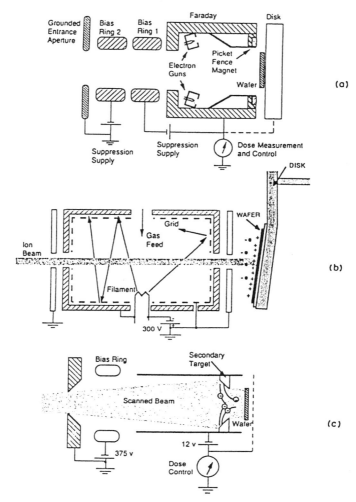

Figure 2.24 *Electron shower apparatus adopted in some industrial implanters: (a) Varian E 1000 batch system, (b) Eaton - NV20 A batch system, (c) Eaton NV 6200/6200 AV serial system (from ref. 2.22).*

V_b and Q_{bd} with the implantation process are related to the wafer charging and to the damage of the oxide.

2.9 Uniformity Control and Mapping

The large scale application of the ion implantation process is based on the accurate control of the dopant over the wafer. The uniformity of the implant over the wafer and the reproducibility of the dose from wafer to wafer and from batch to batch are the main issues in an industrial environment, being directly related to the device yield and reliability.

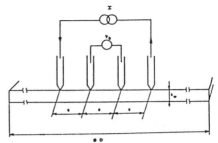

Figure 2.25 *Four-point probe to measure sheet resistivity in a thin sample.*

Several tools have been developed to characterize the implanted wafers, and among them only two will be described briefly in the following: i.e. measurements of sheet resistance on many regions over the wafer after thermal annealing or of the lattice damage in the as-implanted wafer.

The physical principle of the sheet resistance measurement by four point probe is illustrated schematically in Fig. 2.25. Four point contacts are made to the thin implanted doped layer whose conductivity is opposite to that of the substrate. Current is carried through the outer two contacts and the voltage drop is measured across the internal probes. The sheet resistance is given by

$$R_S = K \frac{V}{I} \qquad\qquad 2.11$$

where K is a constant that depends on the configuration, position and orientation of the probes. Considering a common case where the probes are equally spaced, of about 250 μm, in a straight line and the probe spacing is small compared to the dimensions of the layer but large compared to the conductive layer thickness t. In the previous section the thickness was indicated by L to avoid overlap with time t. In this section we prefer to use again t for thickness and ρ for resistivity. The meaning of the symbols is quite straighforward. Under these assumptions the current entering from the contact is spreaded inside the film and the current density J, is constant over the lateral surface of cylinder of radius r, $J = \frac{I}{2\pi r \cdot t}$; the electric field is given by $\mathcal{E} = \frac{J}{\sigma} = \rho J = \frac{\rho I}{2\pi r t}$. The voltage drop between the two inner contacts (2 and 3) is then

$$V_{3,2} = -\int_{Lp}^{2Lp} \mathcal{E} \cdot dr = \frac{\rho}{t} \frac{ln2}{2\pi} \qquad 2.12a$$

to this an equal contribution must be added due to the current (-I) entering from contact 4. The relationship becomes then

$$\frac{\rho}{t} = \frac{V}{I} \frac{\pi}{ln2} = 4.53 \frac{V}{I} \qquad 2.12b$$

The value of K in this case is 4.53 and the ratio ρ/t becomes the sheet resistance R_S. A conducting thin layer of length H (in the direction of the current), width W and thickness t has a resistance R,

$$R = \frac{\rho}{t} \frac{H}{W} \qquad 2.13$$

If $H = W$, i.e. for a square, the resistance is given by $\frac{\rho}{t}$. It is called sheet resistance, has the units of ohms, but it is convenient to refer to is as ohms per square ($\Omega/\square$). The previous relationship assumes a uniform doped layer of resistivity ρ/t; in the case of implanted dopant the value of ρ is given instead by

$$\frac{\rho}{x_j} = \frac{1}{\int_0^{x_i} q\mu(x)c(x)dx} \qquad 2.14$$

where x_j is the junction depth, instead of the thickness t, $c(x)$ the electrically active dopant concentration, μ the carrier mobility depending on the concentration of the implanted species and on the damage. An approximate value of ρ can be obtained by using an average value $\overline{\mu}$ of the mobility. A simple and reasonable criterion consists in selecting a value of mobility corresponding to half the maximum dopant concentration. The sheet resistance becomes

$$R_S = \frac{\rho}{x_j} = \frac{1}{q\overline{\mu} \cdot N_\square} \qquad 2.15$$

where $N_\square$ is the electrical implanted dose. The sheet resistance, for the same implanted dose, decreases with increasing the average mobility value, i.e. with the annealing. The value of R_S ranges usually between $10^3 \Omega/\square$ to $10\Omega/\square$. Automatic apparatus are now available to perform sheet resistance measurement on the whole processed wafers. Sheet resistance probes are used to characterize implant doses ranging from $2 \times 10^{11}/cm^2$ up to $10^{17}/cm^2$.

The low dose regime from $2 \times 10^{11}/cm^2$ to $5 \times 10^{12}/cm^2$ requires a high resistivity substrate and a surface chemical treatment with sulfuric peroxide to reduce the native oxide charges that interfere with the measurement.

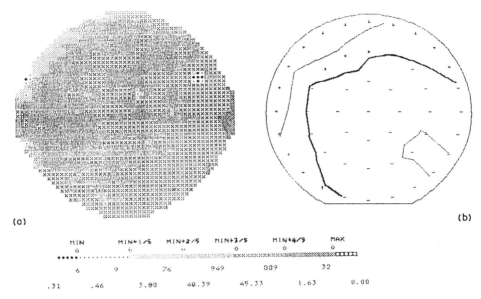

(a)   (b)

Figure 2.26 *Graphical representation of the sheet resistance distribution values by (a) different degree of shades and (b) by contour map. The analysis reported in (a) refers to a 5" wafer implanted with $80keV - 10^{14}/cm^2 B^+$ and annealed at $900^0 C \cdot 1/2hr$. The analysis reported in (b) refers to a 5" wafer implanted with $70keV - 2 \times 10^{15}/cm^2 As^+$ ions and annealed at $900^0 C \cdot 1/2hr$.*

The sensitivity of any implant measurement is defined as the percent change in the measured parameter divided by the percent change in dose. The sensitivity of directly measured sheet resistance is about 0.7 - 1 (see eq. 2.15).

Several graphical representations are used to show the uniformity map in a clear fashion. The range of measured values is divided into a small number of intervals and the values are displayed with a range of shades. An example is shown in Fig. 2.26a for a 5" wafer implanted with 80 keV $\cdot 1 \times 10^{14}/cm^2 B^+$ ions and annealed at $900^0 C \frac{1}{2}$ hr. The sheet resistance values are displayed using five different shades between the minimum and the maximum value. The figure reports below the shades the number of dies measurements. The statistical analysis provides the mean value and the standard deviation. In the case of figure the average value is 471.1 $\Omega/\square$ with a $\sigma \simeq 0.8\%$. A look at the map gives information on the non uniformity and on several problems of the implanters. The other common way for display implantation data is a contour map, where measured data from all the test sites are connected to show the location of the average value (darker line) and the locations of measured values that differ from the average value by a selected scale, in terms of a percent variation either above (+) or below (-) the average. An example is shown in Fig. 2.26b for a 5" wafer implanted with 70 keV - $2 \times 10^{15} cm^2 As^+$ ions and annealed at $900^0 C \frac{1}{2}$ hr. The interval is 1% and the mean is 70.4 $\Omega/\square$ with $\sigma \simeq 0.9\%$

At dose below $2 \times 10^{11} ions/cm^2$ the double implant technique is used [2.33]. The wafer is at first implanted to a dose 50-100 times that of the implant to be measured. The initial implant is followed by an anneal and by four-point-probe measurement. The second implant at low dose damages the sample and the sheet resistance in the initial layer increases. The data before and after the second implant are treated by computer to determine the uniformity of the low dose implant. This technique requires a calibration curve and it suffers from the annealing of defects that can occur even at room temperature and of the interaction of the defects with the preexisting dopant atoms.

The second common technique is based on the optical-thermal response of the layer damaged by the implant [2.34, 2.35]. The technique is contact less, no destructive and does not require thermal treatment. It can map samples implanted over the dose range $10^{10}/cm^2$ to $10^{16}/cm^2$ [2.36]. In the commercial Therma-wave system the change in the reflectance is monitored [2.37]. An Ar ion "pump" laser produces a localized microscopic perturbation in the sample and a He-Ne "probe" laser measures the optical response of the sample to the perturbation (see Fig. 2.27).

The Ar laser is modulated at a frequency of 1-10 MHz and the response is detected in phase. The absorbed energy produces periodic waves of photo-induced heat (thermal-wave) and of plasma-wave (electron-hole pairs) too [2.38]. The thermal and plasma waves propagate several micrometers($\sim 3\mu m$ in Si) beneath the surface and interact with defects or damage. Because the diffusion of thermal and plasma waves is modified by defects, the point to point surface temperature and surface plasma density vary according to the defect sites, and the corresponding variation of the probe laser reflectance is used as a contrast mechanism for imaging

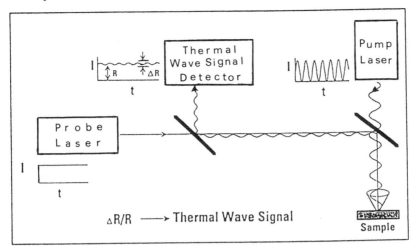

Figure 2.27 *Schematic illustration of a Thermal - wave apparatus.*

the in-depth defects.

The detected signal after processing is reported as a thermal-wave (TW) signal. The change in surface reflectivity is very small ($10^{-5} - 10^{-2}$ range) and requires sensitive detection system. If used to monitor the implant dose the system detects the amount of the crystalline damage which is proportional to the implanted dose.

The spatial resolution depends on the probe spot $\sim 0.8 \mu m$. The measurements are performed and evaluated by computer. The technique needs for each species at each energy a calibration curve to equate damage with dose. The sensitivity ranges from 0.1 to 0.7 depending on the implant damage. Room temperature damage annealing, wafer heating, relaxation and interaction with preexisting dopant must also be considered. The technique can monitore ions not electrically active (as silicon itself in silicon, argon etc.). A series of typical calibration curve are shown in Fig. 2.28a, for boron, arsenic and phosphorus implants and a contour-map is reported in Fig. 2.28b. Usually therma-wave measurements are made for low and medium dose implants up to $10^{15}/cm^2$. Implants at higher doses for heavy ions form a subsurface amorphous layer that causes a nonmonotic behavior of the thermal-wave signal. At high doses the modulated and the direct reflectivity show an interference behavior with increasing amorphous thickness as shown by the peak for *As* and *P* implants in Fig. 2.28a. Recently [2.39] a new method has been adopt: from the optical signal the amorphous thickness is computed by a model and from it, being

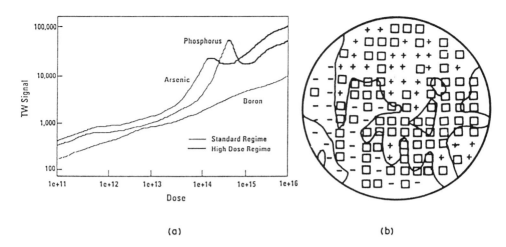

(a)   (b)

Figure 2.28   *(a) Calibration curve of Therma-wave signal for different doses of $50keV B^+$ ions; (b) Therma-wave map of a 8" wafer implanted with $10keV B^+$ at a fluence of $1 \times 10^{12}/cm^2$ (from Therma-Wave Manual).*

the thickness a monotonically increasing function of dose, a simple calibration can be made.

Sheet resistance measurement can be also adopted for standard reference material for ion implantation to calibrate equipment and measurement tools. Among the different species arsenic has been chosen as tested species for its popularity. The following process [2.40] has been proposed: a silicon wafer of 5-10$\Omega \cdot cm$ p-type (100) orientation with a screen oxide 10nm thick is implanted with 80 keV As^+ at a dose of $1 \times 10^{14}/cm^2$. The annealing is performed in nitrogen atmosphere with a 5% dry oxygen at $1000^0 C - \frac{1}{2}$hr. The wafer should have a sheet resistance of 543.1$\Omega/\square$ if measured with a test voltage of 7.5mV.

2.10 Contaminants and Yield

As any other equipment implanters generate particulates that deposited on the wafer surface reduce drastically the yield of devices. Multiple effects occur when a particle comes to rest in an active area of semiconductor. An example is shown schematically in Fig. 2.29 [2.41].

The particle shields the covered portion of the device from the implant. Implantation depths are usually in the 0.05 - 1μm range

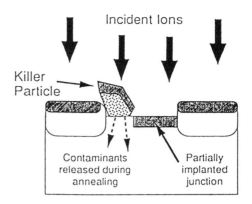

Figure 2.29 *Schematic of a particle acting as a blocking mask and as a contaminant source (from ref. 2.41).*

and the linewidth is also in the 0.5 - 1.0 μm range. A particle of $3\mu m$ (the lowest value that can be measured with reliability today) can block the implant and seriously degrade the performance of the circuit through shorting (or opening) the junction. Secondary effects are due to contaminants released from the particle by sputtering during the implant or by evaporation and diffusion during the following anneal step. These defects are called "killer" because they impaire the device.

With the shrinkage of devices the density of defect must be reduced considerably to maintain a reasonable yield as discussed in the previous chapter in connection of Fig. 1.13 This is true not only for contaminants but also for defects in the silicon and at the silicon substrate - thin layer interface. For yield levels above 50%, in the case of a 4Mbit DRAM, defect densities less than 1 defect $/cm^2$ have to be achieved at the end of the several hundredth process steps [2.42].

Effort is then spent to reduce the amount of particulates introduced by implanters. The goal is 0.05 particles/cm^2 of $0.3\mu m$ or larger to add for a single implant step in the $0.8\mu m$ technology. But for the $0.35\mu m$ technology - 16 M DRAM the target is 0.003 particles/cm^2 of 0.3 μm or larger i.e. one particle per 200 mm wafer diameter per implant step. In a more visible presentation this means a particle of 1 cm on a circle of 6.7 Km diameter.

Several precautions are made to achieve this goal. In the preimplant processing step it is a good practice to eliminate sharp corners where the wafer perimeter and flat meet, and to use implant-masking photoresist more resistant to particle generation. The water vapor must be minimized to avoid their clustering on the photoresist material ejected during implantation. Dirty dummy wafers must be avoided because they are a source of contaminants. Ion damage can release hydrogen from the resist, at relatively moderate implant temperature this hydrogen accumulates as small bubbles that cause exfoliation of the resist material and a shower of particles on the nearby wafer surface.

Particles created in some other parts of the implanter can be transported to the wafer by various means. The process is represented in Fig. 2.30. The arrows pointing toward the ion implanter reservoir identify sources that add particles to it.

Particle Action at Implant

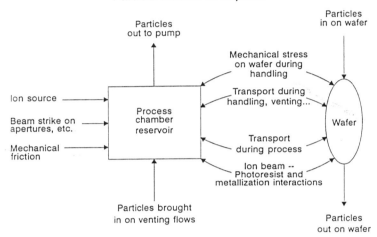

Figure 2.30 *Formation of particle contaminants and their transport to the wafer during an implant.*

Arrows pointing toward the wafer indicate means of transporting particles to the wafer. Probably in the future the miniaturization will require fundamental changes in the implanter control of contaminants. New ways of transporting a beam to reduce contact with implanter surfaces and improvements in the cleaning and handling procedures. As matter of fact beam generated particles in the analyzer flight tube and in the mass slit are the major contributors to ion implanter particulates.

In addition to solid particulates such as dust, silicon particles, resist debris etc. contaminants can be present as heavy metals such as Au, Pt or transition elements. These atoms interact atomically and electrically with silicon. If diffused deep in the substrate they act as recombination- generation center for minority carriers and reduce drastically their lifetime; if migrated at the oxide interface they produce localized charged region, or silicides. Such defects form localized high electric field regions, causing premature breakdown of the oxide under voltage application. Ion implanters are responsible of the largest introduction of heavy metals, expecially Fe up to $10^{15}/cm^3$ [2.43]. These elements are carried by the main beam through sputtering of the stainless steel material that build the implanter. The contaminants are reduced if the beam line is properly designed to avoid that beam hits the walls, or replacing stainless steel with graphite or silicon carbide in the crucial points, or by the use of a thin ($\sim$ 10 - 20 nm) surface oxide layer, so called screen oxide. All of these impurities have a very small range and are trapped in the oxide layer, which can be subsequently removed.

2.11 Plasma Immersion Ion Implantation

A quite different skeme has been proposed recently for the implantation of large area and of shaped material [2.44]. The new doping technique is called plasma immersion ion implantation PIII and has been used initially for metallurgical applications such as tool hardering by nitrogen [2.45], and now in microelectronics [2.46]. At this stage its use is limited to laboratory applications but industrial applications seem feasible through the development and the design of new apparatus.

PIII is similar to a reverse ion sputtering system, being the sputter target replaced by the substrate to be implanted and maintained at tens of kilovolts [2.47]. The PIII apparatus is skematically illustrated in Fig. 2.31a. The substrate is immersed in a high density plasma ($\sim 10^{10} - 10^{11}/cm^3$) produced by efficient ionization sources such as microwave source operating at 21.45 GHz, as indicated in Fig. 2.31b.

The substrate is negatively biased by a direct or by a pulsed voltage. These two operation modes are called dc PIII and pulsed PIII respectively. When an abrupt voltage is applied to the substrate electrons near the surface are repelled, leaving behind a uniform density sheath of positive ions. These ions will be accelerated toward the negatively biased substrate. The ion movement will lower the ion density and the sheath-plasma boundary will expand, other positive ions are extracted until a steady state condition is reached and the current becomes space - charge limited as in a diode (see Fig. 2.31b). The ion current depends then on the time. PIII does not provide ion mass separation and optics for the beam transport, the dose rate can be very large $\sim 10^{16}/cm \cdot s$.

The main use of PIII in microelectronics has been the formation of shallow junctions with depths < 100 nm and with the preservation of ultrathin ($< 10nm$) gate dielectrics [2.48]. These applications are of interest in ultra-large-scale integrated circuit fabrication. In the configuration shown in Fig. 2.31a gaseous

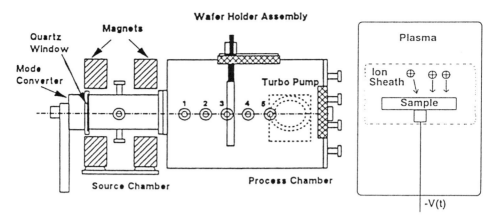

Figure 2.31 *Schematic of the plasma immersion ion implantation apparatus. Ionized gas by radiofrequency source are used for the implant species (from ref. 2.47).*

sources such as Ar, N_2, BF_3 and B_2H_6 can be used to provide the ionization medium. Conformal doping of non-planar device structures such as trenches is also possible by this skeme. The high packing density of devices is made possible by making use of vertical sidewalls for active transistor channels and charge storage elements such as trench capacitor. In a DRAM of last generation to save space the bit information is stored in a capacitor formed by a trench filled with SiO_2, and of high aspect-ratio. The wells of the trench before to be filled must be doped and conventional implantation techniques have focused on multi-step implants with collimated beam at controlled beam incidence angles [2.49]. PIII can replace easily this procedure [2.50]. Several mechanisms have been proposed to explain the conformal doping of vertical structures by PIII; reflection of impinging ions by the internal sidewalls, large angular spreading of the incident ion flux due to ion-neutral scattering, and electrostatic defocusing of ion beams due to protructed surface structures. In the case of shallow junctions sub 100 nm p^+ junction is obtained by a two-step implant. First ions from a SiF_4 plasma are implanted to create a surface layer of amorphous Si. Boron is then implanted using a BF_3 plasma. The two implants can be performed without breaking the vacuum, at extremely low energies (1-6 KV), and with high fluxes.

As a last application we consider the behavior for wafer charging during PIII [2.51]. As discussed in sec. 2.6 the gate oxide during implantation can accumulate enough charge to cause the breakdown of the thin oxide layer. PIII process has a built-in mechanism to neutralize surface positive charge accumulation. Fig. 2.32 illustrates schematically the behaviour of the charge on the wafer during PIII in the pulse mode.

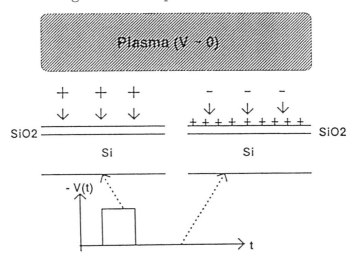

Figure 2.32 *Schematic of the wafer charging sequence with PIII operating in the pulse mode. Electrons are attracted by the surface positive charge on the oxide layer during the off-cycle of pulse (from ref. 2.51).*

The application of a negative voltage to the substrate covered by the thin oxide layer induces a surface negative potential. This negative voltage causes the implant of positive ions that accumulates at the surface of the oxide layer. During the off-cycle of the pulse these positive ions attract the plasma electrons and are neutralized. It is possible also to estimate the duty cycle of the pulse by valuting the maximum positive surface charge, Q_{max}, per unit area acquired for each PIII pulse that induces oxide breakdown.

$$Q_{max} = E_{breakdown}(SiO_2)\epsilon_{0x} \qquad 2.15$$

where $E_{breakdown}(SiO_2)$ and ϵ_{0x} are the breakdown field and dielectric constant of SiO_2 respectively. For a conservative breakdown field of 5MV/cm, Q_{max} is of the order of $10^{13}e/cm^2 \cdot$ pulse, so that at an average dose rate of $10^{15}/cm^2 \cdot s$ a duty-factor less than 1ms should be adopted.

Other applications of PIII are reported in the literature expecially in those areas where the implant depth profile is non critical such as source /drain formation, poly-silicide doping, trench capacitor doping, and multi level metallization.

CHAPTER 3

RANGE DISTRIBUTION

3.1 Introduction

Ions hitting a target penetrate inside, lose their kinetic energy through collisions with the nuclei and with the electrons of the material until they stop. The distribution of the implanted ions depends on several parameters such as the ion mass and energy, the target mass and the beam direction with respect the main axes or planes in the case of a single crystal target as the silicon wafer. The total distance which an ion of mass M_1 travels in coming to rest in a target of atoms of mass M_2 is the range R and is shown in Fig.3.1.

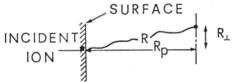

Figure 3.1 *Range R, projected range R_p, and perpendicular distance $R_\perp$ for a single ion incident a target.*

The distance along the axis of incident is called the projected range R_p and the net distance travelled along the axis perpendicular to that of incident is called the perpendicular distance $R_\perp$. Because of the ion scattering, the total path R is greater than the projected range and the following relationship is approximately [3.1] valid between R and R_p for low energy ions $R \simeq [1 + \frac{M_2}{3M_1}]R_p$. The number of collisions experienced by an ion per unit path length and the energy lost per collision are random variables, i.e.

all the ions having the same incident energy do not stop at the same location. There is instead a distribution in space of stopping points which is described by a range distribution function. R and R_p are mean values and the standard deviation, or straggle, in the projected range is ΔR_p. The perpendicular distance from the straight line path or the lateral spreading is characterized by a zero mean value and by a non-zero straggle, $\Delta R_\perp$. The transverse spreading is caused by multiple collisions of the ions and it increases with depth into the target, being a function of the mass ratio M_2/M_1. For $M_2/M_1 \sim 0.1$, projectile mass heavier than the target mass, $\Delta R_\perp \simeq \Delta R_p$ and is some 20% of R_p. At $M_2 = M_1$, $\Delta R_\perp = 0.5\ R_p$ and rises to equal R_p at $M_2/M_1 = 10$, i.e. for projectile ions lighter than the target mass.

The lateral spread is a limiting factor in one of the greatest advantages of ion implantation, i.e. the so called "self-alignment" technique where ions are implanted through a mask that leaves open areas on the wafer. The influence of the lateral spread under a mask is a two-fold, in the amount of parasitic capacitance and in the lower limit of mask opening which determines the maximum device density.

Range calculations require the knowledge of the rate of energy loss of the incident projectiles. In the classical scattering theory [3.2] the interaction of the moving ions with the target atoms is described assuming two separate processes: collisions with the nuclei and collisions with the electrons. The former is due to the coulomb repulsion between the ion and the target nuclei. The collision partners usually approach each other at a distance much less than the interatomic distance in the solid and the probability for three- and more particle collisions is extremely small, thus justifying the validity of the binary collision approximation. Excitations or ionization of electrons are only a source of energy loss, and do not influence the collision geometry. This statement is justified if the energy transferred to the electrons is small compared to the exchange of kinetic energy between the atoms, a condition which is usually fulfilled in ion implantation. The ion is thus deflected by nuclear encounters, and it continuously loses energy to the electrons. Elastic collisions with free electrons, inelastic collisions with bound electrons, inelastic collisions with nuclei, nuclear excitation, nuclear reactions or bremstrahlung radiation and Cerenkov radiation are not significant in the energy and mass regime of ion implantation.

The stopping cross-section or rate of energy loss can be then split into nuclear and electronic stopping

$$-\frac{1}{N}\left(\frac{dE}{dx}\right) = S = -\frac{1}{N}\left(\frac{dE}{dx}\right)_n - \frac{1}{N}\left(\frac{dE}{dx}\right)_e = S_n + S_e \quad 3.1$$

being N the atomic density of the target. It is measured in eV/cm^2.

3.2 Elastic Stopping Power

In a binary collision the interaction potential between a projectile and a target atom used for the scattering problem is usually assumed for practical purposes to be dependent only on the distance r between the too atoms. If the elastic collision occurs with a target atom initially at rest conservation of energy and momentum yields a transferred energy

$$T = \frac{4M_1 M_2 E}{(M_1 + M_2)^2} sin^2 \frac{\theta_c}{2}. \quad 3.2$$

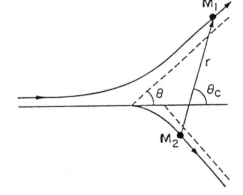

Figure 3.2 *Dynamics of the collision between a projectile of mass M_1 and a target atom of mass M_2.*

E is the initial energy and θ_c the scattering angle in the center of mass (see Fig. 3.2). The maximum energy transfer $T = T_m$ is obtained in a head-on collision ($\theta_c = 180^o$). For istance a 100 keV boron ion transfers to a silicon atom 81 keV in a head-on collision while a 100 keV arsenic ion 79 keV. The scattering angle θ_c is related to the interatomic potential V(r) through [3.3]

$$\theta_c = \pi - 2p \int_{R_{min}}^{\infty} \frac{dr/r^2}{\sqrt{1 - V(r)/E_r - p^2/r^2}} \quad 3.3$$

where p is the impact parameter, R_{min} the minimum distance of approach, and E_r the energy in the center of mass system equal to $M_2 E/(M_1 + M_2)$. The probability for scattering in a direction θ_c is given by the differential cross section $d\sigma(\theta_c)$

$$d\sigma(\theta_c) = 2\pi p dp = -2\pi p \left[\frac{dp}{d\theta_c}\right] d\theta_c = -\frac{p}{sin\theta_c}\left[\frac{dp}{d\theta_c}\right] d\Omega. \quad 3.4$$

where $d\Omega = 2\pi \, sin\theta_c d\theta_c$ is the solid angle and $dp/d\theta_c < 0$. The form of the repulsive potential $V(r)$ is of critical importance in range calculations. The potential used in essentially all the calculations is a screened coulombic potential due to the screening of the bare nuclei of the ion and target atoms by their electron clouds

$$V(r) = \frac{Z_1 Z_2 q^2}{4\pi\epsilon_0 r}\phi\left[\frac{r}{a}\right], \lim_{r\to 0}\phi\left[\frac{r}{a}\right] = 1. \quad 3.5$$

Z_1 and Z_2 are the atomic numbers of the ion and target atoms respectively, q the electronic charge, and a screening distance.

Only few potentials, such as the hard-sphere, the Coulomb and the inverse square potentials, allow the analytical solution of the scattering integral 3.3 [3.4]. The hard-sphere potential is defined by

$V(r) = 0$ for $r > R'$

$V(r) = \infty$ for r< R', where R' is the sum of the radii of the two spheres. The scattering angle is then given by

$$\theta_c = \pi - 2p \int_{R'}^{\infty} \frac{dr}{r^2\sqrt{1 - p^2/r^2}} = \pi - 2\arcsin\frac{p}{R'} \quad 3.6a$$

or

$$\frac{p}{R'} = cos\frac{\theta_c}{2} \quad 3.6b$$

The Coulomb potential

$$V(r) = \frac{Z_1 Z_2 e^2}{4\pi\epsilon_0}\frac{1}{r} = \frac{C}{r} \quad 3.7$$

yields

$$\theta_c = \pi - 2p \int_{R_{min}}^{\infty} \frac{dr}{r^2 \sqrt{1 - \frac{c}{E_r}\frac{1}{r^2} - \frac{p^2}{r^2}}} = 2 \arccos \left(\frac{2pR_{min}}{p^2 + R_{min}^2} \right)$$

3.8

with

$$R_{min}^2 = \frac{C}{E_r} R_{min} + p^2 \qquad\qquad 3.9$$

The inverse square potential represents one of the simplest screened potentials and it is given by [3.5]

$$V(r) = \frac{Z_1 Z_2 e^2}{4\pi\epsilon_o} \frac{a}{r^2} = \frac{A}{r^2} \qquad\qquad 3.10$$

The scattering angle becomes:

$$\theta_c = \pi - 2p \int_{R_{min}}^{\infty} \frac{dr}{r^2 \sqrt{1 - \frac{1}{r^2}\left(\frac{A}{E_r + p^2}\right)}} = \pi\left(1 - \frac{p}{R_{min}}\right) \quad 3.11$$

with

$$R_{min}^2 = \frac{A}{E_r} + p^2 \qquad\qquad 3.12$$

Knowing the dependence of the scattering angle on the impact parameter the nuclear stopping cross-section can be calculated by integration over all possible energy weighted by the differential cross-section, i.e.

$$S_n(E) = \int_0^{T_m} T d\sigma \qquad\qquad 3.13$$

The three potentials previously considered do not describe with the accuracy required by the users of the ion implantation technique the interaction projectile - target atom. The main problem is the determination of the screening function for different impact distances. A useful approximation adopts a screened potential given by a power function, i.e. [3.6]

$$V(r) = \frac{1}{4\pi\epsilon_o} \frac{Z_1 Z_2 e^2}{r^s} \frac{a^{s-1}}{s} (s = 1, 2...) \qquad 3.14$$

The $s = 1$ form corresponds to the simple Coulomb interaction, while $s = 2$ to the inverse square potential. A better approximation is given by the Thomas-Fermi approach in which the two overlapping electron clouds are treated as a free electrons gas in which energy is minimized. The $\phi_o(\frac{r}{a})$ Thomas-Fermi function has been tabulated by Gombas [3.7] and an analytical approximation is given by:

$$\Phi_o\left(\frac{r}{a}\right) = 1 - \frac{r}{a}\left[\left(\frac{r}{a}\right)^2 + 3\right]^{-1/2} \qquad 3.15$$

The screening length is given by $a_{TF} = 0.8853 a_0 \cdot Z_2^{-\frac{1}{3}}$, where a_0 is the Bohs radius. For the Thomas-Fermi potential the screening function is independent of Z_2. The Thomas-Fermi model results in a potential which is too repulsive at large distances; this is due to the fact that this model does not take into account the detailed shell structure of the electronic distributions in the two colliding particles. In the LSS treatment [3.2], so called from the initials of the authors, the scattering cross section is approximated by a function only of one variable, $t^{1/2}$, given by

$$t^{1/2} = \epsilon \sin\frac{\theta_c}{2} = \epsilon\sqrt{\frac{T}{T_m}} \qquad 3.16$$

where ϵ is the energy in reduced units. The treatment of LSS uses reduced energy ϵ and length ρ parameters

$$\epsilon = \epsilon_1 E = \left[\frac{M_2}{M_1 + M_2} \frac{4\pi\epsilon_0 a}{Z_1 Z_2 q^2}\right] E \qquad 3.17a$$

$$\rho = \rho_1 R = N\pi a^2 \frac{4M_1 M_2}{(M_1 + M_2)^2} R \qquad 3.17b$$

The screening radius a is given by $a = 0.885 a_0 (Z_1^{2/3} + Z_2^{2/3})^{-1/2}$, where a_0 is the Bohr radius. For screened coulomb interaction the differential nuclear cross section is

$$d\sigma_n = \pi a^2 \frac{d(t^{1/2})}{t} f(t^{1/2}). \qquad 3.18$$

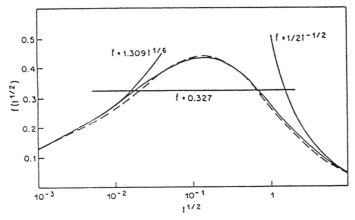

Figure 3.3   *Scattering function $f(t^{1/2})$ for a Thomas-Fermi potential (−) and for different polynomial approximations (solid lines) (from ref. 3.8).*

The function $f(t^{1/2})$ is shown in Fig.3.3 for the Thomas - Fermi and for some potentials of the power form [3.8].
It must be remarked that Eq. 3.18 describes all collisions characterized by the variables E and θ_c by the use of only one parameter: $t^{1/2}$. Furthermore, all ion-target combinations are described by the same cross-section using reduced parameters. For large values of t, that is for large energy transfers in close collisions, this function merges smoothly with the prediction of the Coulomb formula, i.e. with the Rutherford scattering and $f(t^{1/2})$ is equal to $\frac{1}{2}t^{1/2}$ for pure Coulomb interaction.

For the power law potentials, $f(t^{1/2})$ assumes the following relation

$$f(t^{1/2}) = \lambda_s t^{(\frac{1}{2}-\frac{1}{s})}. \qquad .03 < \lambda_s < 1.0 \qquad\qquad 3.19$$

In the case of s=2, inverse square law approximation, $f(t^{1/2})$ is a constant.

In reduced variables the universal stopping cross section becomes

$$\frac{d\epsilon}{d\rho} = -\frac{1}{\epsilon}\int_o^\epsilon f(t^{1/2})d(t^{1/2}) \qquad\qquad 3.20$$

shown in Fig.3.4 as a function of the square root of the reduced energy.

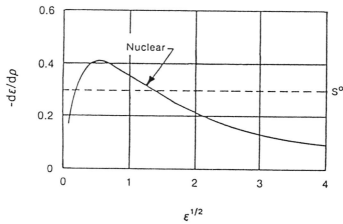

Figure 3.4 $d\epsilon/d\rho$, *nuclear stopping cross section versus* $\epsilon^{1/2}$, *Thomas-Fermi* (————) *and inverse square law (- - -) potential (from ref. 3.8).*

The solid line is found when the Thomas-Fermi model is used while the dashed line results for the inverse square law potential. In this last case the stopping power is energy independent and in the usual variables it becomes

$$-\frac{dE}{dx} = 2.8 \times 10^{-22} N \left[\frac{Z_1 Z_2}{(Z_1^{2/3} + Z_2^{2/3})^{1/2}} \right] \left(\frac{M_1}{M_1 + M_2} \right) \frac{eV}{nm}$$

3.21

and for B and As in Si becomes 94 and $1200 eV/nm$ respectively.

The basic relation between the reduced range and the specific energy loss $d\epsilon/d\rho$ is simply

$$\rho = \int_o^\epsilon \frac{d\epsilon}{d\rho/d\epsilon}$$

3.22

Integration of the previous equation would give the reduced range along the path of the incident ion for a given ϵ if only nuclear stopping mechanisms were to be considered. For instance in the case of a constant $d\epsilon/d\rho$ (square law approximation) the range of 100 keV B in Si must be $\sim 10^3 nm$ while that of 100 keV As 83 nm. The experiments indicate instead a range of 300nm for B and of 50nm for As. Clearly the nuclear stopping power is only a small fraction of the total energy loss rate in the case of B while it is a large fraction in the case of As ions.

The shell structure of the electrons is not included in the statistical models so far discussed. More realistic approximations can be obtained by using Hartree-Fock-Slater atomic charge distributions as done by Ziegler, Biersack and Littmark [3.9]. The atoms of the solid are represented by non overlapping spheres arranged in the proper crystal lattice (muffin-tin atoms). The charge distributions for the atoms is constructed using H.F.S. atomic distributions. Since the range of the atomic charge distribution exceeds the muffin-tin radius a residual charge of about one electronic charge remains outside the atomic spheres. This charge is distributed spherically symmetric in a shell around the muffin-tin atom yielding an effective atomic radius. From the charge distribution the atomic potential is calculated by a simplified quantum-mechanical approach by taking into account the interactions between the nucleus and the electrons of the other atom, the electron-electron interaction, a simplified model for Pauli excitation and exchange energy in the region of overlapping electron clouds. The different combinations projectile-target atoms potentials were fitted by a "universal" screening function Φ_u and the "universal" screening length a_u:

$$\Phi_u(x) = 0.1818e^{-3.2x} + 0.5099e^{-0.9433x} + \cdots$$

$$+0.2802e^{-0.4029x} + 0.02817e^{-0.2016x} \qquad 3.23$$

$$a_u = 0.8853 \cdot a_0 / (Z_1^{0.23} + Z_2^{0.23}) \qquad 3.24$$

3.3 Electronic Energy Loss

The electronic energy loss is defined as the inelastic energy loss of the incident ion as it moves through the cloud of electrons of a target atom. The interaction between the projectile and the electrons is still coulombic. The electrons can be excited to higher discrete energy levels, can be ejected (ionization) or can be excited in the collective motion of plasmons. The prevalence of one contribution over the other depends on the ion velocity, v, with respect that of the target electrons. In the low velocity region, i.e. if $v < \frac{1}{4\pi\epsilon_o}Z_1^{2/3}\frac{e^2}{\hbar}$, the electronic energy loss is proportional to the ion velocity. The interaction arises from scattering of electrons that are no longer attached to any specific atom but form a gas in which the positive charged are embedded. The linear increase of the stopping electronic loss with the velocity (like a viscous force) may be derived from Ohm's law: due to impurity scattering, the small drift velocity of conduction electrons is proportional to the applied electric field. Since the electrons are not accelerated on the average, the force is proportional to velocity, too.

The situation is the same if the projectile (considered as an impurity) moves slowly through the electron gas. For this velocity regime, which corresponds to energies used in most applications of ion implantation, the following formula has been derived [3.10]

$$\frac{d\epsilon}{d\rho} = k_L \epsilon^{1/2} \qquad\qquad 3.25$$

with

$$k_L = \frac{Z_1^{1/6} 0.0793\, Z_1^{1/2} Z_2^{1/2} (M_1 + M_2)^{3/2}}{4\pi\epsilon_o (Z_1^{2/3} + Z_2^{2/3})^{3/4} M_1^{3/2} M_2^{1/2}} \qquad\qquad 3.26$$

It is determined by M_1, M_2, Z_1 and Z_2 and, unlike the nuclear term, cannot be described by a single universal curve. The electronic stopping power is proportional to ion velocity for energies below about 1MeV. It reaches a maximum when the ion velocity is comparable to the average velocity of the outer electrons so the interaction time is the highest possible and the maximum transfer of energy occurs between the projectile and the target electrons. The shape at the peak depends on the detailed structure of the target electrons so that shell effects are more evident in this energy regime [3.11a-3.11b]. This phenomenon is of relevance today due to the use of high energy implanters, and must be considered. The stopping cross section varies non-monotonically with target atomic number (Z_2 oscillations). At higher velocities excitations and ionization of core electrons dominate and collective excitations of the electron gas take place. The cross section falls with energy and the target becomes progressively more "trasparent". If the ion velocity v is higher than $\frac{Z_1 e^2 1}{4\pi\epsilon_o \hbar}$ (velocity of the K shell electrons), it has a high probability to lose all its electrons, and the Bohr approximation becomes valid. In this energy regime the electronic stopping power is inversely proportional to the square of the velocity, in formula [3.12]

$$\frac{dE}{dx} = \frac{Z_1^2 e^4 N}{mv^2 4\pi\epsilon_o} B \qquad\qquad 3.27$$

m, electronic mass, B measures the penetration of the incident ion through the electron shell, given by $Z_2 ln\frac{2mv^2}{I}$ with I average

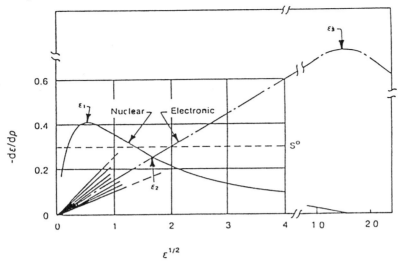

Figure 3.5 *Nuclear and electronic stopping powers versus energy in reduced units (from ref. 3.). For electronic stopping a family of lines (one for each combination of projectile and target) is obtained (from ref. 3.13).*

excitation energy of the electrons of the target electrons, given by $I = I_0 Z_2$ with $I_0 = 10 eV$.

Figure 3.5 provides an overview of the nuclear and electronic stopping power over the entire range of interest in ion bombardment studies [3.13]. It permits a separation into three different regimes:

(i) $\epsilon \leq 10 (Z_1 > 4)$, the velocities of the impinging heavy ions are much smaller than the Thomas-Fermi velocity, both nuclear and electronic losses contribute significantly to the slowing down. This is the tipical regime of low energy-ion implantation.

(ii) $\epsilon >> 10$ and low Z_1. The nuclear energy loss is very low and amounts to about 10^{-3} of the electronic loss. This is the regime of nuclear mycroanalysis.

(iii) $\epsilon > 10$ and medium Z_1. The electronic losses contribute mainly during the first part of the trajectories, nuclear energy loss dominates at the end. This is the regime of high energy implantation ($\sim MeV$).

For some commonly used ions implanted in silicon and gallium arsenide the conversion factors for ϵ and ρ to $E(keV)$ and $R(\mu m)$, the values of k, ϵ_1, ϵ_2 and ϵ_3 are given in Table 3.1. ϵ_1 is the ion energy for the maximum nuclear stopping power, ϵ_2 the energy at which the nuclear and electronic energy loss are equal, and ϵ_3 the energy at the maximum electronic energy loss.

Table 3.1 *The LSS parameters* $\epsilon, \rho, k, \epsilon_1, \epsilon_2$ *and* ϵ_3 *for several dopants in Si and in GaAs.*

Ion	Target	ϵ/E (keV)	ρ/R (μm)	k	ϵ_1 (keV)	ϵ_2 (keV)	ϵ_3 (keV)
B	Si	0.113	32.2	0.22	3	17	3 x 10^3
P	Si	0.021	29.0	0.14	17	140	3 x 10^4
As	Si	0.0048	17.0	0.12	73	800	2 x 10^5
Sb	Si	0.0019	10.7	0.11	180	2 x 10^3	6 x 10^5
Be	GaAs	0.636	9.6	0.55	5	4	2.5 x 10^3
Si	GaAs	0.0163	15.0	0.26	26	145	3 x 10^4
Zn	GaAs	0.0040	15.5	0.17	80	9 x 10^3	1.5 x 10^5
Te	GaAs	0.0014	11.8	0.13	220	4 x 10^3	5.5 x 10^5

In the case of boron implant in silicon, the electronic stopping exceeds nuclear stopping for energies $\geq 10keV$, while for arsenic in silicon it happens for energies $\geq 800keV$.

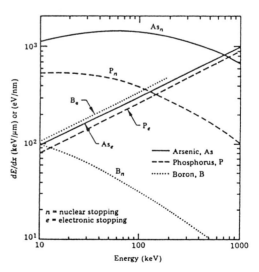

Figure 3.6 *Nuclear and electronic energy loss versus energy for B, P and As ions in Si (from ref. 3.14).*

Due to the relevance of B, P and As as common dopants in *Si* the nuclear and electronic stopping powers are reported in more details in Fig. 3.6.

The Bethe-Bloch and the Lindhard's treatment of electronic stopping power are based on the assumption that the projectile of atomic number Z_1 is a point charge. A projectile can be considered a point charge for high velocities $v_1 >> v_0 Z_1^{2/3}$, at lower velocities the ion may bind electrons and its charge state may be lower than Z_1. Several assumptions are made usually to describe the charge state of an heavy ion and its dependence on the ion-target combination and energy. The ZBL [3.9] treatments adopted in the popular TRIM [3.56] program is based on the stopping power of protons. These are assumed not binded to electrons and the stopping cross section of a fast heavy ion Z_1 is scaled to that of protons, S_H, by the relationship

$$S_{Z_1}(v) = [Z_1^*(v)^2] S_H(v) = [\gamma(v) Z_1]^2 S_H(v) \qquad 3.28$$

γ is defined by $Z_1^* = \gamma Z_1$ and is called the "fractional effective charge" [3.9]. There is no general agreement on the evaluation of the effective charge. The ion charge state is established in dynamic equilibrium, i.e. by capture and loss of electrons. Every change of the projectile charge state generally reduces the projectile's kinetic energy and thus contributes to stopping in a way not included in the effective charge. As a matter of fact the stopping cross sections are predicted not better by 10% in general, in some cases the errors even exceed 30% especially around the maximum and below where shell effects and oscillations in both Z_1 and Z_2 play a relevant role. This is the region of high energy implants and being increased their use a major attention should be payed to the comparison of experimental and theoretical stopping powers.

3.4 Depth Profile of Implanted Ions

The knowledge of the total energy stopping power

$$\frac{d\epsilon}{d\rho} = \left(\frac{d\epsilon}{d\rho}\right)_n + k\epsilon^{1/2} \qquad 3.29$$

allows the calculation of the range ρ by the integral

$$\rho = \int_0^\epsilon \frac{d\epsilon}{(-d\epsilon/d\rho)} \qquad 3.30$$

Statistical fluctuations occur in the energy loss by nuclear collision, than a dispersion or straggling of the ion range is present.

In the case of nuclear stopping power and for power-law potentials one finds

$$\left(\frac{\Delta\rho}{\rho}\right)^2 = \left(\frac{\Delta R}{R}\right)^2 = \frac{s-1}{s(2s-1)}\gamma \qquad 3.31$$

with

$$\gamma = \frac{4M_1 M_2}{(M_1 + M_2)^2}$$

Instead of R and ΔR the measured quantities are R_p and ΔR_p, i.e. the projected range and the standard deviation of the projected range or straggle, along the normal to the sample surface. The evaluation of R_p and ΔR_p requires to follow the energy loss during the ion trajectory until it stops. A Boltzmann transport equation [3.1] is set up to solve the statistical problem of the final ion distribution using the previously found analytical relations.

The transport equations can be solved [3.8] to determine the moments of the distribution and from these, the distribution, i.e. the shape of the concentration versus depth. In comparison with experiments, the projected ranges calculated according to the LSS treatment deviate by a factor two at the lowest energies, while the agreement is good, ($\sim$ 10%), at the higher energies. The deviations are due to the unrealistic long tail of the Thomas-Fermi adopted potential.

The range distribution of implanted impurities is described, as a first approximation, by a symmetrical Gaussian distribution. This distribution is characterized by two moments, the projected range R_p and the straggle, ΔR_p. The implanted profile is given, with the implanted dose $N_\square$, by

$$C(x) = \frac{N_\square}{\sqrt{2\pi}\Delta R_p} \exp\left[\frac{-(x - R_p)^2}{2(\Delta R_p)^2}\right] \qquad 3.32$$

where x is the distance from the target surface, measured along the axis of the beam incidence. The gaussian approximation is useful for quick and simple estimates. The range parameters R_p and ΔR_p are shown in Fig. 3.7 for boron, phosphorus, arsenic and antimony into silicon [3.15].

The peak concentration of $10^{15}/cm^2 - 100keV\,As^+$ in Si amounts to $2 \times 10^{20}/cm^3$ and is located at a distance of $0.05\ \mu m$ from the surface.

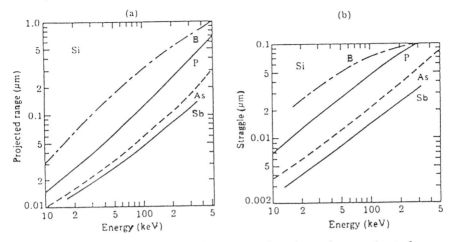

Figure 3.7 *Range distributions showing the projected range R_p(a) and the straggle ΔRp (b) of B, P, As and Sb in Si (from ref. 3.15).*

Table 3.2 reports the normalized ion concentration and the corresponding distance from the peak concentration. $C(x)$ becomes 60% of C_{max} at $R_p \pm \Delta R_p$, and the full width at half maximum, ΔX_p, is given by 2.35 ΔR_p.

Table 3.2 *Normalized ion concentration to the peak at several distances from R_p for a gaussian distribution.*

C(x-Rp)/C(Rp)	(x-Rp)
1	0
0.5	$\pm$ 1.18 ΔR_p
0.1	$\pm$ 2.14 ΔR_p
0.001	$\pm$ 3.04 ΔR_p
0.001	$\pm$ 3.72 ΔR_p

By comparison with experimental data, the gaussian distribution is not a satisfactory fit. The experimental profiles [3.16] of B in Si (see Fig. 3.8) clearly show deviations from a symmetrical distribution and exhibit a substantial skewness[3.17].

The two parameters Rp and ΔRp are not enough to describe the depth distribution and more moments are needed to account for the shape.

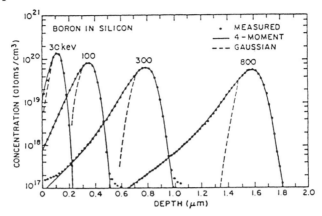

Figure 3.8 *Experimental B profiles into polycrystalline silicon (from ref. 3.16). The four moment distributions are calculated with the following β values (800 keV - 60; 300 keV - 19; 100 keV - 7; 30 keV - 3.6).*

In constructing a profile the moments about the mean depth are more common used. The first four moments are defined in the following:

R_p(average projected range)=

$$\int_{-\infty}^{+\infty} x f(x) dx \qquad\qquad 3.33a$$

ΔR_p (standard deviation or range straggling) $=$

$$\left[\int_{-\infty}^{+\infty} (x - R_p)^2 f(x) dx \right]^{1/2} \qquad\qquad 3.33b$$

$$\gamma(skewness) = \frac{\int_{-\infty}^{+\infty} (x - R_p)^3 f(x) dx}{(\Delta R_p)^3} \qquad\qquad 3.33c$$

$$\beta(kurtosis) = \frac{\int_{-\infty}^{+\infty} (x - R_p)^4 f(x) dx}{(\Delta R_p)^4} \qquad\qquad 3.33d$$

where f(x) is the probability function which satisfies the condition $\int_{-\infty}^{+\infty} f(x)dx = 1$. The real distribution is given by $f(x) \cdot N$, where N is the implanted dose.

The first moment is the average penetration depth and does not influence the shape of the distribution. The second moment or standard deviation, ΔR_p, is a measure of the width of the modified gaussian distribution. The third moment or skewness, γ, is a measure of how much the distribution is cocked or tilted away from symmetry about the mode of the first moment, in one direction (positive γ) or in the opposite direction (negative γ). The fourth moment, β, Kurtosis, is a measure of how pointed or flat topped the distribution is at the peak and consequently how spreaded out it is below the peak. The shape of the distribution is characterized by the combination of the third moment (skewness) and the fourth moment (Kurtosis), γ and β are dimensionless.

The calculated γ and β values are shown in Fig. 3.9 for B and P in Si [3.18].

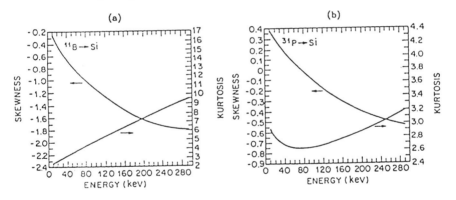

Figure 3.9 *Calculated values of skewness γ and kurtosis, β, for B and P in Si (From ref. 3.18 -* [(c)] *1983 IEEE).*

A Gaussian distribution has $\gamma=0$ and $\beta = 3$. γ is positive when the slowing down of the ions is dominated by nuclear stopping power. It gives rise to a tail extending to the right to relative large depths and places the peak of the distribution farther to the surface than R_p. γ is negative when prevails the electronic energy loss and places the peak of the distribution closer to the surface than R_p. The shape of the profile changes then with the ion energy. In cases where the kurtosis is not available the following relation is adopted.

$$\beta = 2.6 + 2.4\gamma^2 \qquad\qquad 3.34$$

The profiles reported in Fig. 3.10 refer to As implanted Si at several energies and the measurements were performed by Rutherford backscattering [3.19].

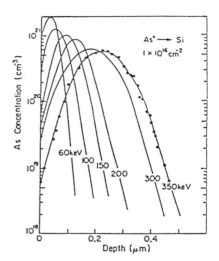

Figure 3.10 *Experimental As profiles in Si for several beam energies (from ref. 3.19). The measurements were performed by Rutherford backscattering.*

The depth resolution is a few hundred angstroms.

In a comparison between B and As profiles in Si (Figs. 3.9 and 3.10), one notes a pronounced skewness toward the surface (γ is negative and the maximum of the distribution occurs at a depth greater than the mean distance R_p) for B and a skewness toward the deep side of the implant profile (γ, is positive) for As. The different behavior of the skewness is associated with the forward momentum distribution for light and heavy ions at comparable energy. Light ions in their encounters with the target atoms will experience a relatively large amount of backward scattering. The distribution will be filled-up on the surface side. Heavy ions on the other hand will experience a large amount of forward scattering and their distribution is filled-up on the deep-side of the profile [3.14].

Several distributions have been used to fit more accurately the experimental ion distribution than by using a Gaussian. Among them we consider only the Pearson IV distribution being used in the SUPREM program for calculating the profiles of implanted boron. This program will be discussed in sect. 3.8.

The distribution is given by the solutions of the following differential equation

$$\frac{df(s)}{ds} = \frac{(s-a)f(s)}{b_o + b_1 s + b_2 s^2} \qquad 3.35$$

where $s = \frac{x - R_p}{\Delta R_p}$. The Pearson coefficients can be written in terms of the first four moments of the distribution [3.20]:

$$a = -\gamma(\beta + 3)/A \qquad 3.36a$$

$$b_o = -(4\beta - 3\gamma^2)/A \qquad 3.36b$$

$$b_1 = a \qquad 3.36c$$

$$b_2 = -(2\beta - 3\gamma^2 - 6)/A \qquad 3.36d$$

$$A = 10\beta - 12\gamma^2 - 18 \qquad 3.36e$$

The Pearson IV distribution applies when the coefficients satisfy the relation $0 < \frac{b_1^2}{4b_0 b_2} < 1$.

The solution has a single maximum at $x = R_p + a$ and decays smoothly to zero on both sides. Integrating eq. 3.33 for a Pearson IV solution gives

$$ln\left[\frac{f(s)}{f_o}\right] = \frac{1}{2b_2}ln(b_0 + b_1 s + b_2 s^2) =$$

$$= \frac{b_1/b_2 + 2b_1}{\sqrt{4b_0 b_2 - b_1^2}} tan^{-1}\left(\frac{2b_2 s + b_2}{\sqrt{4b_0 b_2 - b_1^2}}\right) \qquad 3.37a$$

i.e.

$$f(s) = f_o e^{m(s)} \qquad 3.37b$$

$$m(s) = \frac{1}{2b_2}ln(b_0 + b_1 s + b_2 s^2) -$$

$$- \frac{b_1/b_2 + 2b_1}{\sqrt{4b_0 b_2 - b_1^2}} tan^{-1}\left(\frac{2b_2 s + b_2}{\sqrt{4b_0 b_2 - b_1^2}}\right) \qquad 3.38$$

As said the Pearson IV distribution fits quite well the boron profile in polycrystalline silicon as shown in Fig. 3.8. The projected range R_p and the depth at which the peak distribution is located differ for a Pearson distribution. For 150 keV B in Si, for istance, $R_p = 430nm$, $\Delta R_p = 80nm$, $\gamma = -1.4$ and $\beta = 6$. The Pearson distribution is peaked at $x = R_p + a$, with a=55nm, i.e. at 485nm. The full lines of Fig. 3.8 are Pearson IV distributions.

All the previous treatment has been based on an amorphous target. The depth profiles are then representative of implantation into an amorphous target. In many cases the target is a single crystal and the wafer is usually misaligned by 7° off the axis to give the appearance of a pseudo-amorphous target, although [3.21] it has been shown that a misalignment of 5° is more effective in reducing channeling, which increases the range of the implanted ions. Channeling will be discussed in detail in the next section.

Implants in monocrystalline silicon are characterized by a tail due to feeding-in and channeling phenomena. Therefore a modified Pearson distribution is defined by adding an exponential tail to the shoulder of the standard Pearson distribution function. The tail distribution has the form $A \exp[-\lambda(x - x_o)^2]$. The constants A, λ and x_0 are determined by fitting the value of the first and the second derivative at the matching point of the two distributions.

Several implants are made usually through a thin layer of a dielectric material like SiO_2 or Si_3N_4. For practical cases a simple model can be applied to obtain the distribution in a two layer sample for material whose average atomic numbers and masses are nearly equal. Assuming gaussian profiles, the two parts of the profile in layer 1 of thickness t and substrate 2 are given by [3.22].

$$C_1(x) = \frac{N_\square}{\sqrt{2\pi}\Delta R_{p1}} \exp\left[-\frac{(R_{p_1} - x)^2}{2\Delta R_{p_1}^2}\right] \qquad 0 \leq x \leq t \qquad 3.39a$$

or

$$C_2(x) =$$

$$= \frac{N_\square}{\sqrt{2\pi}\Delta R_{p_2}} \exp\left[-\frac{[t + (R_{p_1} - t)\Delta R_{p_2}/\Delta R_{p_1} - x]^2}{2\Delta R_{p_2}^2}\right] x > t$$

$$3.39b$$

The error in using the previous equation can be

$$\eta = \frac{R_{p2} - R'_{p2}}{R'_{p2}} \qquad 3.40$$

with

$$R'_{p2} = R_{p_1}\frac{\Delta R_{p_2}}{\Delta R_{p_1}} \qquad 3.41$$

A further refinement can be obtained according to the following scheme of calculation: the profile C_1 in the top layer is calculated, and the total number of implanted atoms in this layer ($N_{\square 1}$) is obtained by integration. Then the profile C_2 in the target material 2 is calculated (assuming no masking layer), and the thickness t' which contains $N_{\square 1}$ atoms is determined. The final profile is composed of profile C_1 in material 1 up to t and of profile C_2 starting from t', thus resulting in a profile containing $N_{\square}$ implanted ions.

3.5 Penetration Anomalies

In semiconductor device technology, implantations are usually made through a mask which may be a photoresist or an insulating film (SiO_2, Si_3N_4 etc.). During implantation, the incident ions can knock-on mask atoms giving rise to a recoil implantation of impurity atoms in the masked substrate if the mass of the implanted ions is not too different from the mass of the atoms of the mask. Recoil-implanted impurity profiles are characterized by an extremely shallow distribution with the maximum at the surface. Recoil-implanted impurities, especially oxygen, can have serious detrimental effect on device performance after annealing [3.23].

In addition part of the atoms may be sputtered with a change in the stoichiometry of the layer. This is of relevance when the implants are performed through the thin gate oxide ($\sim 10nm$) of the Mbit memories. To be effective as a mask, the layer should be thick enough to stop almost all the incident ions. The $R_p + 4\Delta R_p$ values are reported in Fig.3.11a for As, P and B in PMMA and in Fig.3.11b for B in several materials [3.24].

Alternatively one can use the transmission coefficent T given for a gaussian profile by

$$T = \frac{1}{2} erfc \left[\frac{t - R_p}{\sqrt{2}\Delta R_p} \right] \qquad 3.42a$$

where t is the mask thickness. The complementary error function can be approximated if the argument is large, thus

$$T \simeq \exp(-\tilde{t}^2)/2(\pi\tilde{t})^{1/2} \qquad 3.42b$$

where the parameter $\tilde{t}$ is given by $\tilde{t} = \frac{t - R_p}{\sqrt{2}\Delta R_p}$. In order to stop 0.9999 of the incident ions, i.e. $T = 10^{-4}$, the minimum thickness t of the mask is $\simeq R_p + 4\Delta R_p$.

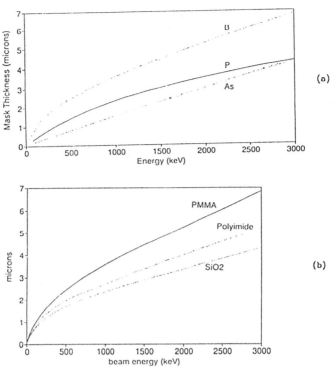

Figure 3.11 *(a) Thickness of PMMA to reduce to 10^{-4} time the implanted dose in Si substrate as a function of B, P, and As energy, (b) thickness of PMMA, polymide and SiO_2 to reduce to 10^{-4} time the implanted B dose in Si substrates (from ref. 3.24).*

The range distribution is modified by several phenomena which occurs during the slowing of the implanted ions. A fraction of the impinging projectiles can be reflected by the surface atoms [3.25]. The backscattered fraction corresponds to the amount of the integral $\int_{-\infty}^{surface} C(x)dx$. When the incident ion is heavier than the target atom, i.e. $Z_1 > Z_2$, the backscattered fraction is negligibly small. However if $Z_1 < Z_2$, the reflection losses become significant and can amount to as much as 30% of the implanted dose, as for low energy implants of boron ions into silicon. This is of relevance for the doping of the walls in the trenches of DRAM.

Another phenomenon of interest which plays a relevant role for low energies and high doses of heavy element implants is the sputtering process. When the collision cascades intersect the surface of the materials, atoms will be ejected. In cases where this proceeds very rapidly, it sets a limit to the number of ions that can be implanted in the target as an equilibrium between doping and surface erosion is reached. The sputtering yield, i.e. the number of ejected target atoms per incident ion, is a function [3.26] of M_1, M_2, E, the angle of beam incidence with respect to the sample surface and the target temperature.

Sputtering yields of Sb, As, B and P projectile ions in Si are shown [3.27] in Fig.3.12 versus the ion energy.

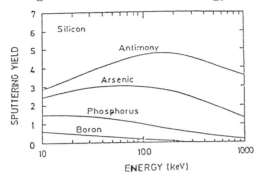

Figure 3.12 *Sputtering yields of Sb, As, P and B ions impinging on Si as a function of the energy (from ref. 3.27).*

The sputtered Si layer thickness, t, is calculated from S and from the ion dose according to $t = \frac{S}{N} N_\square$ with N being the atomic density $(5 \times 10^{22}/cm^3 Si)$. Some data are given in table 3.3:

Table 3.3 *Sputtered thickness t (nm) of Si after different doses of Sb, As, P and B ions.*

Ion	Energy (keV)	Dose (cm^{-2})		
		$10^{15}/cm^2$	$10^{16}/cm^2$	$10^{17}/cm^2$
B	10	0.1nm	1nm	10nm
P	10	0.3nm	3nm	30nm
As	50	0.6nm	6nm	60nm
Sb	100	1.0nm	10nm	100nm

The change of the profile caused by the sputtering can be evaluated in a simple way [3.28] assuming that:

i) the sputtering yield S is constant,
ii) no knock-on takes place,
iii) the volume change due to damage may be neglected,
iv) the distribution is described in absence of sputtering by a gaussian function.

The profile is thus given by [3.28]

$$C(x) = \frac{N}{2S} \left[erf \frac{x - R_p - N_\square \frac{S}{N}}{\sqrt{2}\Delta R_p} - erf \frac{x - R_p}{\sqrt{2}\Delta R_p} \right] \qquad 3.43$$

At saturation, i.e. for $N_\square \to \infty$, the profile is given by

$$C(x) = \frac{N}{2S} erfc \frac{x - R_p}{\sqrt{2}\Delta R_p}. \qquad 3.44$$

The maximum concentration occurs at the surface and is given by

$$C_{max} = \frac{N}{2S} erfc \frac{-R_p}{\sqrt{2}\Delta R_p} \simeq \frac{N}{S} \qquad 3.45$$

for

$$R_p \geq 3\Delta R_p$$

This maximum concentration is independent of the implanted dose, and depends only on the relation between the atomic density and the sputtering. For 60 keV As and with S=3, $C_{max} = 1.5 \times 10^{22}/cm^3$, while for Sb (S=5 at 150 keV) $C_{max} = 10^{22}/cm^3$. These values are higher than the solubility limits of the considered species. At higher impurity concentrations the sputtering yield may change and the near surface concentration may be determined by the preferential sputtering of one of the two atomic species present in the near surface region.

3.6 Channeling Implants

The previous treatment describes the slowing-down of energetic ions in a random medium where the target atoms are located in a random distribution. A silicon target is instead a single crystal and according to the particular orientation, open structures appear either along axes or along planes (see Fig. 3.13). Directional effects are present in the interaction of the projectile ions with the ordered arrangement of the crystalline target. The

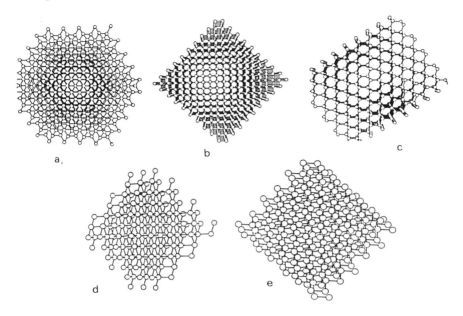

Figure 3.13 *Different atomic arrangements of Si lattice: (a) < 111 > axis, (b) < 100 > axis (c) < 110 > axis, (d) {111} plane, (e) random.*

phenomenon is called channeling. It was discovered almost 30 years ago more or less accidentally by Monte Carlo simulations of the range of heavy ions implanted in solids [3.29] and by profile measurements of keV heavy ions in polycrystalline metal targets [3.30]. These profiles exihibited a small penetrating exponential tail which was subsequently attributed to channeling in those crystallites which happened by chance to be aligned with the incident beam.

Channeling is a simple steering effect resulting from the Coulomb repulsion between the positive charged projectiles and the target atoms along row or planes. If the beam direction is almost parallel to a low-index axis or plane the projectile ions suffer a series of correlated and gentle collisions at small angles with the target atoms as schematically shown in Fig. 3.14.

Close encounters with the target atoms are strongly prevented so that all processes that require small impact parameters are greatly reduced. As a consequence a channeled ion does not displace target atoms and experiences a reduced energy loss not only for elastic nuclear encounters but also for electronic interactions.

A comprehensive theoretical framework for predicting the

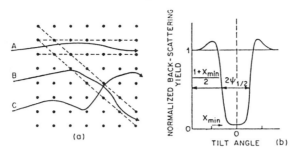

Figure 3.14 *(a) Schematic representation of channeled and non-channeled trajectories in a cubic lattice, (b) backscattering yield around a channeling direction. Yield is minimum when the beam is well aligned with a channel.*

principal experimental parameters was developed in terms of continuous potential for the atomic rows or planes and assuming that a channeled particle conserves its transverse energy $E_\perp = E\psi^2$ during each encounter with an atomic row or plane [3.31]. The minimum distance of closest approach to the row, r_{min}, is given by equating the transverse kinetic energy $E\psi^2$ in the mid-channel region to the transverse potential. As ψ is increased the corresponding value of r_{min} continuously decreases until (for $r_{min} = \rho$, vibrational amplitude) violent atomic collisions become possible and the steered motion breaks down. This condition physically determines the concept of the critical angle, above it the motion is not correlated, below it does. At the surface a small fraction ($\approx$ a few percents) of the incident beam enters the crystal within an area (πa^2) around each aligned row in which the potential energy already exceeds the maximum allowable transverse energy for channeled particles. During its slowing down the channeled particles experience electronic energy loss, multiple scattering by various defects and by vibrating lattice atoms with a gradual increase in $E_\perp$ until they are dechanneled (ungoverned motion). The particles then after see the crystal as a random medium.

A channeled ion samples regions of different electronic density according to its specific trajectories. This means that channeled particles can stop at different distances from the surface being different the electronic stopping power for each of their trajectories.

The best channel in a diamond lattice is the [110] axis (see Fig. 3.13c), it has the lowest electronic density in the middle region and the strongest row potential. In the microelectronic industries the devices are fabricated in (100) oriented wafers and with just few exceptions in (111) oriented wafers. These two orientations have quite similar electronic distributions and row potentials. In the following only the [100] axial channeling and some related planar channeling will be considered The crystal axis of a wafer is not perfectly perpendicular to the wafer surface, with 1^0 of tolerance normally from this condition. So implants at 0^0 tilt angle between the beam direction and the normal to wafer surface are near to the channeling implant condition although they cannot be considered properly channeling implants. With the reduction in device dimensions and with the use of tridimensional structures such as trenches, for capacitors in MOS transistors the tilting of the wafer with respect to the beam direction plays a relevant role. To avoid the asymmetry between left and right caused by shadowing of the mask it is necessary to perform implants at 0^0, i.e. with the beam impinging perpendicular to the wafer surface, or by rotation the wafer at different angles during the implant.

Even if not perfectly aligned with the main axial or planar directions, beam particles can be feeded-in [3.32] the channels either planar or axial and can continue their trajectories as channeled ones. The phenomenon is schematically described in Fig. 3.15.

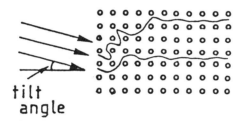

Figure 3.15 *Schematic representation of feeding-in mechanisms.*

Feeding-in cannot be avoided for implants performed in crystals also choosing a direction where the crystal atoms appear randomly distributed. The critical angle of 1 MeV P along the [100] axis is 1.85^0 (see eq. 3.46), but for implant at 3^0 from the axis a huge tail is instead present (see Fig. 1.14) clearly feeding-in is taking place and the misalignment angle will determine the final profile.

As an example of implants performed normally to the wafer surface (0^0) the chemical and the carrier depth profiles of 80 keV $1 \times 10^{14} B^+/cm^2$ implanted in Si are shown in Fig. 3.16 [3.33].

The 0^0 tilt angle profile shows the influence of channeling effect, in fact it is deeper with a peak shift from 0.25 μm to 0.3 μm with respect to the 7^0 tilt angle implant. For comparison the Secondary Ion Mass Spectrometry (SIMS) profile of the same implant

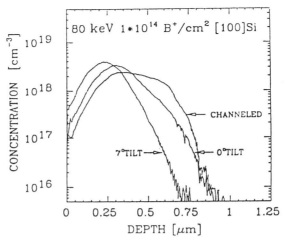

Figure 3.16 *Chemical concentration profiles for 7^0 tilt, 0^0 tilt and along the < 100 > axis implants of $80keV - 1 \times 10^{14}/cm^3$ boron ions (from ref. 3.33).*

but performed along the [100] axis is reported. The boron profile is broader but the maximum penetration value is the same of the 0^0 distribution evidencing that no misalignements are present in the 0^0 geometry but that the beam divergence causes a high number of dechanneled particles. The beam divergence is a difficult parameter to control in high current implanters. The medium current implanters usually have a reduced divergence, industrial implanters are now equipped with sophysticated apparatus for the control and for the measurement of the beam divergency during the implant [3.34]. The sensitivity of the channeled profiles to the misalignment is described quantitatively by the critical angle for channeling Ψ. This represents the half width at half maximum of the yield of all violent collision processes (Rutherford backscattering, nuclear reactions, ion induced X rays ...) by varying the angle of incidence with respect to the axial or planar direction. It is given approximately by the relationship:

$$\Psi = 0.307(Z_1 Z_2/Ed)^{1/2} \qquad (deg.) \qquad 3.46$$

d is the atomic spacing along the atomic rows in $\overset{\circ}{A}$ and E is the incident energy in MeV. The previous relationship does not take into account the narrowing induced by the thermal vibrations. For heavy ions in keV and MeV regimes (for $\Psi > a_{TF}/d$ or $E < E' = 2Z_1 Z_2 e^2 d/a_{TF}^2$) the following expression must be used:

$$\Psi^* = 7.57 \left[(a_{TF}/d)\Psi\right]^{1/2} \qquad\qquad 3.47$$

Table 3.4 reports values of E' in MeV obtained for B, P and As impinging along different silicon axes. Except for boron these values are higher than 1 MeV.

Table 3.4 *Energy E' (MeV) below it the critical angle is given by 3.45.*

	[100]	[110]	[111]
B	0.4	0.3	0.36
P	1.7	1.2	1.5
As	5.0	3.6	4.4

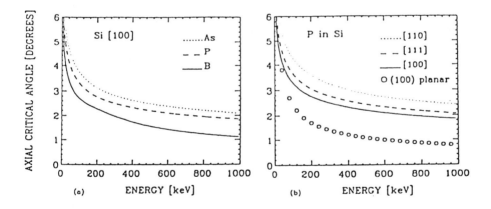

Figure 3.17 *(a) $< 100 >$ axial critical angles for As, P and B in Si; (b) $< 100 >, < 111 >, < 110 >$ and [110] critical angles for P in Si as a function of the beam energy (adapted from ref. 3.35).*

The planar critical angle is given by:

$$\psi_p = \left(\frac{2\pi Z_1 Z_2 e^2 a_{TF} N d_p}{E} \right)^{\frac{1}{2}} \qquad\qquad 3.48$$

being d_p the interplanar distance. The critical angle as a function of the energy is reported in Fig. 3.17a for different ions in silicon and in Fig. 3.17b for phosphorus ions along axial and planar orientations of silicon [3.35]. In Fig. 3.18 the SIMS profiles of $1 \times 10^{14}/cm^2 B$ implanted at different energies along the [100] axis of silicon are reported [3.36].

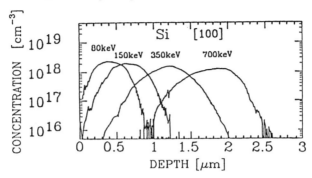

Figure 3.18 *Profiles of $1 \times 10^{14}/cm^2 B$ - implanted along the $< 100 >$ axis at different energies (from ref. 3.36).*

The profiles are characterized by a flat concentration region and by a rapidly decreasing concentration near the maximum depth. The carrier concentration profiles obtained by spreading resistance measurements for $2 \times 10^{13} cm^3 P$ implants along the [100] are reported in Fig. 3.19 [3.37]. Also in these cases the profiles are quite flat up to the maximum range.

The experimental maximum penetration, R_{max} measures the depth at which the concentration reaches the 1% value of the channeled peak. The R_{max} values for B and P implants along the [100] direction are shown in Fig. 3.20a and 3.20b as a function of the square root of energy [3.36-3.37]. The values for both ions follow a linear trend. The projected ranges for implants in amorphous silicon are also reported in the same figures. The electronic stopping power, Se, determined from R_{max} assuming $S_e = k\sqrt{E}$ and neglecting the nuclear stopping power contribution is shown in Fig. 3.21.

The k values depend not only on the adopted projectile-target atoms combination but also on the crystallographic direction.

Channeled ions, move along crystal regions with different electron densities and the dependence of the electronic energy loss on the impact parameter, p, must be taken into account explicitly. In the Oen and Robinson [3.38] approach the local energy loss is given by the product of the LSS stopping power (see eqs. 3.25 and

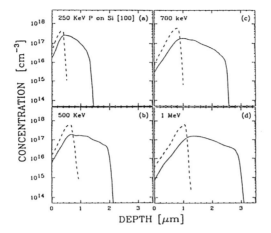

Figure 3.19   *Profiles of* $2 \times 10^{13} P/cm^2$ *implants along the* $< 100 >$ *axis and in an amorphous (dashed lines) Si layer at different beam energies (from ref. 3.37).*

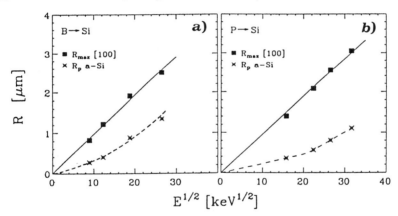

Figure 3.20   *(a) Maximum range,* R_{max}, *and projected range, Rp, as a function of the square root of energy of boron ions implanted along the [100] axis or in amorphous Si (b) same as (a) but for phosphorus ions (from ref. 3.36-3.37).*

3.26) by an exponential function of the impact parameter. The electronic stopping is given by

$$S_c(p, E) = \frac{s^2 k E^{\frac{1}{2}}}{2\pi a^2} exp\left(-\frac{sp}{a}\right)$$ 3.49

where $k = \alpha k_L$, α being a fitting parameter, and k_L the

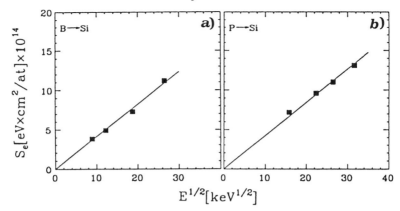

Figure 3.21 *Experimental electronic stopping power values, Se, for well channeled B (a) and P(b) ions entering a Si target along the [100] axis (from ref. 3.36-3.37).*

Lindhard constant (see eq. 3.25). The value of α changes with the ion, it is 1.7 for B and 1.5 for P respectively. The parameter s ranges around 0.3, for P is 0.315 and for B 0.27. The integration of eq. 3.49 all over the impact parameters reproduces the LSS stopping power. Channeled ions have instead a selected values of impact parameters and the well channeled ions, corresponding to those that start and stop as channeled ones, have an impact parameter p slightly changing along their trajectories. Simulation procedures will be described in section 3.8.

So far the influence of damage on the channeled profile has been neglected, channeled ions avoid close collisions with nuclei, and then the damage formation is drastically reduced for these implants. However due to multiple scattering with the thermal vibrations of the atomic rows, with the electrons and with crystal imperfections, the transverse energy of the channeled ions can increase and if it overcomes the maximum allowed value the ions leave the channel and move in a random medium. These ions can collide with target nuclei at small impact parameter and can displace them from the lattice sites. With increasing the implanted dose increases the amount of damage and consequently the dechanneling probability. As a result the increase in dose influences the in-depth ion distribution as shown in Fig. 3.22a for 200 keV B implants in Si [3.39].

The channeled component saturates at a dose of $5 \times 10^{14}/cm^2$. A similar trend is shown by P implants, (see Fig. 3.22b), the channeled fraction saturates again and a peak grows-up at a lower depths.

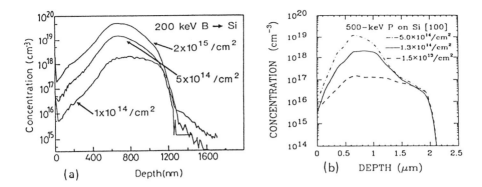

Figure 3.22 *(a) SIMS profiles of 200 keV B implanted along the < 100 > axis of Si at increasing doses (from ref. 3.39). (b) Carrier profiles of 500 keV P implants along the < 100 > axis of Si at different doses (from ref. 3.37).*

At these fluences the overall profiles can be divided into three components: (i) particles entering the crystal in a random direction, (ii) channeled particles that are dechanneled along their trajectories and, (iii) well channeled particles. The random component is originated by the ions that enter the crystal having a small impact parameter with the atomic row and are scattered at large angles. Among ions entering the channel, only a small fraction of them is well channeled, these projectiles move in the channel until they stop. These ions represent the channeled fraction. All the ions that leave the channel after multiple collisions with thermal vibrations, or after scattering with crystal imperfections, contribute to the second component, that dominates with increasing the dose.

The dechanneled fraction can be determined experimentally as shown in Fig. 3.23 [3.36]. The overall profile is given by three components: random A, dechanneled B and channeled C. The random component A is determined scaling the profile in the amorphous target in such a way that the left part coincides with that of the high dose channeled implant. The channeled component C is obtained from the experimental channeled profile performed at low dose. After subtraction of these two components A and C from the whole profile the dechanneled contribution B is determined. For simplicity each component can be fitted by a gaussian function. This is a reasonable and simple method to determine the depth distribution of channeled profiles at increasing doses.

Channeling implants at this stage are not easily introduced in the large scale fabrication of devices. At high dose the damage

Figure 3.23 *Schematic representation of the procedure utilized to obtain the dechanneled component. A is the experimental random fraction given by the profile of an implant in amorphous target. C is the experimental channeled component obtained by the profile of low dose implants in channeling conditions. The component B, i.e. the dechanneled fraction, is given by the subtraction of A and C from the total profile (from ref. 3.36).*

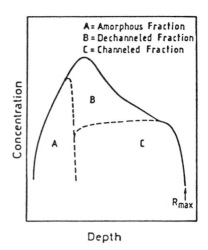

is a limiting factor in the reproducibility of the profile and at low dose it is necessary an accurate control of the wafer orientation with respect to the ion beam direction. For these reasons in almost all the implants directional effects are avoided by tilted the silicon wafer at 7^0 from the normal to simulate a configuration in which the target atoms are seen by the beam particles as randomly distributed. This is not the case, even at this angle some planar feeding-in is still possible. The reason of the 7^0 tilt angle choice is that the [100] axis and the nearest [133] are separated by 14^0. In addition to the tilt angle the rotational angle dependence of the profile has been detailed.

The tilt angle is not enough to specify the wafer orientation with respect the beam incidence, the rotational or twist angle is also needed. The two angles, tilt and twist, are defined according Fig. 3.24 for < 100 > oriented wafer [3.40].

The twist or azimuth is the angle formed between the projection of the incident beam versor onto the wafer and a versor on the wafer surface that is perpendicular to the wafer's primary flat. The major flat in a [100] wafer is aligned to $\pm 1 \div 0^0$ to the [011] direction. The profile depends then also on the rotational angle as shown in Fig. 3.25 that reports the concentration of 15 keV and 80 keV - $5 \times 10^{14} B^+/cm^2$ implants at a fixed angle of 7^0 and for various rotation angles [3.41].

These results clearly indicate the difficulty to avoid channeling and feeding-in.

Table 3.5 reports the optimum combinations of tilt and twist angles to avoid channeling effects for the 100 keV B implants at low

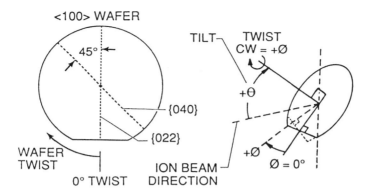

Figure 3.24 *Definition of the wafer orientation with respect beam incidence in terms of twist and tilt angles (from ref. 3.40).*

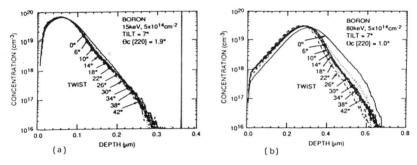

Figure 3.25 *15 keV (a) and 80 keV (b) $5 \times 10^{14}/cm^2 B$ implants at a fixed tilt angle of 7^0 for various rotation angles of the wafer (from ref. 3.41).*

dose.

The ion beam during its scanning on the wafer surface can change orientation with the main axial direction and the penetration of the implanted ions varies. This is illustrated in Fig. 3.26 [3.40]. In the center the ion beam is well aligned with the crystal channel. The beam incidence angle varies as it moves toward the extreme of the wafer diameter and the entrance trajectories of the ions are increasingly misaligned with the channel. These ions will have greatly reduced channeling probability than well aligned ions, and will stop in smaller distances from the crystal surface. Today for medium current implanters parallel beams with a controlled divergence apparatus are available [3.34] , and probably low dose channeling implants might become feasible in the near future if silicon wafers with a well defined orientation ($\pm 0.2^0$) would be

Table 3.5 *Combination of tilt and twist angles to suppress orientation effect for 100 keV B implants in Si (from ref. 3.40).*

Tilt Angles	Optimum Twist Angles
7^0	$25^0 - 33^0$
10^0	$23^0 - 35^0$
12^0	$15^0 - 37^0$
20^0	12^0 or $26^0 - 32^0$
25^0	12^0 or $28^0 - 38^0$
35^0	$28^0 - 38^0$
45^0	$18^0 - 26^0$
55^0	$14^0 - 23^0$ or $32^0 - 38^0$
60^0	$18^0 - 20^0$ or $40^0 - 45^0$

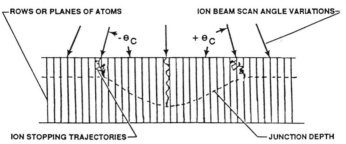

Figure 3.26 *Influence of the non-parallel beam in channeling implants (from ref. 3.40).*

fabricated.

3.7 Lateral Spreading

The lateral spread of the implanted profile under a mask is a limiting factor in VLSI technology. With VLSI and ULSI circuits, dimensions are shrinking into the submicron range for channel length and the lateral straggling must be taken into account. The relevance of the lateral penetration is not limited to low energy implants but also to high energy implants for the formation of p and n-type contiguous wells in CMOS technology. In this case the lateral overlap between the two profiles limits the integration. As previously stated the lateral spread caused by multiple collisions of the incident ions is on the order of the vertical straggle, ΔR_p,

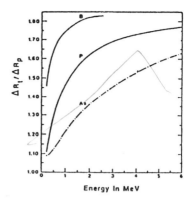

Figure 3.27 *Ratio between* $\Delta R_\perp / \Delta R_p$ *as a function of the beam energy for B, P and As in Si according to TRIM simulation.*

and depends on the ratio M_2/M_1.

The lateral straggle of light ions such as B is greater than of the heavier ions as P or As. The ratio of the lateral to longitudinal spread $\Delta R_\perp / \Delta R_p$ approaches a limit of 0.7 with increasing ion mass at low energy. With increasing the energy the ratio $\Delta R_\perp / \Delta Rp$ increases too, as shown in Fig. 3.27 by the data calculated according to TRIM. This simulation will be described in the next section.

The distribution function for ions incident in (0,0,0) along the axis and stopping at a point x, y, z is given, assuming a gaussian distribution around the y and z coordinates, by

$$f(x,y,z) = C(x) exp \left(\frac{-(y^2+z^2)}{2\Delta R_\perp^2} \right) \frac{1}{2\pi \Delta R_\perp^2} \qquad 3.50$$

In the case of implantation along a line $(0,0,z_0)$ the previous distribution should be integrated over z from $-\infty$ to $+\infty$ and it becomes

$$f'(x,y) = \frac{C(x)}{\sqrt{2\pi}\Delta R_\perp} exp \left(-\frac{y^2}{2\Delta R_\perp^2} \right) \qquad 3.51$$

This two-dimensional distribution is written as the product of vertical $C(x)$ and lateral distributions. When the implant is performed through a mask whose edges are infinitely high and steep the previous distribution must be integrated over the open area where the ion beam can enter. For this mask with an opening from y=-a to y=a, the distribution becomes (see Fig. 3.28a and 3.28b).

$$d(x,y) = \int_{-a}^{a} f'(x, y-y')dy' =$$

$$= \frac{C(x)}{2} \left[erfc \left(\frac{y-a}{\sqrt{2}\Delta R_\perp} \right) - erfc \left(\frac{y+a}{\sqrt{2}\Delta R_\perp} \right) \right] \qquad 3.52$$

Fig. 3.28b shows the equal-ion-concentration contours for 70 keV B implant through a $1\mu m$ wide slit [3.42]. The values are given in terms of the peak concentration.

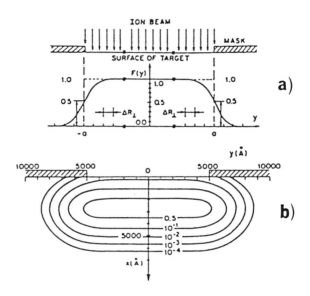

Figure 3.28 *Lateral profiles of ions implanted through a mask of width 2a, where a $>> \Delta R_\perp$, and infinite extension in the Z direction (a), contours of equal-ion concentrations for 70 keV $B^+(R_p = 271nm, \Delta R_p = 82.4nm$ and $\Delta R_\perp = 10nm)$ incident into silicon through $1\mu m$ slit. The concentrations are given in terms of the peak value (from ref. 3.42).*

The lateral doping extends under the mask edges and will modify the geometrical length of a short channel. With decreasing the mask size and increasing the beam energy the influence of the straggling becomes more and more relevant [3.43]. The ratios of the lateral spread of B concentration contours at 10^{-1} and 10^{-3} of the peak value for 10 and 70 keV implant energies to the mask opening size, 2a, are shown in Fig. 3.29 as a function of the mask size. The ratio increases sharply for mask sizes below $0.3\mu m$ for 70 keV and $0.1\mu m$ for 10 keV energy implant respectively. Interactions between dopants can occur in adjacent and separated implanted areas with sizes smaller than the 0.2 - 0.3 μm openings.

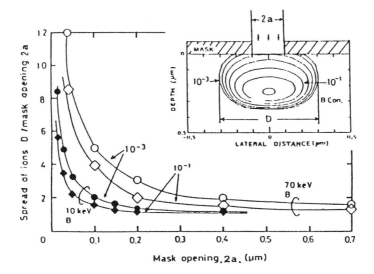

Figure 3.29 *Ratio between the lateral spread at concentration*
10^{-1} *and* 10^{-3} *the peak values and the mask opening size 2a for*
10 and 70 keV B implants (from ref. 3.43).

In many cases masking layers are often tapered at the edge
rather than perfectly abrupt, so that ions are gradually prevented
from entering the silicon. If the masking layers has a thickness
$m_{mask}(y)$, it can be considered as an equivalent silicon films of
thickness $m_{Si}(y)$ and the distribution becomes [3.44-3.45].

$$d(x,y) = \int_{-\infty}^{+\infty} f'(x + m_{s_i}(y'), y - y')dy' \qquad 3.53$$

The previous results were obtained by analytical or numerical
solution of eq.s 3.50 or 3.51 (see the next paragraph) and in some
cases the condition of a constant $\Delta R_\perp$ has been relaxed. The
lateral standard deviation varies with depth in particular for light-
ion profiles such as B or Si. The peak of the lateral distribution
is shifted closer to the surface than the projected range.

Measurement of the lateral profiles is not a simple task in view
of the tied geometrical constraints. Recently improved methods
have been proposed and applied. The preferred chemical etch of
heavily doped layers has been applied to the preparation of cross
section for scanning electron microscopy [3.46] or for transmission
electron microscopy [3.47]. By these methods the contour lines at
different dopant concentrations are delineated with a quite good

lateral and vertical resolution, expecially in the TEM analysis. Several examples will be presented in chapter 6 together with a more detailed description of the used procedure.

Another experimental method of relevance for deep implants is the two dimensional spreading resistance profilometry [3.48]. This technique gives the contour line of the doped region, in particular the isoconcentration curve at the substrate concentration level. The method is based on a magnification of the lateral and in-depth scales using a special beveling technique and a particular measurement configuration (see chapter 6 for the description).

To conclude this section we present a few examples of this latter method relevant to this chapter. The isoconcentration lines at a substrate concentration of $1 \times 10^{15}/cm^3$ for 1 MeV P, $1 \times 10^{14}/cm^2$ implanted along the [100] axis and in a random direction are shown in Fig. 3.30 and those for 400 keV, $B, 1 \times 10^{14}/cm^2$ in Fig. 3.30b [3.49].

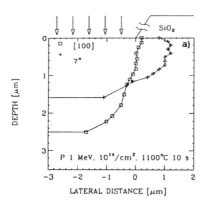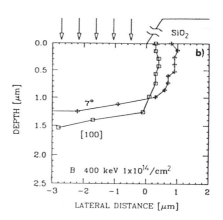

Figure 3.30 *Isoconcentration lines for $1 MeV P$ implants to a dose of $1 \times 10^{14}/cm^2$ in channeling along the [100] axis (□) and in a random direction, 7^0 tilt angle (+) (a). Some curves for 400 keV B, $1 \times 10^{14}/cm^2$ implants (b). The substrate concentration is for both cases $1 \times 10^{15}/cm^3$ (from ref. 3.49).*

The difference between the maximum spreads for random and channeling implants is $1\mu m$. The lateral penetration is higher for the random than for the channeled implants in agreement with the reduced nuclear collision probability experienced by the channeled particles. The experimental maximum lateral penetration is closer to the surface than that predicted by eq. 3.50. This is due to the non perfect steepness of the mask edges. The implant analyzed in Fig. 3.30 were performed with an edge of slope 78^0. The channeling lateral profile is determined by this slope. With increasing dose the difference between lateral spread of channeling and random implants decreases and disappears when the implanted dose is so high to damage the crystal such that the channeled ions see the target as a random medium. The in-depth distribution is instead determined by the ions channeled during the first part of the implant.

3.8 Simulation of Range Distribution

In the previous section the slowing down of the incident ions was described analytically on the basis mainly of the LSS theory and further developments. The to-day requirements of VLSI and ULSI imply a detailed knowledge of the implanted profile in two-dimensions and in the several layers that lay on the silicon substrate. The basic assumptions of the LSS theory do not allow its application to multilayered structures and do not permit an evaluation of the radiation damage. Simulation tools based on new methods have been developed and they are used in all advanced design of ULSI technological processes [3.50-3.51].

The Monte Carlo (MC) and the Boltzmann transport equation (BTE) methods are widely used to simulate ion implantation phenomena in solids. In the BTE approach [3.52] the scattering processes of the ions into the target are described by changes in the statistical momentum distribution.

The calculation of the range, damage distribution is regarded as a transport problem describing the motion of the ions during their slowing down to zero energies [3.53]. Analogous to the kinetic theory of gases the average number of particles scattered into and out of a differential phase space element can be described by a Boltzmann Transport Equation.

Monte Carlo methods are based on the simulation of individual particles trajectory through their successive collisions with target atoms.

There are two main approaches for the detailed calculation of ion bombardment of solids: the Binary Collision Approximation (BCA) and the Molecular Dynamics (MD) [3.54] approach. The molecular dynamics approach studies the movement of atoms in a solid as a function of time and takes the interaction with all neighbouring atoms into account. All moving atoms are followed in small time steps so that their collisions are automatically included. For these reasons MD based programs are used to describe sputtering, the low energies processes and in general all cases where multicollisions must be considered.

In contrast to molecular dynamics approach, in the BCA programs the movement of ions in solid is described by a series of successive binary collisions. It breaks down at low energies, when many-body effects become important.

BCA is normally used to simulate ion implantation. One of the main areas of variation of BCA programs is the target structure. If the target structure can be assumed to be randomized (or structureless) the next collision partner for the incoming ion is found by a random selection process. Because this random selection process is used for each collision throughout the program, these simulations are also called Monte Carlo programs. If the target structure is crystalline, the random selection process has to be applied only at the beginning to find, for example, the impact points of a projectile on the target surface. After the first collision the subsequent procedure is completely deterministic, because the positions of the possible collision partners are fixed and well known, at least within in the thermal amplitude movement.

The final result is based on the summation of the nuclear and the electronic scattering events occurring in a large number N (N>1000) of simulated ion trajectories. By following N histories, distributions for the range parameters of the primary and recoiled ions and the associated damage can be obtained. Each history begins with a given energy, position and direction of the incident ion. The ion is assumed to change trajectory at each elastic collision with the target atom and to move in straight-free path between collisions. The energetic ion loses energy continuously between the elastic collisions by inelastic scattering with electrons. Several programs were developed using different nuclear and electronic stopping power based on different theoretical and semiempirical models. A wide variety of potentials has been used depending on the time of calculation and on the particular problem being studied: the Coulomb potential, the inverse-square potential, the Thomas-Fermi potential, the Bohr potential, the Born-Mayer potential, the Moliere potential, and the Ziegler-Biersack-Littmark potential [3.55]. The most popular program to describe the slowing down of an ion into an amorphous target is TRIM (TRansport of Ions in Matter) code [3.56]. The program and its derivatives have spread-out all over the world, but they can differ quite appreciably so that caution is advised to use any individual version. The entire trajectory of a single ion into an amorphous target can be obtained. The collision cascades determined by the recoiled target atoms can also be calculated. As

an example the trajectories of 10 Si ions incident at 150 keV into
a silicon target are represented in fig. 3.31. In the figure the
path of the incident ion is represented toghether with the path
of recoiled ions. From this kind of simulation the range and the
depth distribution of incident ions can be determined as well as
the displacement in the target.

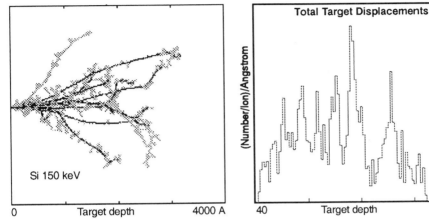

Figure 3.31 *Paths of 10 - 150 keV single Si ion tracks obtained
by TRIM calculation. The relative depth ion distribution is also
reported.*

The first version of TRIM allowed only the calculation of the incident ion trajectory. TRIM.SP (TRIM SPuttering) [3.57] is an
extension of the program TRIM, and is a three dimensional program considering projectiles and recoils. It uses the liquid model
to describe the target structure, i.e. consider a fixed flight distance
(λ) between collisions:

$$\lambda = \lambda_o \quad , \quad \lambda = N^{1/3} \qquad\qquad 3.54$$

with N target number density. The scattering angle is determined
by a randomly chosen impact parameter (p) so that $p = p_{max}\sqrt{R}$
being R the random number while the maximum impact parameter is calculated considering a cylindrical volume with radius p_{max}
and lengh λ_o:

$$\pi p_{max}^2 \lambda_o = N^{-1} \qquad\qquad 3.55$$

$$p_{max} = (\pi \lambda_o N)^{-1/2} \qquad\qquad 3.56$$

at low energies the maximum impact parameter, $p_{max} = \pi^{-1/2}N^{-1/3}$, is too small to take into account appreciable elastic energy losses with more distant atoms. The problem is solved considering simultaneous collisions, one in a cylindrical volume and additional ones in annular volumes.

TRIM.SP uses exclusively elastic kinematics. It proceeds in the following way. After the imput data have been read the projectile moves a distance λ, starting from outside the target at $x_o = -2p_{max}$. The x direction is normal to the target and x is positive inside the target. The target surface is at x=0. Then a collision partner is determined by a stochastic choice of an impact parameter p $0 < p < p_{max}$ as previously defined, and an azimuthal angle Φ, $0 < \Phi < 2\pi$. The scattering angle and the recoil angle are determined in the centre of mass system considering the elastic collision. The energy of the projectile is reduced by the energy transferred to the knock-on atom. The knock atom is followed only if its energy is larger than the surface binding energy, E_s, and if its distance from the surface is smaller than its predetermined range. It can set the target atoms in motion, which will be called secondary knock-on atoms. The ions will be followed until their energy is smaller than the energy E_s. Then the projectile proceeds to the next collision. The projectile and the recoils also lose energy anelastically between subsequent collisions by a continuous loss or by local loss. If the projectile has reached an energy below the cut off energy E_i or if the projectile is backscattered or transmitted, then the next projectile is started. In a multiatomic target, such as a compound semiconductor, the collision partners have to be chosen randomly according to the target compositions, and the change in composition induced by ion bombardment must be taken into account.

Sputtering and reflection will therefore depend on the projectile fluence. The program discussed above simulates only the low fluences or equilibrium case. To study fluence dependence or atomic mixing it is necessary to includes composition changes. This is done in some Monte Carlo codes, as for istance , EVOLVE [3.58], TRIDYN [3.59], TRIPOS [3.60], and HIDOS [3.61-3.62].

New effects are continuously introduced in TRIM to extend the application to different materials or to describe different phenomena. The capability to simulate ion implantation in silicon or microelectronic materials is however already satisfying.

The simulation of ion trajectory in crystalline targets requires the knowledge of the atom locations. These positions can be constructed by three translation vectors starting from a basis of one or more atoms according to the elementary crystal structure. One of the most common programs considering the target structure is MARLOWE code [3.63]. Recently a new version of the TRIM code (CTRIM) has been developed to consider the crystal structure [3.64]. The simulation of channeling implants requires the geometrical description of the target structure and that of the thermal vibrations. They are usually described in terms of uncorrelated Gaussian displacements of the lattice atoms, with a mean square displacement determined by the Debye model. The collision partner search procedure in this case is determined by the initial conditions. The parameters describing the crystal structure are given as input. The initial incident ion direction with respect to the crystal is also an input parameter. The incident ions however are considered to hit the crystal surface on an area (no less than that of the elementary cell of the crystal). The single point of incident is determined by a random number. It may happen that more than one target atom meets the established criteria for the partner collision of the incident ion. In this case simultaneous collisions are considered and the momenta of all the included target atoms are calculated. Energy conservation is achieved by scaling the squares of the final momenta of all particles envolved.

The large angle scatterings of the ions with the target atoms are of great importance for the lateral penetration of the implanted ions under a mask. They are due to the elastic collisions, and in MARLOWE they can be described by different potentials. The most used is the Moliere potential. In this case fitting parameters for different combination of incident ions and target atoms are not necessary.

For the electronic stopping MARLOWE considers local and non local treatments. In a non local treatment a constant dE/dx is considered. The importance of using local treatments was recently shown [3.37]. In the Oen and Robinson approach [3.38] the local energy loss is given by the Lindhard, Scharff, and Schiott [3.2] stopping power time an exponential function of the impact parameter (see eq. 3.47). However no dependence on the channel size is considered. This must be included, indead, to obtain a better description of the experimental data [3.37].

MARLOWE calculations provide the final position of the incident ion determining the three coordinates: depth (z), lateral (x), and height (y) deviation from the point of impingement at the surface of the crystal. Three dimensional distributions can be obtained by counting, for a given number of incident ions, the ions rested in a cube of $z + \Delta z$, $x + \Delta x$, and $y + \Delta y$ edges. The resulting profiles are usually presented in terms of the depth and the lateral distribution under an hypothetical mask. A typical two

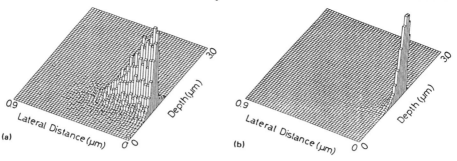

Figure 3.32   *Two dimensional distribution of 400 keV B⁺ ions implanted in silicon at a random incidence (a) or along the [100] axis (b) (from ref. 3.49), as obtained by MARLOWE code.*

dimensional distribution is reported in Fig. 3.32 [3.49]. It refers to a 400 keV B implant in a random direction a) or along the [100] axis b) of a Si target. The 10000 ions are distributed over an area of 1 μm^2, i.e. to a dose of $1 \times 10^{11} at/cm^2$. The simulation can be scaled with a good approximation also to higher doses. Not only the in-depth but also the lateral distribution differs drastically in the two cases.

In Fig. 3.33 simulated and experimental profiles are shown for $2 \times 10^{13} cm^2$ 1 MeV P implants along different directions of a *Si* target.

The agreement between calculations and experiments is quite good for the amorphous silicon target and for implants along a crystalline random direction. This proves the ability of Monte Carlo simulations to follow the trajectories of particles fed into channels. In the channeling implants the maximum penetration is well reproduced, but not the shape. For these implants the escape of the ions from the channel is strongly influenced by the presence of the implant damage. MARLOWE code considers the recoils and follows their trajectories explicitly within a single cascade, but it does not accumulate the damage created by an ion to that created by the next ion. The displacement cascade produced by an ion is then reset before the subsequent ion is implanted. Recently some attempts [3.63-3.66] have been made to model the influence of the damage on the channeled ion trajectories. The number of created couples interstitial-vacancy can be easily determined considering the energy for a *Si* displacement. Then a large number of the generated vacancies recombine with interstitials in sites farther than first neighbouring sites. Taking into account this dynamic annealing is therefore a crucial aspect. A modified version of MARLOWE, UT-MARLOWE, was recently

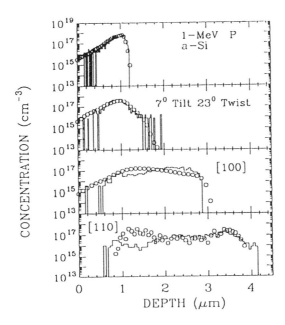

Figure 3.33 *Experimental profiles (open circles) of* $2 \times 10^{13} P/cm^2$ *1 MeV ions implanted into silicon for different configurations from the top to the bottom: amorphous layer; random condition; along the [100] axis and along the [110] axis. These profiles are compared with the calculated distributions obtained by MARLOWE code (histograms) considering 1000 incident ions (from ref. 3.37).*

developed [3.67]. In this version both an improved electronic stopping description and a cumulative damage model was introduced. In fig. 3.34 [3.68] the simulations obtained by UT-MARLOWE for 15 keV B implants in silicon at different incident angles are compared with experimental results. The MARLOWE based model can be applied to any orientation but the defect density might exceed the atomic density of silicon if the capture radius (the distance at which vacancies and Si interstitials recombine) is reduced below a certain value between $0.43a$ and $0.22a$ where a is the Si lattice constant. To account for the dynamic annealing of point defects during the implantation a detailed vacancy-interstitial recombination procedure has been included which associates pairs by minimum distance criteria. In some cases the recombination parameters were fitted with the experimental data to determine the correct damage distribution at different implant temperatures.

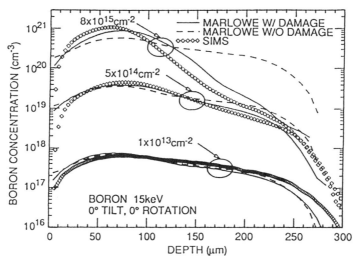

Figure 3.34   *Comparison of experimental SIM profiles of 15 keV boron implanted into silicon at doses of $1 \times 10^{13}/cm^2$, $5 \times 10^{14}/cm^2$ and $8 \times 10^{15}/cm^2$, with profiles simulated with and without the cumulative damage model (from ref. 3.68).*

The processing technique in ULSI is becoming more and more complex and the number of required steps is significantly increasing. Ion implantation is only one of the steps, so that 2D more complete process simulators have been developed [3.69]. They include ion implantation as one of the processes and both analytical and Monte Carlo approaches are included. The complexity of the structure has to be considered to give the 2D distributions of implanted impurities in all the part of the device including different layers of various materials and forms. Typical examples are threshold-adjust implants, channel stop, and source/drain, implants into gate and field oxide regions, which may be covered by Si_3N_4. The existence of multilayered structures results in implant profile discontinuities at the interface between different layers. Additionally atoms from surface layers may be knocked into deeper layers by impinging ions. This recoil effect might degrade the electrical performance of the finished device. To consider all these effects the recent 2D process simulators include Monte Carlo codes for the implantation step. One of the most used is SUPREM IV [3.70]. The acronym comes from Stanford University Process Engineering Modeling Program [3.71].

It can describe all the processes used in the microelectronic industry for the fabrication of devices. As an example let'us consider the doping by ion implantation of the trench walls in a DRAM.

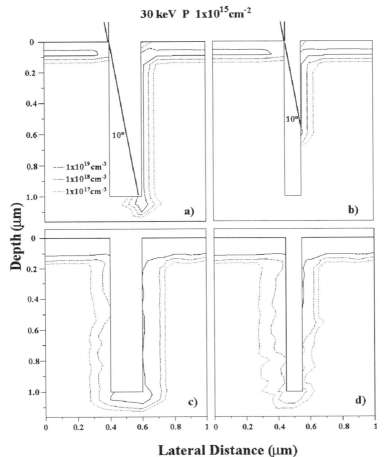

Figure 3.35 *Equiconcentration contours in 1μm - deep trenches implanted with 30 keV P - 1 × 10^{15} at 10°. Analytical simulation for 0.2μm (a) and for 0.1μm (b) wide trench respectively; Monte Carlo simulation for 0.2μm (c) and for 0.1μm (b) wide trench respectively. (Courtesy of M.Saggio)*

The trenches form the capacitors in a MOSFET memory and store the electrical charge as bit of information (see sect. 7.4). The simulations reported in Fig. 3.35 refer to two trenches 1μm deep but of 0.2 and 0.1μm width respectively. These trenches are implanted with 30 keV P, $1 \times 10^{15}/cm^2$ at an angle of 10° to dope the walls. The equiconcentration contours obtained by analytical methods are shown in Fig. 3.35a and 3.35b respectively. The implant into the 0.2μm wide trench is able to dope both the bottom and the

right-hand side (3.35a). In the $0.1\mu m$ wide trench the bottom
instead remains undoped because of the shadowing effect by the
upper part of the trenchs (3.35b). Reflection of ions from the
side wall of the trench and possible scattering into the bottom
is not included in the analytical simulations. To take them into
account Monte Carlo approach must be adopted and the obtained
simulations are shown in Fig. 3.35c and 3.35d for the same cases
previously considered. The dopant distribution is quite different
and ion reflection allows the doping of both sides of the wall.

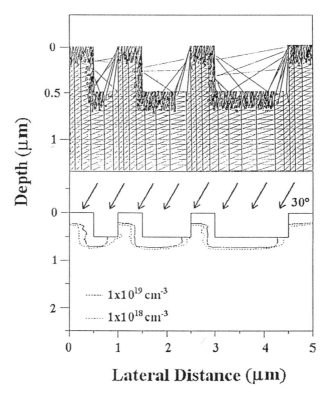

Lateral Distance (μm)

Figure 3.36 *(a) Mesh size distribution adopted to simulate the
implant at 30^0 of $40\ keV$ - $1 \times 10^{15}/cm^3 P$ ions through different
apertures (1.5 μm, 1.0 and 0.5μm). The ion trajectories and
the rest positions are shown. (b) Equiconcentration contour lines.
(Courtesy of M.Saggio)*

   In all the 2D simulation programs the fundamental step is
the definition of a network describing the device structure. This
network is characterized by many nodes and their density must be

carefully determined in all the regions of the structure. In the following steps of the simulation the equations describing the single process (ion implantation, diffusion, etc.) will be calculated only in these nodes. A maximum number of nodes is usually allowed by the program and by the computer characteristics. So they must be distributed in such a way to cover with a fine definition all the regions of better important. As an example the mesh adopted for the simulation of an implant of 40 keV P at 30^0 for different widths of the mask is shown in Fig. 3.36a. The ion trajectories and their rest positions as obtained by Monte Carlo routine, are also reported. The mesh size is lower in the critical regions of the structure allowing a much better resolution of the ion distribution. The equiconcentration contours are plotted in Fig. 3.36b. The shadow effect of the mask edge is clearly evidenced by the asymmetries between left and right-hand side profiles.

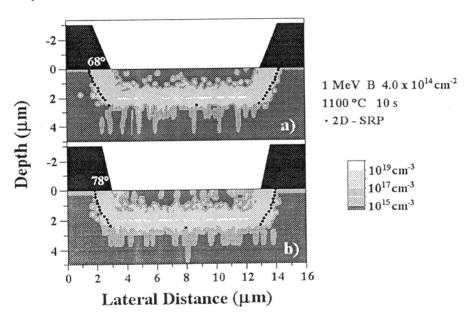

Figure 3.37 *Monte Carlo simulation of 1 MeV B - implant at normal incidence through a mask slope of 78^0 (a) and of 68^0 (b) respectively. The shades indicate different dopant concentration ranges and the dots experimental values at a concentration of $10^{14}/cm^3$ obtained by two-dimensional spreading resistance measurements. (Courtesy of M.Saggio and V.Privitera).*

The influence of the mask edge slope on the as-implanted profile can also be determined by Monte Carlo programs and a simple case is illustrated in Fig. 3.37. Two different mask shapes are considered, one at an angle of 78^0 (Fig. 3.37a) and the other at an angle of 68^0 (Fig. 3.37b). The 1 MeV B implant is performed at normal incidence in both cases. The shades indicate different concentration values. The dots represent experimental data obtained by two-dimensional spreading resistance technique (see sect. 6.3). The points refer to a dopant concentration of $10^{14}/cm^3$ and follow quite well the calculated values, indicating the feasibility of these simulators. Mask sloping increases the lateral penetration of the ions and it adds to the collisional lateral spreading.

CHAPTER 4

RADIATION DAMAGE

4.1 Introduction

Ion implantation is a violent process, a large amount of the projectile kinetic energy is transferred to the atoms of the target displacing them from the lattice sites. The knowledge of the damage created by the implanted ions is of extreme relevance for the planning of the subsequent thermal annealing. It is necessary to eliminate or to reduce drastically the presence of defects in the active zones of the device before the implementation of the implantation process in the industrial fabrication of the devices. This chapter will treat and describe the damage resulting just after the implant. The evolution and the annealing of extended defects will be described in the next chapter, although a region of overlap exists between them. Implants are usually performed at room temperature and as seen in chapter 2, in spite of the adopted precautions, temperature rise is possible during the implant itself. Even at room themperature some defects, in particular point defects, are mobile and may interact among themselves or with preexisting damage giving rise to a variety of extended defects only in part related to the distribution of damage for a OK substrate implant. This part will be considered at the end of this chapter. The next paragraphs describe the slowing down of the projectile, the damage distribution and the related defect distribution under particular assumptions. For clarity a brief description of the most common defects in silicon is also presented.

4.2 Collision Cascade

An ion, during its slowing down, interacts anelastically with electrons and elastically with the other target atoms. If the kinetic energy, E', transfered to the host atom is higher than a certain value, E_d, the displacement threshold energy ($\sim 15eV$ for Si), [4.1] the knock-on atom leaves its lattice site and according to the residual kinetic energy, $E' - E_d$, can move for a certain path length. These atoms (primary collisions) recoil and collide with other atoms (secondary collisions) giving rise to higher generations of collisions. The later generation collisions produce many low-energy recoils, which induce small displacements in nearly random directions. This sequence of collisions and of displaced atom moltiplication is often called collision cascade and it lasts $\sim 10^{-13}s$, [4.2] i.e. the ion range divided by the average ion velocity. For a 200 keV As ion with a speed $\sim 10^8 cm/s$ and a range in Si of $2 \times 10^{-5} cm$, the cascade is completed in less than $10^{-12}s$.

This prompt regime is followed by a redistribution of the energy into the surrounding material by both lattice and electron conduction, a process lasting an additional time interval of $10^{-11} - 10^{-10}$ s. In the following $10^{-9}s$ the unstable disorder relaxes and some ordering occurs by a local diffusion process [4.3]. Later on clustering of lattice damage results in the formation of complex defects. This last process is governed as said before by the target temperature and by the presence of impurities, if any. The time evolution of the processes taking place during the ion slowing down is shown in Fig. 4.1.

The total amount of disorder in a collision cascade is determined by the component of the incident ion energy ending up in recoil processes. This component is somewhat smaller than that calculated on the nuclear stopping power of the projectile because the recoils lose part of their kinetic in ionizing events. In terms of the Lindhard's dimensionless energy ϵ, the energy ending up in atomic collisions, ν, is 0.8ϵ for $\epsilon < 1$; at $\epsilon > 1$ the ν/ϵ ratio falls of rapidly and ν approaches the saturation value 6 in the very high - ϵ regime [4.4]. The size and the character of the collision cascade depend on several parameters such as the mass and the energy of the target.

For light ions, as B in Si, the mean-free-path between successive elastic collisions is larger than the interatomic distance, and thus the collision cascade will result in a dilute distribution of defects. Heavy ions at low energy can have a mean free path comparable with the interatomic distance and a dense cascade will be generated. These two cases are shown in Fig. 4.2(a) and 4.2(b) in

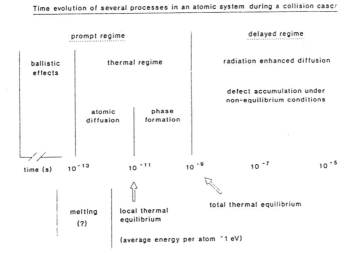

Time evolution of several processes in an atomic system during a collision cascade

Figure 4.1   *Time evolution of a collision cascade.*

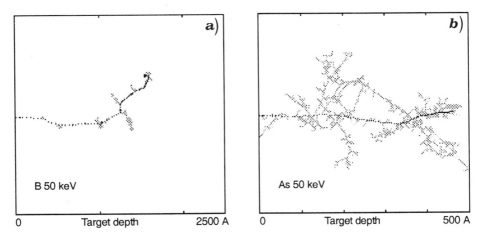

Figure 4.2   *(a) Trajectory of a 50 keV B ion and (b) of a 50 keV As ion in Si evaluated by TRIM program.*

terms of single-ion trajectories obtained by TRIM simulation. A large number of silicon atoms is displaced by the projectiles in the near surface region and the recoiled atoms transfer and deposit energy to a greater depth. The final damage profile is produced then mainly by the recoiled atoms and is a replica of the recoil range distribution. In the heavy ion implants the deposited energy density within the cascade becomes large enough that the assumption of isolated binary collisions breaks down and several atoms are set simultaneously in motion sharing a reasonable high kinetic energy.

The atoms transfer rapidly their high vibrational excitations to those in the surrounding material, and in only few picoseconds the system cools. The material is then subjected to a rapid thermal quench of the order of $10^{14} - 10^{15} K/s$ [4.5]. A different phase can result and a small volume of "hot" Si atoms may, become a small amorphous region surrounded by a crystalline matrix. Low doses ($\leq 10^{12}/cm^2$) of heavy ions create individual isolated damage regions around each ion track with negligible probability of cascade overlap.

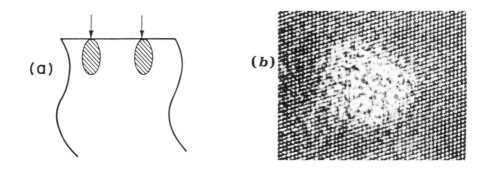

Figure 4.3 *Schematic of low-dose heavy ion implant (a), High resolution transmission electron micrograph cross section of a Si sample implanted with 100 keV Bi. Each bright dot is a column of single Si atoms in the < 110 > projection and their periodicity is 3.1 Å (b) (from ref. 4.6).*

This case is represented schematically in Fig. 4.3a and by high resolution transmission electron micrography of *Si* implanted with 100 keV *Bi* ions in Fig. 4.3 [4.6]. The amorphous silicon has lost the long range order characteristic of crystalline solid and a confuse image in the center of the photograph appears. A *Bi* ion displaces approximately $10^4 - 10^5$ atoms. Section 6.6 will consider in more details the high resolution transmission electron microscopy technique. It must be pointed out that a lattice image is formed by interference of diffracted beams and details of the true atomic structure are provided [4.7]. The bright dots represent columns of single *Si* atoms in the chosen projection and their periodicity is few angstroms.

Even if the quench does not produce a new phase as the amorphous one, it creates in the target, a non-equilibrium atomic distribution containing many defects. In the cascade process atoms are ejected from the center of the cascade leaving a zone rich in vacancies and a surrounding shell rich in interstitials.

At high dose ($\geq 10^{14}/cm^2$) one gets complete overlap when the average separation between cascades becomes comparable to the cascade dimensions and intercascade effects start to occur. This second case is represented schematically in Fig. 4.4a, the corresponding TEM analysis is shown in Fig. 4.4b for 100 keV - $10^{15}/cm^2$ As ions. The amorphous layer $.1\mu m$ thick is continuous all over the implanted area. In the transition region between the amorphous and the crystalline substrate isolated damage zones are still present, and they are responsible of the formation of extended defects, termed end of range damage, in the subsequent annealing stage.

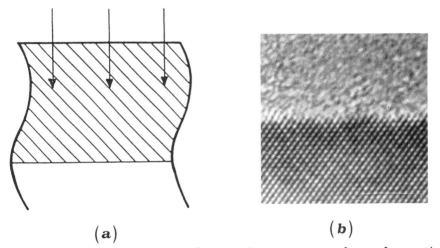

(a)   **(b)**

Figure 4.4 *Schematic of a continuous amorphous formation by high dose heavy ion implant (a), cross section of a Si sample implanted with 100 keV As · $10^{15}/cm^2$ ions(b) (Courtesy of C.Spinella)*

4.3 Damage Distribution

In the ballistic description the depth distribution of the energy deposited into nuclear collisions can be obtained by several analytical or Monte Carlo approachs. One of the more useful treatment consists in the solution of the transport equation similar to that adopted for the range distributions. Binary collisions between the projectile and the target atoms, and between the recoils and the target atoms are described by a suitable potential. The transport equation is solved by taking the spatial moments of distribution and solving the equation for the moments [4.8, 4.9]. The distribution is obtained from the moments following standard methods. An alternative approach is the Monte Carlo method, individual collision cascades are simulated and then the results are averaged from several hundred cascades, as discussed in sect. 3.8.

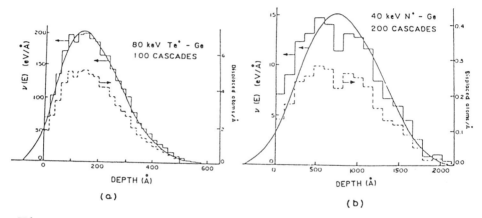

Figure 4.5   *Comparison of the deposited energy distribution, $\nu(E, z)$ derived from analytical (solid curve) and Monte Carlo (histogram) methods. The displaced atom density is also included: (a) 80 keV $Te^+ \rightarrow Ge$, (b) 40keV $N^+ \rightarrow Ge$ (from ref. 4.10).*

Two deposited energy distributions derived from the analytical (solid curve) and Monte Carlo (histogram) approach are shown in Fig. 4.5 for 80 keV $Te^+ \rightarrow Ge$ and for 40 keV $N^+ \rightarrow Ge$ [4.10]. The curves indicate that the heavier ion (Te^+) is producing a much larger deposited energy density. The energy transported by the recoils is relevant for projectile masses heavier than the host and is negligible in the opposite case because the range of a light ion is much greater than the range of the average recoil.

The depth distribution of the energy loss given to the recoil and that lost in ionization events are shown in Fig. 4.6 for 40 keV B in Si and for 600 keV Kr in Si. The distributions are derived from TRIM. Fig. 4.6a and Fig. 4.6c show the computed energy transfer to recoil Si target atoms from 40 keV B and 600 keV Kr ions respectively. Part of this energy is lost by the recoils

in anelastic events, ionization, and is shown as dashed lines in
Fig. 4.6b and Fig. 4.6d respectively. The ionization losses of
the projectiles are reported in the same figures as full lines. For
boron the ionization rate is nearly constant up to the end of the
trajectory while for Kr it reaches the maximum in the initial part
and then decreases with the depth.

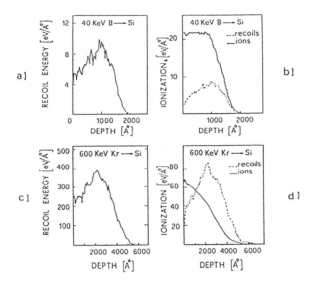

Figure 4.6 *Depth distribution of the energy loss to recoil (a, c)
and that of ionization (b, d) for 40 keV B (a, b) and 600 keV Kr
(c, d) in Si. The distributions are computed by TRIM programme.*

The distribution of the deposited energy density into nuclear
collisions can be converted into a damage distribution, assuming,
for instance, that only those recoils receiving an energy greater
than E_d are displaced. The energy transfers below E_d results
in the creation of lattice vibration or phonons. The number of
displaced atoms per single ion vs depth can be estimated from
$\nu(E,x)$ via the modified Kinchin-Pease [4.11] relationship

$$n_d(x) = \frac{0.8\nu(E,x)}{2E_d}.$$
4.1

with ν(E,x) the total energy density deposited in atomic motion. The factor 0.8 is a correction factor that depends weakly
on the form of the screened scattering potential [4.12]. The total
number of displacements in the implanted volume is

$$N_d = \frac{0.8 \int_0^\infty (E, x) dx}{2E_d}. \qquad\qquad 4.2$$

A commonly used unit of damage is the number of displacements per atom (DPA). A unit of 1 DPA means that, on the average, every atom in the affected volume has been displaced from its equilibrium lattice site once. The dependence of DPA vs. depth is given by

$$DPA(x) = \frac{0.8\nu(E, x)}{2E_d \cdot N} \phi. \qquad\qquad 4.3$$

where N is the atomic density (atoms/cm^3) and ϕ is the ion dose in units of at/cm^2.

The formula 4.2 is based only on ballistic effects. As an example the number of displaced atoms, in terms of vacancies, as calculated by Kinchin - Pease theory is shown in Fig. 4.7a and 4.7b for 20 keV B and 100 keV As in Si respectively. For comparison the corresponding range distributions are reported in Fig. 4.7c and 4.7d respectively, note the different depth distribution for displaced atoms and for stopped ions. In the case of 100 keV As the 12% of the energy is lost by the ions in ionization events, .2% in the creation of vacancies and 0.3% in phonons, while for the recoils the corresponding values are 26%, 3.6% and 58% respectively, i.e. most of the energy is transfered to the target as lattice vibrations, or heat. A 20 keV B ion instead loses in ionization 57%, .4% in vacancies and 1% in phonons, the Si recoils 3.0%, 1.8% and 31.4% respectively.

The measured total number of displaced atoms, N_d, per cascade is shown in Fig. 4.8 for several ions implanted into Si at 35K to reduce thermal effects and to freeze the defects [4.13]. The high energy portion of each curve bends over to approximately the same slope as the N_{kp} line, i.e. the Kinchin-Pease relation, indicating that the rate of damage creation at high energy agrees well with collision cascade theory and that the enhanced value of N_d is entirely due to the low-energy portion of each ion track, where the deposited energy density becomes extremely high. The reciprocal slope $(d\nu(E)/dN_d)$ of the curves in Fig. 4.8 measures the effective energy required to displace target atoms. At high $\nu(E)$ and high Z_1 it falls rapidly and approaches the heat melting value ($\sim 0.8eV$) for Si in agreement with the previous discussion on the formation of small amorphous regions in terms of a fast quench of a "molten zone".

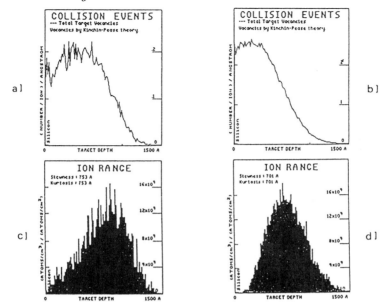

Figure 4.7 *Depth distribution of the vacancies created by 20 keV B (a) and 100 keV As(b) in Si. The calculations are performed by the TRIM program on the basis of the Kinchin-Pease theory. For comparison the corresponding range distributions are shown in (c) and (d) respectively.*

The previous estimates refer to an amorphous target. If an ion impinges along a low index axis or plane it suffers correlated collisions at small angles from the atoms in the row or in the plane. Close impact collisions, responsible of lattice displacements, are prevented to channeled ions, so that a large reduction in damage occurs for channeled implants. As for the ion distribution, the simulation of the energy deposited into nuclear collisions can be performed by the MARLOWE code for any direction of incidence with respect the crystal orientation. As an example the experimental damage distributions, are shown in Fig. 4.8a for 80 keV $B - 1 \times 10^{15}/cm^2$ impinging along the $< 100 >$ axis of the Si wafer and at 7^0 from it. The data are extracted from channeling analysis (see 6.4) and then represent the density of off-lattice site silicon atoms. The calculated vacancy concentration, are shown in Fig. 4.9b for the two angles of incidence [4.14].

The distribution at 0^0 is broaden and its integral lower than the corresponding ones for the 7^0 implant. The calculated profiles include all the generated vacancies without taking into account their

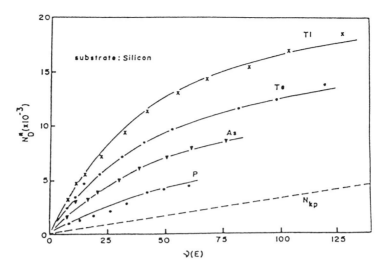

Figure 4.8   *Total number of displaced atoms per cascade versus $\nu(E)$ for several ions implanted in Si at 35 K. The dashed line N_{KP} represents the values predicted by the Kinchin-Pease formula (from ref. 4.13).*

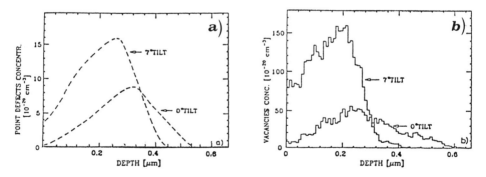

Figure 4.9   *(a) Experimental defect distributions after 80 keV - $1 \times 10^{15}/cm^2 B^+$ implants along the [100] Si axis and at 7^0 from it. (b) Vacancy distributions computed by the Marlowe code for the two implant conditions (from ref. 4.14).*

close pairs recombination with interstitials. The damage build-up is also not included in the computation for the 0^0 tilt implant. The comparison between the experimental and the calculated distributions of displaced atoms can be then performed on the basis of their shape only. It is also implicit in the above discussion that the interstitial distribution copies the vacancy distribution.

4.4 Crystalline Defects

A typical cross section of an implanted sample, if analyzed by transmission electron microscopy, reveals a variety of extended defects: amorphous zone, stacking faults, dislocation loops, twins, clusters etc. The visibility is limited to defects of dimensions above 1-2 nm. Point defects like interstitials, vacancies, and small clusters cannot be detected, although they play a relevant role in the formation of the secondary defects after the annealing steps. Consider now a brief description of the more common types of defects seen in silicon.

The two basic types of intrinsic point defects, are vacancies (V) and self- interstitials (I) (see Fig. 4.10) [4.15]. Vacancies are empty sites where atoms are missing from their normal positions in the crystal lattice. The interstitial is an "extra" atom, located in a position not normally occupied by an atom in the perfect crystal. The vacancy-interstitial pair is called Frenkel defect, and the vacancy a Schottky defect. A third type of native defect is the interstitialcy that consists of two atoms in non substitutional positions configured about a single lattice site. Usually a distinction between an interstitial and an interstitialcy is not made and both are identified as I in the literature. Any crystal will contain a certain concentration of intrinsic point defects that lower the Gibbs free energy of the system as compared to a defect- free crystal. The equilibrium concentration, in atomic fraction, is given by (x=V, or x=I)

$$C_x(T) = A exp(-G_x^f/kT) \qquad 4.4$$

where k is Boltzmann's constant, T the absolute temperature and G_x^f the Gibbs free energy of formation of the intrinsic defect x. The dimensionless factor A accounts for the number of possible defect sites per lattice atom and is unity for vacancies. The Gibbs free energy G_x^f is the sum of two terms $H_x^f - T\Delta S_x^f$, with H_x^f the activation enthalpy of formation and ΔS_x^f the corresponding entropy of formation associated with configuration and usually attributed to lattice vibrations. In spite of the extremely large number of experimental and theoretical investigations no experiment has definitively measured the equilibrium concentrations of vacancies or interstitials in silicon, or even the activation enthalpies of formation [4.16-4.17]. A detailed discussion of this topic is outside the scope of the book, and just to give an order of magnitude we quote at high temperature the following values:

H_x^f 2.4 eV and 3.8 eV for V and I while ΔS_x^f is $1.1k$ and $2k$ for V and I respectively. At 1400 K the concentration of vacancies and of interstitials amounts to 3.4×10^{14} and to $7 \times 10^{19}/cm^3$ respectively.

In semiconductors point defects often introduce electronic states in the band gap and can exist in different electronic configurations. A vacancy in Si can be either neutral or negatively or positively charged, an interstitial can be neutral or positively charged. The thermal equilibrium concentration of charged point defects depends on the doping type, i.e. on the Fermi level and on its energy level position in the energy band gap. Point defects form complexes among themselves, like divacancy V-V, di-interstitial I-I, or with impurities like V-O, As-I etc.

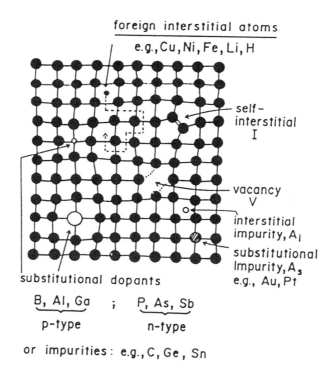

Figure 4.10 *Various types of intrinsic and extrinsic point defects in an elemental semiconductor crystal such as silicon (from ref. 4.15).*

The number and type of possible complexes or clusters is extremely large in view of the possible combinations. As an example several possible configurations for vacancies, interstitials and interstitialcies in a diamond lattice are shown in Fig. 4.11 [4.18].

The role of intrinsic point defects in determining the behaviour of microelectronic devices has been well recognized and these defects have been studied in details [4.19]. In addition to the four silicon self-interstitial localized configurations an extended and fully relaxed atomic configuration with no constraint posed on the location of the extra atom in the simulation all has been considered. Atomistic calculation [4.20] indicates that the migration of interstitial through tetrahedral→ hexagonal→tetrahedral positions can be obtained by the extended interstitial with an activation energy for migration of 1.2 eV.

The other defects present in semiconductors as a result of annealing an implanted layer are classified in line defects as dislocations, plane defects as stacking faults, twins and grain boundaries, and volume defects as precipitates and voids [4.21].

Dislocations are line defects in an otherwise perfect crystal. In contrast to point defects, dislocations are not equilibrium defects and are formed under thermal and mechanical process. The energy is of about 10 eV per atomic plane. Dislocations are classified in two types - the screw and the edge (see Fig. 4.12a and 4.12b).

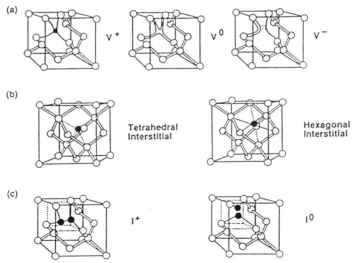

Figure 4.11 *Configurations of point native defects in a diamond lattice. (a) vacancy in the +, 0 and - charge states, darkened bonds indicate orbitals with unpaired spins, (b) tethraedral and hexagonal interstitial site, (c) singly positively charged < 100 > and electrically neutral self-interstitialcy (from ref. 4.18).*

A screw dislocation can be created by cutting the crystal along any plane, then shift the crystal on one side of the cut by a vector parallel to the cut surface relative to the other half and then join the atoms on either side of the cut. If we were to follow a crystallographic plane one revolution around the axis on which the crystal was skewed, we would finish one atom spacing below our starting point. The vector required to complete the loop is called the Burgers vector **b** that is parallel to the screw dislocation. An edge dislocation can be formed by cut the crystal along any plane, spreading the crystal apart and partly filling the cut with an extra plane of atoms.

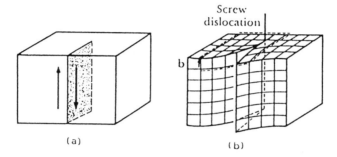

Figure 4.12   *The perfect crystal is cut (a) and sheared one atom spacing (b) , the line along which sharing occurs is a screw dislocation (from ref. 4.22).*

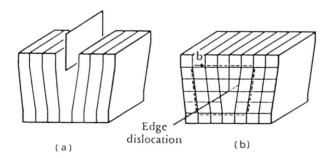

Figure 4.13   *The perfect crystal is cut and extra plane is inserted (a). The bottom edge of the extra plane is an edge dislocation (b) (from ref. 4.22).*

The bottom edge of the inserted plane represents the edge dislocation. The Burgers vector **b** is required to close a loop of equal

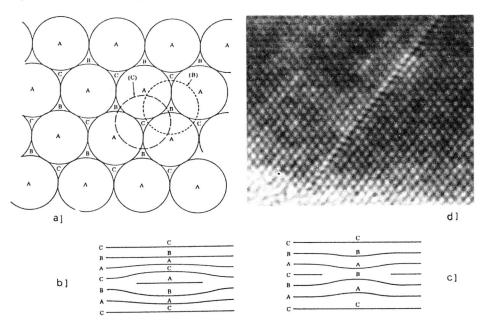

Figure 4.14   *Stacking sequence of spheres in close packed ABC(a). Extrinsic stacking fault: a {111} plane is inserted between a {111}-B plane and a {111}-C plane (b). Intrinsic stacking fault: a{111}-C plane is removed between a {111}-B plane and a {111}-A plane (c). High resolution transmission electron micrograph of a stacking-fault (d), each dot is a column of single Si atoms in the < 110 > projection and their periodicity is 3.1 Å (Courtesy of C.Spinella).*

spacing atom spacing around the dislocation. It is normal to the dislocation line. As a result of the implantation or of the subsequent annealing dislocation network and dislocation loops can be obtained. They can impare the electrical behaviour of devices if located in the electrical active regions.

The single dangling electron at each atomic site along an edge dislocation in Si is usually considered as an acceptor. Thus the central region of the dislocation, the core, may be negatively charged and repels electrons, thus forming a positively charged space which obstructs the transport of free electrons in the material. In addition to the space-charge effects, dislocations may act as genertion - recombination centers for carriers, and may getter metallic impurities along the core, thus forming a conductive path [4.23].

The arrangement of atoms in a crystal lattice can be visualized considering the stacking of hard spheres. The sequence for a close packed lattice is reported in Fig. 4.14 a. When one close-packed layer has been completed (the A layer), the next layer can go into either of the two sets of hollows (B or C) on the first layer. The normal sequence of [111] plans in diamond and in face center cubic lattice is ABCABC. If the sequence has been changed to ABACA with the introduction of an extra plane in between B and C an extrinsic stacking-fault has been formed (Fig. 4.14b). If a {111}-C plane is removed between a {111}-B plane and a {111}-A plane from the ABC sequence, the stacking sequence is changed to CABABC and an intrinsic stacking fault has been formed (Fig. 4.14c). A high resolution TEM showing a stacking fault is reported in Fig. 4.14d.

The extrinsic stacking fault results from the clustering of self-interstitials, while the intrinsic is associated to the clustering of vacancies. The stacking faults is surrounded by dislocation line whose Burgers vector is normal to the defect.

Twins boundaries are planar defects. The boundary is a plane across which there is a spacial mirror image misorientation of the lattice structure (see Fig. 4.15a). The layer limited by two boundaries is called twins. These defects can be formed during the regrowth of amorphous layer on {111} oriented silicon substrates: see 5.2 . A high resolution transmission electron micrograph showing a dislocation and a twin boundary is shown in Fig. 4.15b. Another surface defect is a grain boundary, i.e. a region separating two crystalline grains of different orientation.

Volume defects are mainly associated to the precipitation of a second phase. In ion implanted silicon containing impurities above the solid solubility limit precipitation occurs easily where the crystal is damaged through heterogeneous nucleation and growth process. The driving force for the precipitation reaction is the supersaturation of the solute atoms or the chemical reaction as in the formation of compound like SiO_2. Other than oxygen precipitates, metallic precipitates of Cu, Co, Au are observed to form and to serve as nucleation sites for stacking faults during epitaxial growth of amorphous layers [4.24].

4.5 Primary Defects

The exact nature of isolated defects and of defect complexes produced by ion implantation is quite hard to characterize. Models to simulate the first stage of defect agglomeration have been

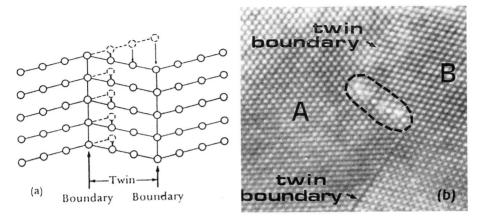

Figure 4.15 *Schematic of a twin layer surrounded by two boundaries (a). High resolution transmission electron microscopy showing a dislocation in the region surrounded by a dashed line and a twinned region B(b), each bright dot is a column of single Si atoms in the < 110 > projection and their periodicity is 3.1 Å (Courtesy of C.Spinella).*

proposed, but the large variety of involved parameters prevent a simple description. Molecular dynamics calculations have provided useful information on the evolution of the collision cascade in metals. A similar approach to semiconductors will be of relevance for the understanding of the defect formation. At this stage only a qualitative description can be given of the primary damage. After implantation several different damage configurations are present in the lattice: isolated point defects or point defect clusters in essentially crystalline silicon (light ions), local amorphous zones in an otherwise crystalline material (low dose, heavy ions) and continuous amorphous layer (high dose, heavy ions). These three different types of primary damage require as we will see in the next chapter different annealing strategies.

Let us consider now the influence of implant temperature and dopant on the damage. The temperature during the irradiation influences the amount of residual damage even for heavy ions. As an example let us consider the damaged created by 150 keV Au at low dose [4.25]. The disorder results in isolated amorphous zones surrounded by defect-free crystal material, similar to that shown in Fig. 4.16a.

The amount of Si atoms displaced from the lattice sites, per single *Au* ion as determined by channeling analysis, changes roughly

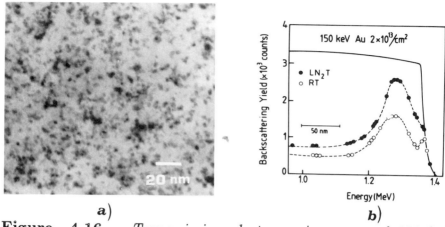

a)   b)

Figure   4.16   *Transmission electron microscopy of 150 keV*
$Au-2\times10^{13}/cm^2$ *implanted in Si showing small amorphous zones*
(a). Amount of damage as determined by channeling analysis of
150 keV $Au-2\times10^{13}/cm^2$ *implants in Si at room temperature and*
at liquid nitrogen temperature respectively (b) (from ref. 4.25).

of a factor two for implants at room (RT) and at liquid nitrogen (LN_2T) temperature. It is 4500 at RT and 7000 at LN_2T
respectively. Annealing is then taking place already at room temperature and several possible mechanisms may operate: i) isolated
point defects (as V and I) can migrate and recombine or vacancies
can move out of the core of the amorphous region thus shrinking
their radius, ii) defects created by the ion collision cascade interact with the already existing damage. This last process usually
called dynamical annealing will be considered in the section on
"hot implants".

    In the domain of temperature where defect migration occurs,
the presence of impurities influences also the amount and type of
primary damage. Formation of impurity-defects complexes either
mobile or immobile can influence the mobility and the recombination mechanisms of point defects. The effect of As dopant on
the defect production is shown in Fig. 4.17a [4.26]. The channeling spectra of undoped and As doped samples implanted with 400
keV Ge at a substrate temperature of 250^0C are reported. The
shape of the spectra indicates the presence of a large concentration of extended defects, mainly dislocation loops as confirmed by
TEM. The data indicate a considerable reduction in the residual
disorder by increasing the As concentration of the substrate. A
concentration of $1\times10^{19}As/cm^3$ reduces the amount of damage by

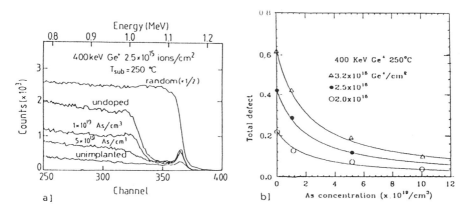

Figure 4.17 *Channeling spectra of As-doped Si samples implanted with Ge at 250°C (a). Amount of damage left by Ge implants at 250°C as a function of the As concentration (b) (from ref. 4.26).*

a factor of 2. The data of a detailed investigation are summarized in Fig. 4.17b where the amount of extended defects is reported vs the As concentration for different ion fluences. The reduction of residual damage in doped samples is related to the mechanism of defect agglomeration during the implant. The rate of extended defects formation is proportional to the average point defect concentration created during the implant. Since the aggregation and clustering of point defects is necessary for the formation of such extended defects, it will depend on the point defects recombination mechanisms. Arsenic atoms form with vacancies As-V pairs that act as recombination centers for self-interstitials through the following reactions [4.26]

$$As + V \leftrightarrow (As - V)_c; \quad (As - V)_c + I \leftrightarrow As + O \qquad 4.5$$

O is an occupied lattice site. In As doped samples the annihilation rate of the point defects is therefore modified by the presence of the dopant. In this model it is assumed that all the As atoms form pairs whose lifetime is longer than the lifetime of point defects. Therefore an As-V pair can assist vacancy-interstitial recombination before annealing out.

Usually implants are performed at room temperature and heavy ions forms amorphous zones. The amorphous formation during heavy ion implantation has been represented schematically in Fig. 4.3a. The model assumes that each ion produces a roughly cylindrical track of amorphous material along its path and that the conversion of the whole bombarded layer occurs when the cylinders overlap to include all the available volume ("track overlap"). As an example of the adopted procedure consider an implant of 40 keV phosphorus ions in Si. The number, N_d, of displaced atoms/cm^2 is given by

$$N_d = \phi N_{d_0} \qquad\qquad 4.6$$

where ϕ is the dose, and N_{d_0} the number of displaced atoms per incident ions. If $dE/dx|_n$ is constant and independent of the ion energy the concentration of displaced atoms/cm^3 is given by

$$n_d = \frac{\phi N_{d_0}}{R_p} = \phi \frac{dE}{dx}|_n \frac{1}{2E_d} \qquad\qquad 4.7$$

For 40 keV P in Si, $dE/dx|_n = 500 eV/nm$, $E_d \simeq 15 eV$, so that at an ion dose of $5 \times 10^{14}/cm^3$ the displaced density n_d equals the silicon density $5 \times 10^{22}/cm^3$. Experiments agree with this value, i.e. the amorphous is formed when the density of displaced atoms equals that of silicon.

Amorphous region can be formed also by light ion implantation at suitable target temperature and dose. The defect density in the single collision cascade is quite low and probably no amorphous region is present within the spatial extension of the collision. Amorphization occurs in this case by damage accumulation in the material until a critical threshold in the defect density is reached (4.27). Ions must strike region which had previously been damaged. The phase transition involves heterogeneous nucleation for heavy ions and homogeneous nucleation for light ions. The layer becomes amorphous when the free energy of the amorphous region equals that of the defect-rich region.

In terms of energy deposition it has been found that amorphous formation requires an energy deposition of $6 \times 10^{20} keV/cm^3$ [4.2] at low temperature and low energy, or that the energy deposition rate should be at least ~ 12 eV per atomic plane. This is about 100 times the energy released in the crystallization of amorphous Si (0.12 eV per atom). In the case of B at low energy with $\frac{dE}{dx}|_n \simeq \frac{0.6 eV}{nm}$ a fluence of $1.6 \times 10^{15}/cm^2$ is required for the amorphization. These estimates are valid at such low temperatures that defects are immobile. The threshold dose for amorphization increases for any ion with the target temperature up to a critical value above which no amorphous can be formed. The critical temperature depends on the ion mass and energy, i.e. on the energy

density deposited into nuclear collisions. Experimental results are
shown in Fig. 4.18 [4.28].

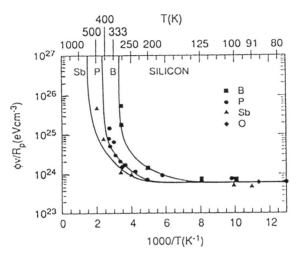

Figure 4.18 *Amorphization dose as a function of 1/T for several species implanted in Si at low energy (from ref. 4.28).*

This figure shows, for ions of various mass implanted at different temperatures, the energy density that must be deposited per unit volume through nuclear collision processes to produce a continuous amorphous layer. At low temperatures the energy density for amorphization is nearly independent of ion mass and temperature. For a light ion such as B implanted at room or higher temperatures, a continuous amorphous layer is never formed. Qualitatively the temperature dependence of the threshold dose, for heavy ions is accounted for assuming [4.29] that cylindrical amorphous tracks shrink in radius by annealing during the time of bombardment, so that more tracks are needed to completely fill the bombarded volume. In the case of light ions due to the annealing during the bombardment at high temperatures, the defect density never builds up to a high enough value to produce the amorphous phase. The real picture is probably much more complex than this simple description, due to the variety of mobile species that can interact among themselves or with preexting damage.

In the model it has been assumed that the formation of the early amorphous regions does not affect the formation of the following $\alpha - Si$ regions. However, once a $\alpha \cdot Si$ zone begin to nucleate in the irradiated volume, the large interfacial region between amorphous and crystalline Si is likely to act as sinks for ion beam generated defects. This will in turn modify strongly the balance of all the defect reactions occuring in the material and therefore the rate of the amorphization process. In addition the amorphous nucleates where the energy density exceeds the threshold value and even at low-temperature once nucleated the subsequent growth depends on the ion energy. Temperature is the main parameter to determine the evolution of the damage, and if defects are mobile then impurities can play also a relevant role as already shown by the data reported in Fig. 4.17.

4.6 Hot Implants

An ion during its slowing-down in the lattice, can deposit into nuclear collision an energy in excess of the threshold value for the amorphization ($\sim 12eV/at$) and a buried amorphous layer is formed. With increasing dose the total energy increases too and the amorphous layer enlarges. The growth sequence requires low dose rate or low-temperatures, at high dose rate and in the absence of a good cooling system the wafer temperature rises (Fig. 2.20) and dynamical annealing takes place. The defects created by the beam are now mobile and can annihilate or agglomerate. At this stage, defect production competes with dynamic defect annealing (see the next section in determining the resultant disorder structure. At high current densities, therefore, low doses produce an amorphous layer (the implantation time is too small to have an appreciable rise in temperature), high doses induce instead a dynamic annealing of the damage previously formed. This process is usually called self-annealing, [4.30] since the damage produced by ion implantation is subsequently annealed out by the ion beam itself. In the data of Fig. 4.18 the rise in temperature shifts the threshold energy density for amorphization toward higher and higher value.

The resulting defect structure is quite complex: a deep region of defect clusters, overlayed by a buried amorphous layer and by a thin crystalline surface layer. An example of self-annealing during 120 $keV P^+$ implantation onto a < 100 > Si substrate at a current density of $9.0\mu A/cm^2$ is shown in Fig. 4.19 [4.31].

The figure reports $2.0MeV He^+$ Rutherford backscattering spectra in the channeling configurations after different doses (see for a description of the analysis technique chapter 6). At low doses ($\sim 1 \times 10^{15}/cm^2$), $\sim 200nm$ thick $\alpha - Si$ layer is formed, at a dose of $7.5 \times 10^{15}/cm^2$ a single crystal, highly defective, is obtained. The wafer reached, according to calculations, a tempera-

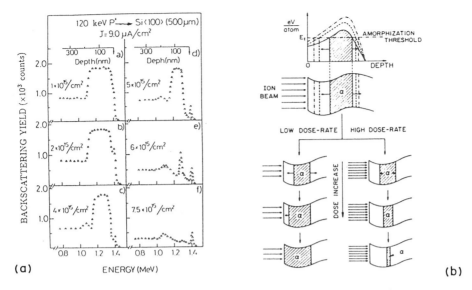

Figure 4.19 *(a) Channeling analysis of 120 keV P^+ implants at a current density of $9.0\mu A/cm^2$ for several doses in Si wafers without a thermal sink; (b) schematic of the disordering-ordering mechanism induced by ion bombardment in Si (from ref. 4.31).*

ture of 480^0C. A schematic illustration of the disordering-ordering mechanism induced by ion-beam bombardment is reported in Fig. 4.20(b).

As before said in the region in which the deposited energy overcomes the threshold value for the amorphization a buried amorphous layer is formed. With increasing the dose the amorphous layer enlarges. However at high dose rates, i.e. at high current densities, and in the absence of a good thermal contact with the sample holder, the substrate temperature increases and the competitive phenomenon of the recrystallization of the α-layer already formed takes place.

The as-implanted layers contain a lot of extended defects, such as twins and hexagonal silicon phase regions [4.32]. The evolution and the concentration of twins are shown in the TEM images and diffraction patterns of Fig. 4.20. The sequence a, b, c, refers to $< 100 > Si$ implanted with 1×10^{15}, 7.5×10^{15} and $1 \times 10^{16}/cm^2$ fluences respectively. The sequence d, c, f, refers to $< 111 > Si$ implanted with 120 keV P^+ at 4×10^{15}, 1×10^{16} and $1.2 \times 10^{16}/cm^2$ fluences respectively. All the micrographs except d are dark-field images obtained with $< 111 >$ twin spots.

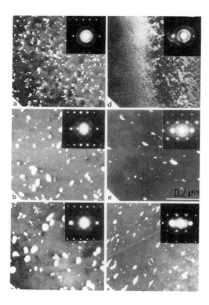

Figure 4.20 *TEM micrographs and electron diffraction patterns of 120 keV* P^+ *implanted* $< 100 >$ *(a)-(c) and* $< 111 >$ *(d)-(f) Si oriented substrates. (a)*$4 \times 10^{15}/cm^2$*, (b)* $7.5 \times 10^{15}/cm^2$*, (c)*$1 \times 10^{16}/cm^2$*, (d)* $4 \times 10^{15}/cm^2$*, (e)* $1 \times 10^{16}/cm^2$*, and (f)* $1.2 \times 10^{16}/cm^2$*. The micrographs are dark-field images from Si* $< 111 >$ *twin spots but (d) is a bright field image. The white particles are twinned regions (from ref. 4.32).*

The holes in the diffraction patterns indicate the presence of an amorphous layer as, for example after a fluence of $1 \times 10^{16}/cm^2$ into $< 111 > Si$ (Fig. 4.20c). Additional spots are found in the diffraction patterns as the fluence increases and are associated to twins and hexagonal silicon particles [4.33].

The defect formation is quite complex in these experiments and a much simpler experimental condition is achieved performing the implants at different substrate temperature and at different ion energies to investigate the nucleation and growth of the $\alpha - Si$ zones by varying the defect mobility (temperature), and energy density (collision cascade). As an example consider implants of Ge at 60 and 400 keV energies in Si for substrate temperatures ranging from 77 to 423 K. The amorphous fraction, as measured by RBS and channeling, increases with ion fluence almost linearly at 60 keV and more than linearly at 400 keV (see Fig. 4.21) [4.34].

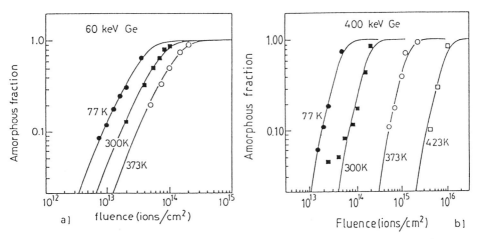

Figure 4.21 *Amorphous fraction created by 60 keV Ge⁺ implants in Si at different temperatures as a function of the fluence (a), same as (a) but for 400 keV Ge⁺ (from ref. 4.34).*

The amorphization fluence (i.e. the dose required to produce an amorphous fractions of 90%) strongly depends on the temperature as shown in Fig. 4.22 for the two energies. The experimental points follow the same trend for both energy implants. At low-temperatures the slight variation in the amorphization fluence for high and low energy implant is due to the difference in the energy density deposited into atomic processes. The average energy loss is 700 eV/nm for 60 keV Ge and 600 eV/nm for 400 keV Ge. The energy density for the formation of the amorphous phase is almost the same in both cases. The critical temperature is 450 K for 400 keV Ge and about 600 K for 60 keV ion. The difference is associated to the details of the defect density within the single collision cascade. In the low energy case the energy is deposited uniformly throughout the cascade volume, while in the high energy case it is deposited only in a small fraction of the cascade volume.

These regions are called subcascades and are relevant at high beam energy. Simulated ion trajectories for 60 and 400 keV Ge are represented in Fig. 4.23, as obtained by TRIM program.

This shape justifies the previous considerations. According to the dynamical annealing process the transition temperature between damage build-up and damage annealing increases with decreasing the size of the disordered regions.

At temperature where defects are mobile the disorder depends

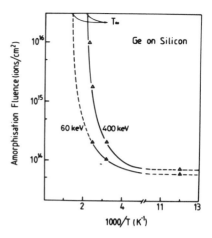

Figure 4.22 *Amorphization fluence for 60 and 400 keV Ge^+ implants in Si as a function of 1/T (from ref. 4.34).*

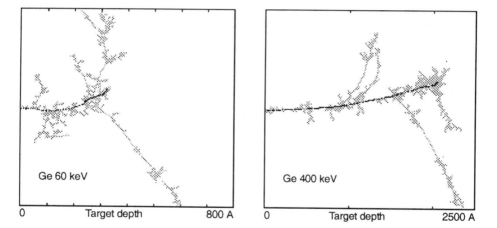

Figure 4.23 *TRIM trajectories for 60 and 400 keV Ge^+ ions in Si.*

also on ion dose rate. Fig. 4.24 reports the displaced atoms vs dose-rate at 373 K for low and high energy Ge implants.

The disorder is slightly dose-rate dependent at 60 keV, while it does at 400 keV. As shown in Fig. 4.23b a large volume of the 400 keV collision cascade contains only point defects and small defect clusters. With increasing the dose-rate, rises the probability of encounters of these defects to form large clusters instead to

Figure 4.24 *Dependence of the damage on dose rate for 60 and 400 keV - $2\times 10^{14}/cm^2$ implants in Si at 373 K (from ref. 4.34).*

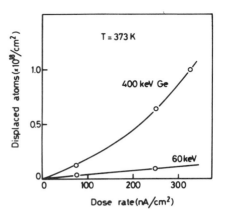

annihilate at sink. Inter collision cascade effects are then present. At low energy instead the disorder extends all over the collision cascade volume and then the dose rate does not influence considerably the amount of damage. The two models of amorphization formation described in par. 4.5 are unable to explain the curves of Fig. 4.21. If heterogeneous nucleation is assumed, the diameter of the amorphous region along the single ion track should be less than $1nm$, which is an unphysical value. Considering a more realistic diameter of 10 nm, the overlapping of more than 4 ions in the same region is needed to produce amorphous which is also quite unlikely to occur.

The quantitative description of the homogeneous amorphization is quite complex and recently another description has been presented [4.35]. It is based on the following considerations:

i) an energetic ion crossing an amorphous crystalline ($\alpha - c$) interface can induce amorphization or crystallization depending on the substrate temperature and ion dose-rate.

ii) similar behaviour, i.e. growth or shrink, is followed by small and isolated amorphous clusters,

iii) the growth and shrink rates, i.e. amorphization or crystallization, are dopant dependent,

iv) The lifetime of ion beam induced defects is of the order of seconds at 250^0C and of days at room temperature.

Not only the very fast and energetic early stages of the collision cascade ($10^{-12}s$, Fig. 4.1) but also the long-living defects must be considered when describing these processes. Once amorphous islands are nucleated in the irradiated volume, the long living defects interact with the interfaces between these islands and the crystal and can induce their growth or shrinkage depending on the irradiation temperature and on the ion flux. According to this assumption the kinetics of amorphization growth can be described by the Avrami-Johnson-Mehl [4.36] equation for the phase transition with no composition change and where nucleation and growth occur simultaneously. The growth velocity u of the amorphous zones depends on several parameters (temperature, dose rate, dopant, energy density) etc. that do not change during the implant. If V_ϕ is the amorphized volume after a dose ϕ, the amorphous fraction F_ϕ is V_ϕ/V_0, being V_0 the initial volume to be transformed. The change with dose of the amorphous fraction becomes then

$$\frac{dV_\phi}{V_0} = dF_\alpha \cdot Id\phi(1 - F_\phi) \qquad 4.8$$

V_α is the volume of the amorphous region that was nucleated at dose ϕ', for a spherical shape $V_\alpha = \frac{4}{3}u^3(\phi - \phi')^3$, I is the nucleation frequency, i.e. the number of nuclei forming per unit volume of untrasformed material and per unit dose.

Integrating equation, 4.8 with I and u constant, the amorphous fraction becomes

$$F_\phi = 1 - exp\left[-\frac{\pi}{3}u^3 I\phi^4\right] \qquad 4.9a$$

or in more general form

$$F_\phi = 1 - exp[-k\phi^n] \qquad 4.9b$$

with $k = Iu_\alpha^{n-1}$, n is an exponent which for three-dimensional growth lies in the range 3-4. Several experimental curves of the amorphization fraction vs dose are fitted by n=3.5, that is a reasonable value. In the following discussion n is assumed 3.5 so that the fitting parameter is now k which contains the growth rate and the nucleation frequency. These two quantities can be separated if one can measure the growth rate in the absence of nucleation. This is possible by performing irradiation on a preexting amorphous layer or on isolated amorphous regions. The schematic arrangement is shown in Fig. 4.25.

At temperatures below a critical value, dose rate and dopant dependent, the amorphous material grows layer by layer with a shift of the planar amorphous/single crystal interface or with an increase in the size of the isolated damaged region [4.37].

At temperature above the critical value a layer by layer crystallization of the amorphous region or shrinkage of the isolated zones takes place instead. The crystallization and amorphization rates

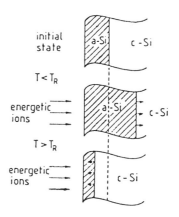

Figure 4.25 *Schematic of the crystallization and amorphization growth by irradiation of a preexisting amorphous layer. At a temperature above T_R the amorphous layer shrinks, at temperatures below it grows.*

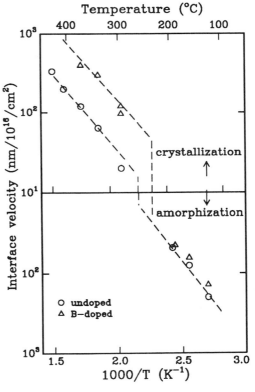

Figure 4.26 *Crystallization and amorphization rates as a function of temperature for undoped and B-doped Si wafers irradiated with 600 keV Kr (from ref. 4.38).*

as determined by experiments [4.38] are reported in Fig. 4.26, as a function of the reciprocal temperature for undoped and B-doped samples. The upper part refers to crystallization regime at higher temperature substrate. The boron concentration of the doped samples was $\sim 1 \times 10^{20}/cm^3$.

Ion flux is a critical parameter in controlling amorphization. This is illustrated in Fig. 4.27 where the interface displacement rate is reported as a function of the reciprocal temperature from several dose rates [4.39].

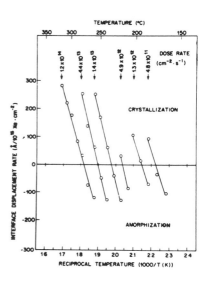

Figure 4.27 *Displacement rate of the c-a interface versus the reciprocal temperature for irradiations with a 1.5 MeV Xe beam at different dose rates. Upper part crystallization regime, lower part amorphization regime (from ref. 4.39).*

With increasing dose rate the critical or the the reversal temperature, at which the interface displacement is zero, is shifted towards higher values.

The critical ion flux for amorphization is plotted as a function of temperature for 1.5 MeV Xe to a fixed fluence of $9 \times 10^{15}/cm^2$ in Fig. 4.28 [4.39].

At this dose, conditions to the right of the line result in the formation of a buried amorphous layer whereas those to the left do not produce amorphization. The experimental points follow an Arrhenius dependence with a slope of 1.8 eV. This value has been associated to the activation energy of formation and dissociation of the Si divacancy. The knowledge of the amorphization rate allows the determination of the nucleation rate in equation 4.9a through a fit of the experimental data. As an example in Fig. 4.29 the amorphous fraction is reported as a function of ion fluence for 400 keV Ge ions impinging on undoped and B-doped silicon at $100^0 C$ [4.40].

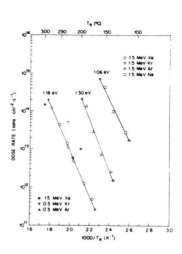

Figure 4.28 *Arrhenius plot of the reversal temperature for a given dose rate (from ref. 4.39).*

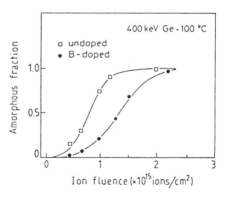

Figure 4.29 *Amorphous fraction versus ion fluence for undoped and B-doped < 100 > Si samples irradiated at 100^0C with $400\,keV$ Ge ions. The continuous lines are fits to the data by means of the model described by Eq. 4.9 (From ref. 4.40).*

The solid lines represent the fits to the experimental data by means of Eq. 4.9b and using n=3.5 for all the curves. The nucleation rate I can be then derived. It decreases from ~ 5 nuclei/ion at room temperature to 0.04 nuclei/ion at $150^0 C$ for 400 keV Ge ions. The value larger than one may be related to the number of dense subcascades in the single collision cascade (see Fig. 4.23). The presence of boron enhances the crystallization rate and depresses then the net amorphization rate. Small change in u influences considerably the amorphous fraction dependence on the dopant, being u in the exponential to a power of 2.5 (eq. 4.9a).

The nucleation rate evaluated in the doped samples coincides within 20% with that of the undoped samples. The nucleation of amorphous region is probably associated to the prompt part of the collision cascade that is independent of the dopant. Doping effect is less pronounced when amorphization is made at room temperature and is totally absent at 77K, in agreement with the decrease of defect mobility with temperature. The influence of the dopant is then associated to the delayed part of the collision cascade.

The residual damage in so-called hot implants results in a variety of extended defects: dislocation lines, loops, stacking-faults, twins, hexagonal silicon phase. This last defect is metastable with respect to the diamond structure, but is stabilized by the presence of internal stresses [4.33]. As discussed with reference to Fig. 4.20, hexagonal regions are found also after high current and high dose implants [4.32]. In implants performed at a well defined temperatures it has been found that hexagonal silicon particles are formed at target temperatures above $350^0 C$ and at fluences higher than $2 \times 10^{15}/cm^2$. By increasing the wafer temperature during the implant the threshold fluence increases slightly [4.41].

The mechanism for a diamond cubic to a hexagonal silicon phase transformation has been associated with the effect of a large uniaxial compressive stress along the $< 100 >$ direction. This transformation is accomplished by a rearrangement of bonds between atoms and a subsequent relaxation of the rearranged lattice in an orderly manner [4.42]. The energy required for the transition from diamond cubic to hexagonal phase is of the order of $10^9 erg/cm^2$ [4.33].

In the case of ion implantation this energy may be supplied by the high energy density deposited by the ion in the cascade volume or by a direct transfer of momentum for the impinging ions to the atoms. The critical temperature of about $350^0 C$ may be related to the local and short distance migration of atoms.

The hexagonal phase is stabilized by the stress induced by other defects which are gradually created during the implantation process. With increasing temperature the probability of defect recovery increases and a higher fluence is then required to stabilize the phase. The hexagonal phase regions anneal out at temperatures above 700^0C. In this temperature range the dissociation of interstitial clusters of the easy glide of dislocations can relieve the stress and destabilize the phase. The temperature range for the existence of the hexagonal phase in hot implants follows closely that found by indentation.

4.7 Ion Beam Induced Enhanced Crystallization

Although the crystallization under ion beam is related to the pure thermal counterpart, we prefer to discuss it in this chapter and devolve the next completely to the influence of thermal annealing on the evolution of defects left by the implant. Studies on ion-beam-induced epitaxial crystallization are performed by heating a pre-existing amorphous layer onto a single- crystal substrate at a fixed temperature and by irradiating it with ion beams having low current densities, in order to avoid further heating [4.43]. The adopted beam energies is such that the projected range of the irradiating ions is well beyond the original c-a interface. This allows one to discriminate between the damage clustering, which is typically produced at the end of range, and the effects of a passing beam on a pre-existing amorphous layer, i.e. the interaction of point defects created by the beam with an α -layer. The regrowth is followed by RBS channeling measurements (see chapter 6) or in situ by transient reflectivity of a polarized laser beam that impinges on the sample during the irradiation [4.44]. The reflectivity oscillates with decreasing the amorphous thickness due to successive constructive and destructive interferences occurring between the light reflected from the surface and from the α-c interface.

The first and main result is the large enhancement of the crystallization kinetics induced by the ion beam irradiation. It is possible to regrow at temperature as low as 250^0C with a rate of 0.007 nm/s under 600 keV $\cdot 1 \times 10^{12} Kr/cm^2 \cdot s$ irradiations. At this temperature the pure thermal regrowth rate is negligible ($\sim 10^{-11} nm/s$). The beam induced regrowth depends on the energy density deposited into nuclear collisions as shown in Fig. 4.30 by the comparison of the growth rate with the number of vacancies generated by the 600 keV Kr beam irradiation and calculated by the TRIM code.

The experimental rates closely follow the shapes of the generated vacancies. This trend suggests that the ion-induced growth rate is associated with the production of point defects, or in a more

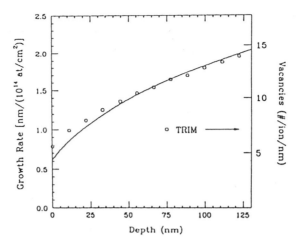

Figure 4.30 *Growth rate versus depth for an* $\alpha - Si$ *layer recrystallized at* 350^0C *by 600 Kr ions (from ref. 4.45).*

general way, with the energy lost into elastic collisions at the α-c interface by the impinging ions. Long-range diffusion of ion-induced defects toward the α-c interface is inconsistent with these experimental results. The primary role played by the α-c interface in IBIEC (Ion Beam Induced Epitaxial Crystallization) is further demonstrated by the presence of an orientation dependence. In Fig. 4.31 the growth rates for $\alpha - Si$ layers onto $< 100 >$ and $< 111 >$ oriented substrates irradiated at 350^0C with a 600 keV Kr beam are shown [4.45].

Figure 4.31 *Growth rate versus depth for* $\alpha - Si$ *layers onto* $< 100 >$ *and* $< 111 >$ *oriented substrates and recrystallized at* 350^0C *by Kr ions (from ref. 4.45).*

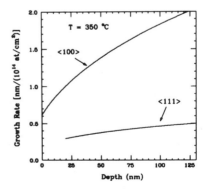

The rate is much lower (almost a factor 4) on the $< 111 >$ substrate with respect to the $< 100 >$ substrate. As we will show in the next chapter the dependence of interface velocity on substrate orientation is well established in pure thermal annealing processes. The thermal regrowth onto < 111 > oriented substrate is, ~ 25 times lower, than on < 100 > substrates. These results suggest that probably the same interfacial defect responsible for thermal regrowth is thought to promote IBIEC, the role of the ion beam being that of changing the average defect concentration. The dependence of the growth rate on temperature is illustrated in Fig. 4.32, for both IBIEC and thermal annealing [4.46].

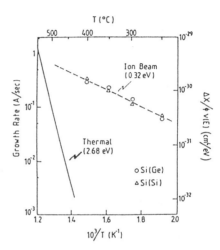

Figure 4.32 *Ion - induced growth rate vs reciprocal temperature for* α *- Si layers recrystallized by irradiation with 600 keV Kr beam.*

The growth rate is reported both in nm/s (left-hand side), taking into account that the ion flux was $1 \times 10^{12} Kr/cm^2 \cdot s$, and in cm^4/eV (right-hand side). This latter scale represents the growth rate in the form $\Delta X/\phi\nu(E)$, ΔX being the extent of the regrowth, ϕ the ion fluence and $\nu(E)$ the energy deposited into displacement production at the $\alpha \cdot c$ interface. The data indicate an activation energy of 0.32 eV to be compared with 2.68 eV for the pure thermal process. The meaning of this so low value is not clear, and probably it represents only an apparent activation energy. The linear dependence in a logaritmic plot comes, instead, from a balance between crystallization and amorphization, as we will discuss briefly at the end of this section.

The presence of dopants like B, P and As enhances the IBIEC, although in a less manner than the pure thermal process. Fig. 4.33 reports the ion-induced growth rates, deduced from an analysis of the reflectivity signals, versus the depth. The data are shown for undoped and boron doped samples. In the same figure the B profile is reported also in a logarithmic scale. At a B concentration of $8 \times 10^{20}/cm^3$ the rate is enhanced by more than a

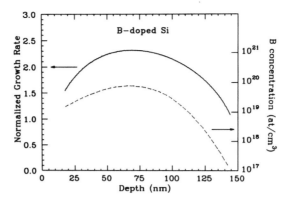

Figure 4.33   *Dependence of the ion-induced growth rate on the B concentration reported in a logarithmic scale (right-hand side). The irradiation was performed with 350 keV Kr ions (from ref. 4.46).*

factor 2. The shape of the curve shows strong similarities with the B profile plotted in a logarithmic scale and it is suggested then a logarithmic relationship between the ion induced growth rate and B concentration. P doping produces rate enhancements similar to those presented for B and also in this case a logarithmic concentration dependence has been observed [4.47].

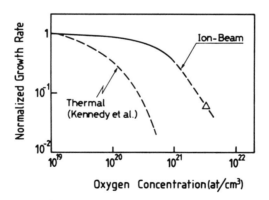

Figure 4.34   *Dependence of the normalized growth rate on O concentration for both IBIEC and pure thermal growth (from ref. 4.46).*

Impurities like oxygen depress instead the purely thermal epitaxial growth even at concentrations of $10^{20}/cm^3$ [4.48]. During

IBIEC a similar, but less severe, retardation is observed. The dependence of the ion-induced normalized growth rate (i.e. the rate of O-implanted sample divided by that of the impurity free sample) is shown in Fig. 4.34 (600 keV Kr - 450^0C). For comparison the data relative to pure thermal annealing at 550^0C are also reported. Probably O forms strong bonds with the Si atoms (SiO_2), at the $\alpha-c$ interface. These strong bonds hinder the bond breaking required for crystallization and retard it. During ion irradiation some of these bonds may be broken and the regrowth can take place.

Crystallization under ion beam has been also obtained for deposited silicon layers in presence of a thin native oxide layer at the interface [4.49]. The same sample cannot be epitaxially crystallized by a pure thermal treatment and a polycrystalline structure results. During ion irradiation the interfacial oxygen layer is broadened by ions beam mixing and the oxygen concentration decreases up to a such value to allow the epitaxial regrowth. The intermixing is shown in the TEM cross section of Fig. 4.35 [4.50].

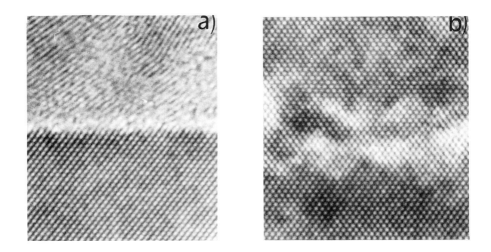

Figure 4.35 *High resolution transmission electron micrograph of the deposited polycrystalline Si layer - Si single crystal substrate before (a) and after $600 keV Kr^+$ ions irradiation (b) (from ref. 4.50).*

The process is not limited to $\alpha - Si/Si$ structures and has been applied for the fabrication of silicon on insulator structures, of

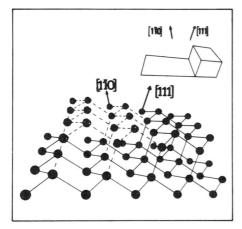

Figure 4.36 *Schematic representation of the c − α interface. Closed spheres represent crystalline atoms while open spheres represent amorphous atoms. The interface is composed by (111) planes connected by [1̄10] ledges*

Si/Ge heterostructures [4.51] and of epitaxial silicides like $NiSi_2$ [4.52]. The ion beam regrowth has also been applied to damaged compound semiconductors such as $GaAs$ [4.53] and InP [4.54]. In this last compound twin formation is reduced drastically during ion beam regrowth of the amorphous layer. These results are a clear indication of the large potentialities offered by this novel technique in crystal growth.

Ion beam induced crystallization and amorphization is a quite complex process and there have been several attempts to model it by an atomistic description. A quantitative phenomenological model has been elaborated by Jackson [4.55] and subsequently modified to include the dopant and the orientation dependent. In this description each impinging ion converts, within a single collision cascade, a small volume of crystal at the $\alpha - c$ interface to the amorphous state and at the same time, creates defects that promote crystallization. These defects can diffuse only small distances from this generation site, cannot escape away from the volume of the collision cascade and are assumed to annihilate in pairs. The net motion of $c - Si/\alpha - Si$ interface is then a consequence of a balance between two distinct terms: amorphization and temperature dependent crystallization

$$R = R_c - R_a = \frac{a < N > \Lambda}{\phi \tau_j} - V_\alpha exp \left(\frac{E_\alpha}{kT} \right) \qquad 4.10$$

Λ is the volume recrystallized at the $\alpha - c$ interface by a single defect jump, a is the lattice parameter, $< N >$ is the average defect density, ϕ is the dose rate and τ is the defect jump period. This period is temperature dependent with an exponential form of the type $\frac{1}{\tau_j} = \frac{1}{\tau_0} \left(exp - \frac{E_j}{KT} \right)$ and $E_j \simeq 1.2 eV$ The amorphization term is characterized by an activation energy E_α. By assuming that the defects promoting crystallization annihilate in pairs the average defect density $< N >$ can be evaluated and the net rate results:

$$R = \frac{\Lambda \pi r_0^2}{\sigma^2} ln \left[1 + \frac{\gamma}{2} \left(1 + \sqrt{1 + \frac{4}{\gamma}} \right) \right] - V_\alpha exp \left(\frac{E_\alpha}{kT} \right) \qquad 4.11$$

where

$$\gamma = \frac{N_0 \sigma^2 a \tau_0}{\tau_j} \qquad 4.12$$

r_0 is the radius of the collision cascade, σ^2 is the cross section for defect annihilation and τ_0 is the average time between the arrival of two consecutive ions in the same region. The influence of dopants on the crystallization can be explained assuming that defects exist in charged states with different mobilities [4.56]. As a consequence the average mobility becomes:

$$\frac{1}{\tau_j} = \frac{1}{\tau_j^0} \frac{N_0 - fN_d}{N_0} + \frac{k}{\tau_j^0} \frac{fN_d}{N_0} = \frac{1}{\tau_j^0} \left[1 + \frac{f(k-1)}{N_0} N_d \right] \qquad 4.13$$

where τ_j^0 is the jump period of the neutral defects, k is the ratio between the mobility of charged and neutral defects and f is the fraction of dopant active in the charging process.

The term γ is modified as follows

$$\gamma = \gamma_0 (1 + \beta N_d) \qquad 4.14$$

with $\beta = \frac{F(k-1)}{N_0}$. The dependence on the dopant is then related to the logaritmic of the concentration, in agreement with the experimental results reported in Fig. 4.33.

The orientation dependence of the recrystallization growth and the temperature dependence of the amorphization rate require a knowledge on the interfacial defects responsible of these two processes. Recently a detailed investigation on the dependence of the growth rate on substrate orientation has provided crucial informations for atomistic growth models [4.57].

The $a - Si/c - Si$ interface consists of (111) terraces separated by $[01\bar{1}]$ ledges to minimize the number of unsatisfied crystal bonds. To be considered part of the crystal an atom must have bonds to two other crystal atoms, immediately restricting crystallization to the ledges. Dangling bond generation and propagation on the ledges, or defect-assisted motion of kinks along the ledges result in the crystallization. The geometry of the interface is shown in Fig. 4.36, while the dynamic of the kink motion is represented schematically in Fig. 4.37a and Fig. 4.37b [4.50]. The orientation dependence of the growth rate is determined by the areal density of ledges. During IBIEC the defect flux generated by the beam interacts with interface bonds and catalyzes a limited number of crystallization events per defect.

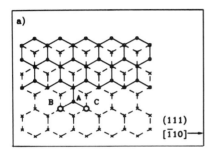 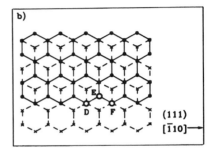

Figure 4.37   *Section of the c − a interface along a {111} terrace, the crosses represent atoms in the lower plane, while closed circles are atoms in the top plane. In (a) the recrystallization of the atom lebelled A generates two kink sites (B and C). These are the next amorphous atoms to recrystallize. In (b) the two kinks lebelled D and F are going to annihilate each other. Recrystallization of atoms O, F and E will produce the annihilation. Kinks are generated in pairs (from ref. 4.56).*

Defects must break bonds on ledges to cause crystallization, thus the controlling factor will be the ratio of interface bonds where crystallization can occur compared to the total number of interface bonds. The other effect caused by the interaction is a change of the interfacial bonding configuration that in turn may affect the chance of the next arriving defect causing growth. The phenomenological model previously described determines how many defects are near the interface, and the atomistic description determines their effect on the interface. For what concern the identity of the defect, the most likely candidate is a dangling bond in the amorphous region [4.58]. If a dangling bond encounters the $\alpha - Si/c - Si$ interface it is plausible that it would lead to bond rearrangements. Furthermore, dangling bonds recombine in pairs and thus meet the requirements of both the phenomenological and atomistic models of IBIEC. Defects produced in the crystalline side (such as Frenkel pairs) are considered to be responsible of amorphization. Ion generated interstitials may agglomerate at the pre-existing c-a interface producing a preferential site for amorphization. The rate of this process is inversely proportional to the recombination rate of Frenkel pairs. Therefore, decreasing the temperature the condensation probability increases with an exponential law, where E_α (see eq. 4.10) may represent the activation energy for recombination of vacancy- interstitial pairs. This description agrees with the experimental value of $E_\alpha \sim 3.0\text{eV}$. The absence of an orientation dependence suggests that the terraced interface structure present during crystallization is probably lost when the interface motion is reversed. Therefore the c-a interface does not contain preferential amorphization sites and has a similar structure regardless of the substrate orientation.

4.8 Ion Implantation into Localized Si Areas

The reduction in junction size requires the determination of the two-dimensional distributions of primary and secondary damage as well as impurities in the implanted and annealed layers. In the previous chapter we have considered the influence of mask-pattern dimension on the lateral straggling of implanted ions, in this chapter the influence of mask and lateral straggling on the primary damage will be considered, while in the next chapter we add the influence of the thermal annealing on the final aspect of the secondary damage.

The distribution of primary damage in terms of Frenkel pairs can be obtained by Monte Carlo simulation using the Kinchin-Pease equation. Fig. 4.38 shows a typical example of calculations in terms of two-dimensional Frenkel pair density contours normalized by the peak concentration for 70 keV B implant. The size of the mask varies between 0.04 and 0.4 μm. The decrease of the mask size shifts toward the surface the depth at which the maximum concentration of Frenkel pairs occurs [4.59].

The results of the simulation show that Frenkel pairs move toward the surface with a decrease in the mask size below a threshold value characteristic of the implant. The threshold size decreases with the ion mass. The lateral straggling of the impinging ions decreases with the ion mass, being reduced the probability of collisions at large angle for heavy ions. A decrease in beam energy is also accompanied by a corresponding decrease of the threshold mask size value, again in agreement with the dependence of the lateral straggling on the ion energy.

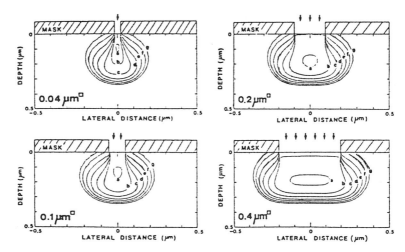

Figure 4.38 *Monte Carlo simulation of Frenkel pair density contours for 70keV B implants through a mask of different size. The density contour lines are normalized to the peak concentration: (a) 9×10^{-1}; (b) 3×10^{-1}; (c) 10^{-1}; (d) 3×10^{-2}; (f) 3×10^{-3}; (g) 1×10^{-3} (from ref. 4.59).*

Not only the mask size influence the location at which the maximum concentration of Frenkel pair occurs, but also the peak concentration decreases with a decrease in the mask size below the threshold value. The critical size is about $0.4\mu m$ for B, $0.2\mu m$ for P and $0.1\mu m$ for As. As a consenquence if the implant is performed on the same wafer with masks of different sizes the dopant concentration will vary from one area to another. From the point of view of device yield the distribution and type of secondary defects is a key parameter.

CHAPTER 5

ANNEALING and SECONDARY DEFECTS

5.1 Introduction

The major drawbacks of the ion implantation process are the secondary defects that form during the thermal treatment required to activate electrically the dopant and to restore the crystalline order. Defects are detrimental for the electrical behaviour of the devices if located in the active regions and limit drastically the yield as shown in Fig. 1.14. Annealing is the key step of the implant and most of the work has been performed in this field to understand the evolution of defects, their aggregation, the influence of the ambient in which the annealing is performed, of the thermal procedure etc. Prolonged anneal at high temperature could result in a perfect crystal, but other related effects such as the diffusion of dopant may not be compatible with the device characteristics.

The aim of the annealing is the production of defect-free silicon or nearly defect-free silicon with the minimum amount of thermal budget. The trade-off between electrical profile required by the device characteristics and the annealing procedure should be considered in details for any specific application. There are no unique recipies for the annealing, each of them depends on the device, and on its structure. e.g. the mask breaks the usual planar symmetry. In the following we will describe the annealing and the residual defect for several kinds of damage: amorphous layers, isolated amorphous zones, and heavily disordered crystalline regions. The residual damage results mainly in extended defects, like dislocation lines, dislocation loops, intrinsic and extrinsic stacking-faults, twins etc. The chapter is dedicated to silicon, in view of

the different behaviour the anneal procedure of implanted compound semiconductors such as *GaAs* and *InP* will be considered
in chapter 8. A section is also devoted to the description of rapid
thermal systems in view of their relevance as heating source in
several process steps of VLSI and ULSI.

5.2 Solid Phase Epitaxial Growth of Amorphous Silicon

Heavy ions implanted at room temperature and at doses
above $10^{14}/cm^2$ form an amorphous surface layer whose thickness
depends mainly on the beam energy. At high energy the amorphous layer is buried and two interfaces with the single-crystal
substrate are present. When the implanted sample is annealed
in a furnace at a fixed temperature above $550^0 C$ the amorphous
layer recrystallizes by a movement of the planar single crystal-
amorphous interface with a velocity which depends on temperature, substrate orientation and doping [5.1].

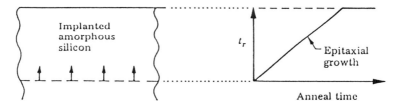

t_r = regrown layer thickness

Figure 5.1   *Solid-phase epitaxial regrowth versus anneal time
for an amorphous implanted layer on $< 100 > Si$*

The process is called Solid Phase Epitaxial Growth (SPEG). As
shown in Fig. 5.1 for $< 100 >$ oriented silicon wafers the thickness
of the regrown layer increases linearly with time, thus indicating
a constant growth velocity. At $550^0 C$ the interface velocity is 0.12
nm/s and measurements of the growth velocity are shown in Fig.
5.2 in an Arrhenius plot for layers amorphized by multiple energy
Si implants at low temperature [5.2].
The measured velocities extend over nearly 10 orders of magnitude
and are characterized by a single activation energy $E_A = 2.76eV$.
The regrowth velocity, v_g, is given then by

$$v_g = v_o \ exp(-E_A/kT) \qquad\qquad 5.1$$

where the preexponential factor $v_0 = 3.68 \times 10^8 cm/s$. The value
of the activation energy is about half that observed in diffusion of

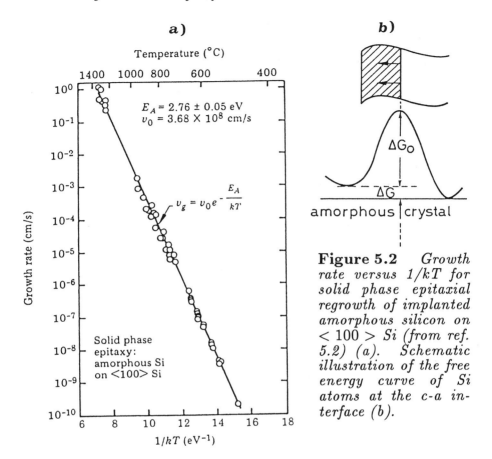

a)

Temperature (°C)

$E_A = 2.76 \pm 0.05$ eV
$v_0 = 3.68 \times 10^8$ cm/s

$$v_g = v_0 e^{-\frac{E_A}{kT}}$$

Solid phase epitaxy: amorphous Si on <100> Si

Growth rate (cm/s)

$1/kT$ (eV^{-1})

b)

ΔG_0

ΔG

amorphous | crystal

Figure 5.2 *Growth rate versus $1/kT$ for solid phase epitaxial regrowth of implanted amorphous silicon on $< 100 > Si$ (from ref. 5.2) (a). Schematic illustration of the free energy curve of Si atoms at the c-a interface (b).*

dopants in Si. Consequently regrowth of the implanted amorphous layer can be carried out at times short compared to those required for appreciable dopant diffusion.

During the epitaxial growth process implanted dopants move onto substitutional lattice sites as the crystal-amorphous interface crosses their location. Substitutional concentrations of group III and V dopants can exceed the equilibrium solubility limits, as the dopants are frozen at these relatively low regrowth temperatures.

The driving force for the amorphous to single crystal ($\alpha \rightarrow xtl$) transition is the lowering of the free energy: in the amorphous phase [5.3-5.4] it is higher of the crystalline one by about 0.12 eV/atom. The interface is then pushed toward the amorphous part. The energy diagram is shown in the inset of Fig. 5.2 for an atom at the two contiguous phases separated by the interface. The transition rates of atoms from $a \rightarrow b$ and $b \rightarrow a$ at a given temperature T are given by

$$r_{a \to b} = K_{a \to b} \cdot exp \left[\frac{-\Delta G_o + \Delta G_i(T)}{kT} \right] \qquad 5.2a$$

$$r_b \to_a = K_{b \to a} \cdot exp \left[\frac{-\Delta G_0}{kT} \right] \qquad 5.2b$$

the net rate is then

$$r = r_{b \to a} - r_{a \to b} \simeq K \; exp \left[\frac{-\Delta G_0}{kT} \right] [1 - exp \left(-\Delta G_i/kT \right)] \qquad 5.3$$

assuming $K_{a \to b} \simeq K_{b \to a}$.

The driving force $\Delta G_i (\sim 0.1 eV)$ is higher than kT and is approximately temperature independent, then

$$r = K \; exp \left[-\frac{\Delta G_0}{kT} \right] \qquad 5.4$$

In terms of energy and entropy of activation the previous expression coincides with eq. 5.1. The details of the atomic defect structure responsible of the growth are still under investigations as discussed in the previous chapter on the ion beam induced epitaxial regrowth of amorphous layer.

Recrystallization of the amorphous layer is strongly orientation -dependent, < 100 > oriented wafers recrystallize faster than < 111 >-oriented wafers as shown in Fig. 5.3 [5.1-5.2].

All orientations exhibit the same activation energy, but < 100 > samples regrow about 3 times faster than < 110 > samples and about 25 times faster than the initial growth rate for < 111 > substrates. The growth of < 111 > oriented silicon is not linear with annealing time and after a regrowth of about 100nm, it exibits pronounced twinning along (111) planes, the twins occupying about 30-40% of the overall volume of the regrown layer [5.6]. The orientation dependence of the growth velocity is summarized in Fig. 5.4 [5.1].

The regrowth seems governed by cooperative interfacial bond breaking and local rearrangement rather than long range transport of silicon atoms [5.7]. The sites at the interface responsible for the regrowth have been associated to kinks at the [110] ledges connecting two consecutive (111) planes at the $\alpha - xtl$ interface. (see also chapter 4, and the description of the IBIEC mechanism)

Figure 5.3 *Arrhenius plot for isochronal anneal of amorphous layers on Si substrates of different orientation. The amorphization was performed at 77K by multiple energy implants of Si²⁸ (from ref. 5.1).*

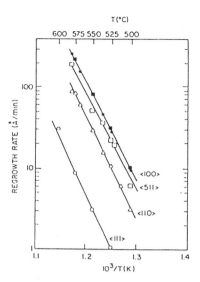

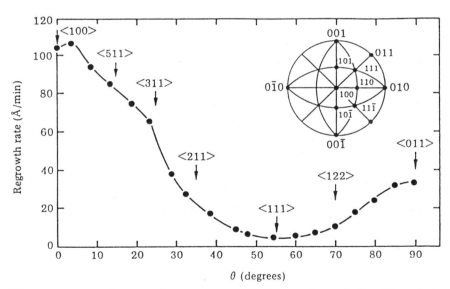

Figure 5.4 *Regrowth rate versus orientation of the Si substrate for implanted amorphous Si annealed at 550°C (from ref. 5.1).*

Kinks represent a structural deformation of the $\alpha - xtl$ interface. They generated there, diffuse there and annihilate there. The regrowth proceeds via a bond breaking process and a subsequent rearrangement along the $[1\bar{1}0]$ ledges as illustrated in Fig. 5.5

[5.8]. Since the number of the [1$\bar{1}$0] ledges onto (111) oriented terraces strongly depends an crystal orientation this description is able to qualitatively explain the orientation dependence as reported in Fig. 5.4.

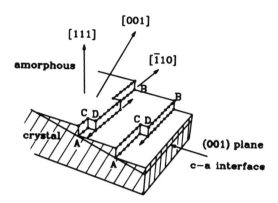

Figure 5.5 *Schematic description of the amorphous-crystal interface, showing the movement of kinks in the epitaxial growth of amorphous Si on (100)Si (from ref. 5.8).*

The growth velocity is maximum when the crystallographic planes containing the ledges are perpendicular to the surface and decreases when the ledges are inclined to the surface. On (111) oriented wafer nucleation of ledge is required, this gives rise to twinned material.

Small concentration of implanted impurities ($\leq$ 1 atom percent) can increase (B, P, As) [5.9] or decrease (O, N, C, Ar) [5.10] the regrowth rate of < 100 > oriented silicon substrates. This effect is shown in Fig. 5.6 for P, O and $P + B$ implants into < 100 > Si [5.9-5.10]. The regrowth rates are normalized to that of the impurity-free case.

The regrowth rate increases for P, decreases for O and remains constant for the P plus B double implants. The latter effect occurs for the simultaneous presence of p- and n-type dopants [5.11]. With increasing the dopant concentration the regrowth rate reaches a maximum and then slowes down appreciably at a concentration close to or exceeding the equilibrium solid solubility. SPEG in some cases can be completely inhibited by the presence of impurities such as Cu and Pb [5.12]. The phenomenos is illustrated schematically in Fig. 5.7.

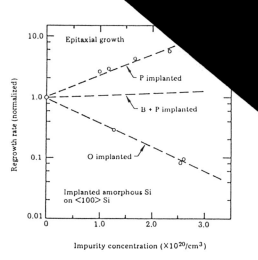

Figure 5.6 *Growth rate versus impurity concentration for* $< 100 >$ *silicon implanted with P, O and B ions. The data are normalized to the free-impurity case (from ref. 5.9-5.10).*

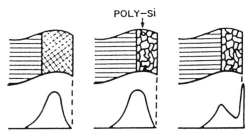

Figure 5.7 *Schematic illustration of impeded solid phase epitaxial growth and the initiation of polycrystalline growth processes in a region with high impurity concentration (from ref. 5.12).*

For moderate and high dose implants, the silicon is amorphized at a depth well beyond the peak of the implanted distribution. During the subsequent thermal annealing, the epitaxial regrowth occurs in a region of the tail of the dopant distribution, but as it proceeds toward the high concentration region it is impeded. As epitaxy is retarded, due to either local strain or impurity precipitation effects, polycrystalline nucleation and grain growth within the amorphous layer may well dominate the latter stages of recrystallization. The poly nucleation is favored in cases where (i) low solubility species, as rare gases, fast diffusing species such as Cu or Ag atoms precipitate in the amorphous phase prior to regrowth, thus providing nucleation sites for competing silicon growth processes or (ii) if implant segregation and precipitation

179

ng regrowth when the concentrations exceed

)endence of the regrowth rate, of more than
.tude, on impurity concentrations as low as 0.5
)lies nucleation growth sites whose concentra-
ty depend on the charge state i.e. on the Fermi
oping. Additionally cooperative bond breaking
)e important in providing multi- atom reorder-
ing of the ... :e in the vicinity of an impurity atom or of an
impurity - generated defects.

So far we have assumed a sharp interface between the amorphous and the crystalline regions. End of range damage results instead in isolated amorphous or partially amorphous regions [5.13]. The transition region contains then a variety of primary defects, that evolve after annealing at 550^0 C into small extrinsic dislocation loops of 15-30 nm in diameter. The loops are formed normally by agglomeration of excess silicon interstitials. The end of range region is supersaturated with interstitials or become so upon annealing. The end of range damage requires usually annealing at high temperature ($900^0 C - 1000^0 C$) to be removed [5.14]. The shrinkage of extended defects, like dislocation is controlled by an activation energy of 5.6± 0.5 eV which is greater than the activation energy for common dopant diffusion but similar to that of silicon self-diffusion [5.15].

The formation of extrinsic dislocation loops is energetically favorable to the formation of a large cluster of interstitials. This is attributed to the large reduction in strain energy associated with the formation of a plate. The strain field around the dislocation loop can act as gettering site for impurities and implanted dopants. Detrimental aspects include pipe diffusion and higher leakage current if the defects occur near a p-n junction. End of range damage is reduced if the $\alpha - xtl$ interface is quite abrupt so that the transition region, where defects are nucleated, is narrowed. This is accomplished by low temperature implants to reduce the threshold dose for amorphization.

Other defects associated to the imperfect regrowth of the amorphous layer produced during implantation are hairp-in dislocation, twins [5.16-5.17]. Hairpin dislocations nucleate when the regrowing $\alpha - c$ interface encounters small microcrystalline regions sligthly misoriented with respect to the bulk crystalline material. V shaped dislocations are formed easily from $\alpha - c$ broad interface which contains a large number of microcrystallites. For this reason Ge ions are used instead of Si for the preamorphization step in the process of shallow junction formation (see chap. 7 -

sect. 4). The interface structure is quite dependent on the details of the implant. Even a slight temperature rise during implant can induce dynamic annealing and the formation of hairpin disloca-tion. Very low temperature annealing $\sim 250^0 C - 450^0 C$ prior to solid phase regrowth [5.18] reduces the concentration of misori-ented crystallites with a narrowing of the $\alpha - c$ transition region.

In (111) oriented substrates microtwins are formed upon SPEG [5.6-5.19]. In the (111) oriented wafer the simultaneous attachment of three adjacent atoms from the amorphous phase to a crystalline atom is required for the nucleation of a growth step. These three atoms can add in the correct position or with a twin orientation. Upon annealing at high temperature $(900^0 C)$ the mi-crotwins evolve into a tangle of dislocations which are difficult to dissolve even with a furnace annealing at $1100^0 C$ [5.20].

The previous treatment has considered the regrowth of amor-phous layer with only one interface with the crystalline material. In the case of high energy implants of heavy ions at high doses a buried amorphous layer is formed [5.21-5.22](e.g. few MeV As at dose above $10^{15}/cm^2$). The SPEG occurs in this case from the two outer c-a interfaces, towards the center of the amorphous layer. These regrowth modes follow the previous scheme, are strongly dependent on the orientation and on the dopant. For istance in Czokralsky grown silicon substrates due to the presence of oxygen in concentration above $10^{18}/cm^2$ the regrowth velocity is lower than in floating-zone grown wafers or in epitaxial layer grown in a reactor by chemical vapor deposition [5.23]. In the recrystal-lized zones, hairpin dislocations and microtwins remain in (100) and (111) substrates, respectively. The formation of dislocation is strongly connected with the broader a-c transition regions which incorporate misoriented microcrystallites.

The regrowth velocity of the two interfaces in the buried layer, is the same if it has been formed by self-ion implantation. In the case instead of amorphous formation by dopant species other than Si the two growth rates can differ slightly. This difference may be associated to the non symmetrical distribution of dopants along the thickness of the buried layer. Higher dopant concentration implies higher SPEG rate. Another reason may be due to the difference in abruptness of the two interfaces; the deeper is usually more abrupt than the outer.

In addition to the defects formed by the imperfect regrowth of the amorphous layer, another category of defects is present af-ter the SPEG of buried layer. This defect forms where the two advancing interfaces meet [5.24]. After a subsequent annealing at high temperature $(\sim 950^0 C)$ these defects evolve in large dislocation loops ($\sim$ 100 nm in diameter) of both faulted, i.e. with a stacking fault inside, and perfect types respectively. These defects have very different morphologies depending upon the wafer orientation. The defects appear to arise from either a slight displacement of the two-intersecting α-c interfaces or the coalescence of excess interstitials rejected by the advancing α -c interfaces.

At high energy and high dose implants dynamic annealing is more pronounced due to the relatively large power delivered to the wafer. In many cases, expecially in the absence of a suitable cooling system, buried amorphous layers are produced only with heavy ions as As. The structure of the secondary defects is quite complex and it depends crucially on the implant conditions. End of range damage, hairpin dislocations, microtwins and interstitial loops when the two interfaces meet are common in these cases.

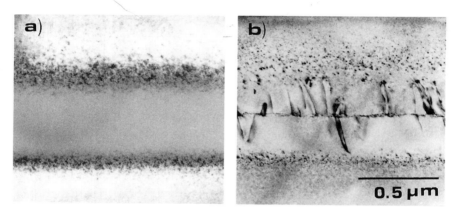

Figure 5.8   *Cross section - TEM of a Si sample implanted with 1.7 MeV Si - 2 $\times$ 10^{15}/cm^2 (a) and annealed at 650^{0}C - 5 min. (b) (Courtesy of C.Spinella).*

As an example the TEM cross sections of a Si sample implanted at room temperature with 1.7 MeV ^{29}Si 2 $\times$ 10^{15}/cm^2 and annealed at 650^0C for 5 min. are reported in Fig. 5.8. The implant causes the formation of a buried amorphous layer 0.4μm thick. End of range damage is already detected at both amorphous-single crystal interfaces. The annealing induces the epitaxial regrowth of the buried layer. Hairpin dislocations are clearly visible and a band of defects is formed when the two interfaces meet.

5.3 Annealing of Low-Dose Heavy - Ion Implant

At dose below amorphization heavy ions form small damage clusters surrounded by defect-free crystal material. These amorphized regions anneal out partially at low temperature such as $250^0 C$ [5.25]. Longer treatments at the same temperature, however, do not produce further annealing as seen in Fig. 5.9 for a sample implanted with $2 \times 10^{13}/cm^2 150 keV Au^+$ ions. It is necessary to increase the temperature to anneal further the residual damage.

Figure 5.9 *Recrystallized fraction vs annealing time for samples damaged at LN_2T (circles) and at RT (squares) with 2×10^{13} /cm^2 - 150 keV Au and for two different temperatures (523 K, 623 K) (from ref. 5.25).*

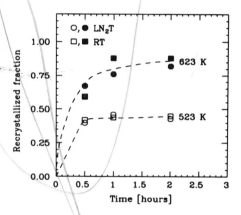

A temperature of 650 K anneals-out the damage but a few small dislocation loops remain in the implanted layer.

These results seem incompatible with the idea that these damage clusters are composed mainly of amorphous material, being this temperature too low to produce any appreciable recrystallization in continuous planar amorphous layers. The interface of these partially amorphized region is wavy and kinks and other growth sites, can be already present at the boundary, thus facilitating recrystallization. As an example two different possible c-a interface configurations are shown in Fig. 5.10 for a planar and non planar interface. A kink site at a planar terraced c-a interface is depicted in Fig. 5.10 a.

The (111) planes are as usually connected by $[1\bar{1}0]$ ledges. The atom A, unlike the others in the amorphous state, forms two bonds with crystalline atoms. This atom (kink) is therefore the next to recrystallize. As soon as A is crystallized the atom B becomes the next kink. Regrowth will then proceed by a kink motion along the $[1\bar{1}0]$ ledge until the whole row is completed.

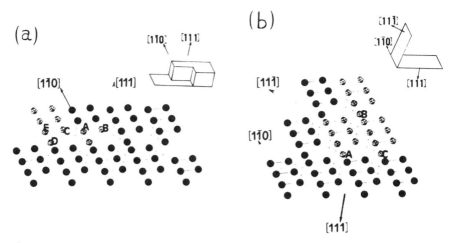

Figure 5.10 *Isometric drawing showing different possible c-a interface configurations. Dark spheres refer to crystalline atoms, bright spheres to amorphous ones. On the upper side of each atomic structure the relative interface configuration is schematically sketched. In (a) a boundary having a terraced structure composed of (111) planes connected by [1$\bar{1}$0] ledges is shown, the presence of a kink-like growth site can also be observed; in (b) an interface bounded by (111) and (1$\bar{1}$1) planes at an angle of 70,5⁰ is reported (from ref. 5.25).*

When a kink is already present at the interface the kink migration energy is therefore enough to produce the crystallization of an entire row. To reinitiate the process, however, an atom on the successive [1$\bar{1}$0] row (such as C) should be recrystallized in order to produce new kink sites in the positions D and E. At this stage, therefore, regrowth is limited by kink nucleation.

Different situations may be found at nonplanar c-a interfaces, as for istance in Fig. 5.10b, that depicts an interface with a sharp edge composed by (111) and (1$\bar{1}$1) planes encountering at an angle of 70 · 5⁰. Every other atom along the [1$\bar{1}$0] ledge (such as A) presents two bonds with the crystal. These atoms are therefore in the same situation as a kink site. Kinks are already present in this structure and should not be nucleated. The crystallization of atoms in the [1$\bar{1}$0] ledge containing A results in the formation of new identical growth sites along two different [1$\bar{1}$0] ledges. In particular atoms such as B and C will find themselves in this situation. In turn recrystallization along these two rows generates preferential sites along three different [1$\bar{1}$0] ledges, and so on. In contrast with planar growth in this configuration reordering is not limited by kink nucleation; the process is self-sustained and speeds-up with time. At low temperatures the annealing process is mainly due to the reordering of the most favorable configurations. As soon as they complete the reordering, a saturation effect is observed. Crystallization can then proceed only by a further increase in the annealing temperature which stimulates the nucleation and migration of new kinks. Similar kind of damage is present at the end of the range damage even if the surface layer is amorphous. A low temperature pretreatment is adopted to sharp the c-a interface by a mechanism similar to that discussed.

5.4 Regrowth of Amorphous Layer Under a Mask

So far the annealing of continuous amorphous layers and of partially amorphized regions has been described for a planar geometry. In device fabrication selected regions are doped by the use of masks. High-dose As ion implantation through thin SiO_2 films is generally used for source and drain formation in MOS-FET devices and the gate is masked by a thick SiO_2 layer or photoresist.

The dose is above $10^{15}/cm^2$ and the implant energy is around 50 keV. At this condition a continuous a-Si forms from the surface to a depth of about 100nm. The a-Si shapes under the mask edge have two-dimensional depth distributions. The c-a interface is not located on a well defined crystallographic plane but changes plane along the mask edge. The regrowth velocity depends strongly on the orientation of the plane so that regions of the same initial a-c interface moves at different velocities [5.26]. The situation is schematically represented in Fig. 5.11.

The SPEG proceeds vertically along the $< 100 >$ direction from the bottom of a-Si to the surface and the front under the mask regrows from the envelope of different and slower directions. The net result is the formation of secondary defects under the mask edge.

The sequence of two-dimensional annealing behaviour for $5 \times 10^{15}/cm^2$ - 80 keV As implantation through a $1.5\mu m$ wide mask is reported in Fig. 5.12 [5.27]. After annealing at $550^0 C$ for 7 min. the $\alpha - Si$ layer regrows halfway. The solid phase epitaxial regrowth proceeds mainly from two directions: vertically along the $< 100 >$ direction from the bottom of the $\alpha - Si$ to the surface and along the $< 110 >$ direction parallel to the surface under the mask. The vertical regrown layer thickness is about three times

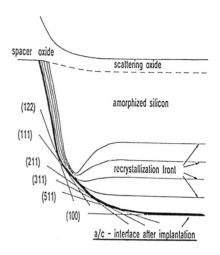

Figure 5.11 *Schematic of regrowth from an amorphous - layer under a mask (from ref. 5.26).*

the lateral one in agreement with the data shown in Fig. 5.4. The residual defects are classified into three different groups schematically represented in Fig. 5.12(e). Type - 1 defects formed the end of range damage just beneath the interface between the original a-Si layer and the underlying Si substrate. Type - 2 defects are located near the projected range of As and are associated to the precipitation of the dopant in excess of the solubility limit at the annealing temperature. Type - 3 defects are observed under the mask edges, are not implantation induced defects but are formed during the recrystallization of the amorphous layer. They remain even after annealing at high temperature such as $1000^0 C$. Type - 1 and 2 defects anneal out at this temperature. Type - 3 defects show dislocation - like contrast and they are dislocation lines with different characters in different parts running parallel to the mask edges. Under compressive stress these defects may penetrate into unimplanted areas under the mask edges and may degrade the electrical performance of the devices.

The influence of submicron mask patterns on high - dose P implant is shown in Fig. 5.13. The cross sectional TEM micrographs show the two-dimensional distributions of defects generated in submicron areas with different mask sizes by $5 \times 10^{15}/cm^2 - 50 keV P^+$. In as-implanted layers, continuous $\alpha - Si$ layers are formed from the surface to a depth of 107 nm in each implanted layer, regardless of the mask size. After annealing at $300^0 C$, type 1 dislocation loop forms just below the original α/c interface for all of the implanted layers.

Type 2 defects related to the precipitation of dopants in ex-

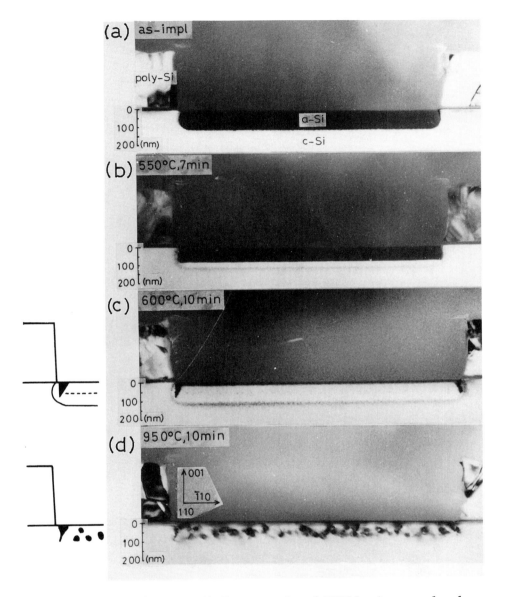

Figure 5.12 *(a, b, c, d) Cross sectional TEM micrographs showing the annealing sequence of two-dimensional* $\alpha - Si$ *in* Si *implanted through* $1.5 \mu m$ *wide mask with* $5 \times 10^{15}/cm^2 - 80 keV As^+$ *(from ref. 5.27).*

cess of the solid solubility limit are seen in samples with mask

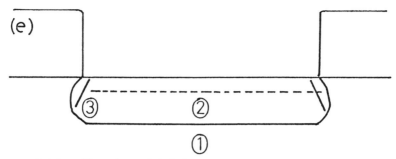

Figure 5.12 *Continued (e) Schematic representation of residual defect classification after solid phase regrowth (from ref. 5.27).*

$$50 \, keV, \, P, \, 5 \times 10^{15} \, ions/cm^2$$

Figure 5.13 *Cross sectional TEM micrographs showing two-dimensional distributions of defects observed in $50keV - 5 \times 10^{15}/cm^2 P^+/cm^2$ implanted submicron areas (from ref. 5.27).*

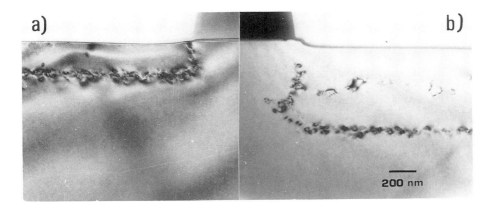

Figure 5.14 *Cross sectional TEM micrographs showing two-dimensional distributions of defects observed in* $2 \times 10^{15}/cm^2 -$ *$50keV B^+$ - annealed at $900^0 C$ - 15 min (a) and in $10^{14}/cm^2 -$ $400keV P^+$ annealed $1000^0 C$ - 10 s. (Courtesy of C.Spinella and V.Raineri).*

opening size above $0.1 \mu m$. Below $0.1 \mu m$ these defects are absent, probably due to the sligth decrease of the maximum impurity concentration as discussed in chapter 3. Type 3 defects just under the mask corners remained after annealing at $1000^0 C$. Below $0.1 \mu m$ mask these defects are not any more separated. These defects form in a similar way as those in *As* implants. The generation of such defects depends on the corner shape of the amorphous region. If the SPEG occurs only in one dimension they are not generated. One method to prevent the mask-edge defects is to form amorphous region with arc-shaped corners. This requires less steep implantation sidewalls. If the implant does not produce amorphous layer the residual defect shape follow the damage distribution, and no new defects are observed. This is illustrated in Fig. 5.14(a) for $2 \times 10^{15}/cm^2 - 50keV B^+$ annealed at $900^0 C$-15 min and in Fig. 5.14(b) for $1MeV P^+ - 10^{14}/cm^2$ annealed at $1000^0 C$ - 10 sec. The dislocation loops follow the as-implanted damage distribution in both cases. The mask influences the distribution, bringing defects near by the surface in proximity of the oxide gate where they may deteriorate its electrical behaviour, by decreasing the breakdown voltage [5.28-5.29].

5.5 Annealing of Heavily Disordered Regions

Low ·temperature $\sim 77K$ implants of high dose boron ions in the range of $10^{15}/cm^2$ result in the formation of a continuous amorphous layer. During solid phase epitaxial regrowth the implanted boron is electrically activated at the recrystallization temperature as shown in Fig. 5.15 where the ratio between the areal density of the carriers and the dose is reported for different implantation temperature as a function of the isochronal annealing temperature [5.30].

Figure 5.15 *Isochronal annealing curves for 50 keV - $10^{15}/cm^2 B^+$ implants in Si at several substrate temperatures (from ref. 5.30).*

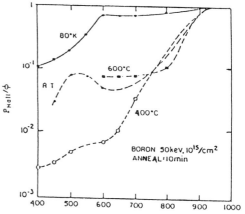

At low implant temperature the activation is quite abrupt and sharp at the recrystallization temperature around 500^0C. With increasing implant temperature even at room temperature boron ions do not form amorphous layer. The annealing behaviour is quite complex as illustrated in Fig. 5.16 where typical isochronal treatments of 150 keV B implanted Si are shown [5.31] together with the percentage of boron atoms which occupy substitutional lattice site for the $2 \times 10^{15}/cm^2$ dose [5.32].

Three annealing regions are indicated as I, II and III. The low dose implant, $8 \times 10^{12}/cm^2$, shows a monotonic increase of activity with temperature. The other two implants, $2.5 \times 10^{14}/cm^2$ and $2 \times 10^{15}/cm^2$ respectively, are characterized by a reverse annealing in the temperature range $500^0C - 600^0C$, i.e. in the region II. With increasing temperature decreases the fraction of active boron dopant. The complete activation requires high temperature annealing in excess of 900^0C, region III. The different behaviour has been associated to the defects. Region I is characterized by point defect disorder that dominates the electrical free-carrier concentration. In non-annealed sample the electrical behaviour is as-

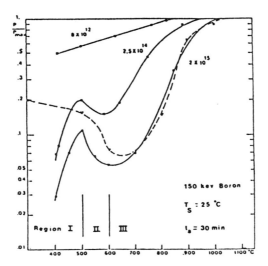

Figure 5.16 *Isochronal annealing behaviour of boron. The ratio of the free-carrier to dose is plotted versus the anneal temperature for three doses of boron (from ref. 5.31). The dashed line represents the fraction of boron atoms located in substitutional lattice sites for the $2 \times 10^{15}/cm^2$ dose (from ref. 5.32).*

sociated with deep - level traps created by the large number of displaced Si atoms, with temperature rise annealing of compensating damage clusters takes place.

Small agglomerate defects dissolve and the released mobile defects recombine with the formation of extended defects including also boron atoms. Substitutional boron is pushed off-lattice sites and the free carrier concentration decreases. In this region II TEM studies indicate the formation of a dislocation structure in which boron atoms have precipitated. Road - shaped defects elongated along the $< 110 >$ directions are formed. It is not clear if these defects contain B or are boron precipitates [5.33]. In region III at higher annealing temperature ($900^0C - 1000^0C$) the boron substitutional concentration increases and the process is governed by an activation energy of about 5.0 eV. This energy corresponds to the generation and migration of a silicon self-vacancy at elevated temperature. These defects interact with non-substitutional boron atoms and cause precipitates to dissolve. In spite of a complete electrical activation of the boron dopant, extended defects in high density are still present in the sample. The defect annealing in the $800^0C - 100^0C$ range causes the formation of high concentration 15-20 nm diameter vacancy loops. A prolonged anneal transforms these defects into linear defects located at the peak of the B range profile. At 900^0C the linear defects are replaced by large loops.

As an illustration of complexity plan view TEM micrographs of 80 keV - $2 \times 10^{15}/cm^2 B$ implanted Si are shown in Fig. 5.17. The sequence refers to 900^0C (a, b), 1000^0C(c, d) and 1100^0C(e, f) annealing temperature respectively and times of 1 hr about. A large density of dislocation loops is seen after 900^0C and 1000^0C annealing for 0.5hr. Some of them are extrinsic in nature and contain a stacking-fault as shown by the interference fringes, Fig. 5.17(d). A prolonged anneal transforms these defects first into linear defects which tend to shrink and to disappear. An anneal at 1150^0C - 10 min causes the complete disappearance of dislocations visible by TEM analysis. As already pointed out a complete elimination of the defects is in some cases incompatible with the final profile of the dopant. Prolonged anneal causes the diffusion of the impurities. The 1100^0C - 1hr anneal of 80 keV - $2 \times 10^{15}/cm^2 B$ shifts the junction at a substrate concentration of $10^{15}/cm^3$ to a depth of three microns. A decrease in the dose reduces the thermal budget for the anneal of the damage.

TEM analysis is limited to a density of $\sim 10^6$ defects$/cm^2$. Lower density cannot be detected and other analytical tools are used, e.g. optical microscope. Residual defects are enlarged in size by the injection of self- interstitials coming from the $SiO_2 - Si$ interface, during oxidation at high temperature [5.34-5.35]. The driving force for self-interstitial injection is the volume expansion of about a factor of two associated with the SiO_2 formation at the $SiO_2 - Si$ interface. These self-interstitials interact with small defects and give rise to the nucleation and growth of interstitial-type dislocation loops containing a stacking fault, termed oxidation - induced - stacking fault (OSF) (see Fig. 5.18a). The lenght of the stacking-fault reaches the maximum at a temperature of 1200^0C for (100) oriented wafer and 1150^0C for 5^0 off (100) oriented wafer respectively [5.35]. The analysis by optical microscope requires a chemical etch to evidence the damaged region. For oxidation stacking-fault the following solution is adopted: one part of 75g CrO_3 in 1000 ml H_2O mixed to two parts of 48% HF [5.36]. The solution is applicable to (100) and (111) silicon wafer orientation and the etch rate is approximately 1 $\mu m/min$. As an example Fig. 5.18 reports the optical micrographs of a sample implanted with $80 keV - 10^{15}/cm^2 B^+$ ions annealed in a nitrogen ambient at 1150^0C for two minuts. The sample was oxidized at 1100^0C for 1.5hr in a dry oxygen, ambient and then chemically etched.

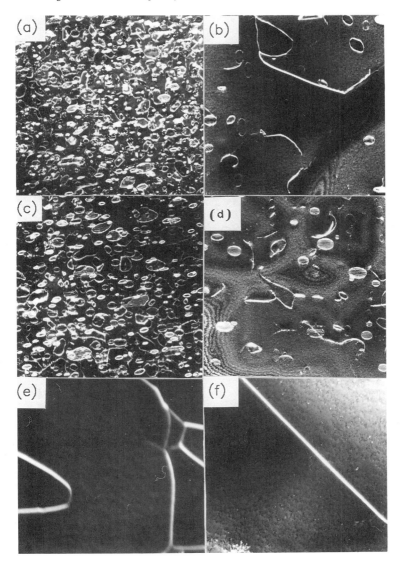

0.2 μm

Figure 5.17 *Plan-view TEM micrographs of $80keV - 2 \times 10^{15}/cm^2 B^+$ implanted $< 100 > Si$ wafer annealed at 900^0C - 0.5 hr (a), 900^0C - 3 hr (b) , 1000^0C - 0.5hr (c), 1000^0C - 1 hr (d), 1100^0C - 5 hr (e), and 1100^0C - 1 hr (f). (Courtesy of C.Spinella).*

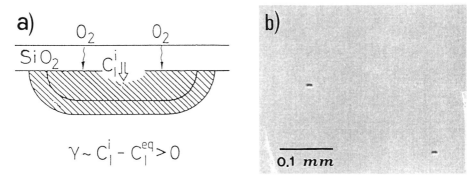

Figure 5.18 *(a) Schematic illustration of the effect of near-surface point defect concentration on growth of oxidation stacking faults. C_I is the interstitial concentration while superscripts i and eq represent values at the interface for OSF equilibrium, respectively. (b) Optical micrograph of a sample implanted with 80 keV - $10^{15}/cm^2$ B, annealed at 1150^0C for 2 min, and then oxidized in dry O_2 at 1100^0C for 1.5 hr.*

The stacking fault density is of $8 \times 10^3/cm^2$. The grown oxide layer is 200 nm thick. It must be pointed that even if by TEM defects are not seen the subsequent oxidation can show the presence of stacking faults. These grow, as said, by the addition of self- interstitial silicon to small clusters of point defects not detected by TEM. On the other hand the oxidation process should be performed after a suitable anneal in a free-oxygen ambient to reduce considerable the nuclei for stacking-faults. The optical micrographs of Fig. 5.19 was taken on a 80 keV - $10^{15}/cm^2 B$ implanted Si sample annealed directly at 1100^0C in dry oxygen atmosphere. The stacking fault density is quite high indicating the necessity to perform the first step of the annealing in a non-oxidizing ambient.

Oxidation stacking-fault is a classic example of the influence of a process on the evolution of defects. Oxidation enhances also the diffusivity of species that move by interstitial mechanism [5.37], like B, retard the diffusivity of species that move by vacancy mechanism such as Sb [5.38]. For these reasons the annealing conditions should be optimized during the drive-in. In many cases the implanted dopant should be diffused in the substrate and an oxide layer is also required for subsequent step-process. Attention must be paid to the simultaneous occurrence of oxidation and diffusion in a layer with still residual defects. Other processes

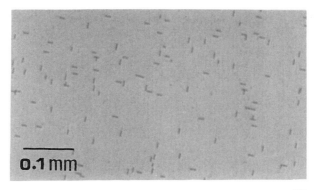

Figure 5.19 *Optical micrograph of* $80keV - 10^{15}/cm^2 B$ *implanted Si wafer annealed at* $1100^0 C$ *in a dry oxygen ambient.*

such as nitridation or silicidation [5.39] are instead associated to the injection of vacancies that may shrink the dislocation loops by dislocation climb or dissolve small clusters of interstitials.

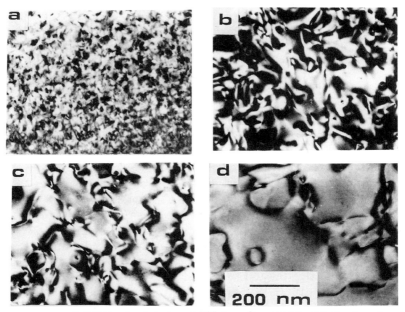

Figure 5.20 *TEM of: (a)* $4 \times 10^{15}/cm^2 - 600keV Ge$ *implanted Si at* $250^0 C$; *(b)* 1100^0 , *2s; (c)* $1100^0 C$, *5s and (d)* $1100^0 C$, *30 s annealed samples (from ref. 5.40).*

The evolution of defect structure by annealing of boron implants is common to heavily disordered materials where no-amorphous layer is present as in hot implants. As an example the TEM analysis of $600keV - 4 \times 10^{15}Ge^{+}/cm^{2}$ implanted silicon samples at $250^{0}C$ and subsequently annealed at $1100^{0}C$ for different times is shown in Fig. 5.20 [5.40].

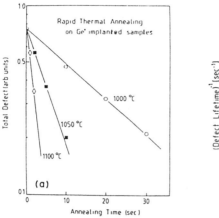
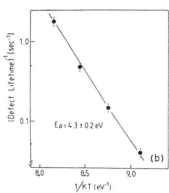

Figure 5.21   *(a) Total defect length as measured by channeling analysis vs annealing time for Si implanted at $600keV$ with $4 \times 10^{15}/cm^{2}$ and annealed at different temperatures, (b) Arrhenius plot of the reciprocal of the defect lifetime from the curves of (a). (From ref. 5.40).*

The total length of dislocation loops decreases exponentially with the annealing time as indicated by the straigth line fit of the experimental data (see Fig. 5.21(a)). The quantitative analysis was performed by channeling measurement technique to be discussed in chapter 6. The annealing behavior is then described by $N(t) = N_0 exp(-t/\tau)$ where N(t) is the length at time t, N_0 the dislocation length at t=0 and τ the defect lifetime. The dependence of $1/_\tau$ on temperature is reported in Fig. 5.21(b). The slope of the inverse of the defect lifetime in the Arrhenius plot gives the activation energy of the process, which results 4.4eV. Other measurements performed by TEM analysis [5.41] on the minimum time for the complete disappearence of dislocations gives a slightly higher value, ~5.0eV. In any case the shrinkage of the loops is associated to climb which is controlled by the self-diffusion activation energy of Si. In the case of non-amorphized layers the nucleation of secondary defects takes place at peak of the damage profile if the overall damage level is sufficiently high.

For a systematic analysis of defects in ion implanted silicon see also the reference 5.42.

5.6 Rapid Thermal Processing

The annealing of extended defect requires high temperature ($\sim 1000^0C$) and long times. During this thermal treatment dopants, like boron, can diffuse over considerable distances and the final profile differ from the as-implanted, vanishing in some cases the advantages of ion implantation in locating at a well determined depth the impurities. Diffusion is a serious drawback expecially for the formation of shallow junctions, so that new methods have been employed to anneal out the damage but with a limited dopant diffusion. The point defects involved in the shrinkage and removal of extended defects as previously shown are usually self-interstitial or vacancies with an activation energy for formation and migration of ~ 5 eV, that is ~ 1 eV higher than that for impurity diffusion.

The data reported in Fig. 5.22a [5.41] indicate experimentally what stated before. The dislocation removal rate is governed by an activation energy of ~ 5.0 eV while the broadening of ths *As* implanted profile is characterized by an activation energy of ~ 4.0 eV.

Due to the different slopes, high temperature depresses diffusivity with respect to the concentration of point defects. It is then advantageous to heat the wafer at high temperatures but for short time, conventional furnace cannot be used of course. For each dopant is possible the optimization of the thermal process to anneal out defects and to limit diffusion. Fig. 5.22b reports the different time-temperature regions where diffusion of *As* is limited to a broadening lower than 40nm and defect anneal takes place. The optimum annealing region is determined from the overlap of the regions.

Furnace annealing has several other disadvantages. The tubes must be dedicated to a single process in order to prevent contamination that occurs because the hot furnace wall traps and then releases contaminants to which it has been exposed.

Heating of the wafers from their edge inward also causes thermal gradients and stress unless very slow heating and cooling rates are used. A temperature difference of ΔT between two regions of the same wafer induces a thermal stress S given by

$$S = \alpha E \Delta T \qquad\qquad 5.5$$

with α the thermal expansion coefficient, E the Young's modulus.

If this stress exceed the yield strength of the material dislocation will form. In silicon the yield stress decreases with temperature and with the amount of oxygen precipitates. It ranges

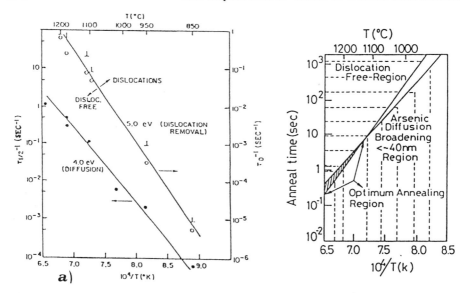

Figure 5.22 *(a) Dislocation removal rate* (τ_D^{-1}) *measured by transmission electron microscopy in As implanted silicon as a function of the reciprocal temperature (open circles). The bar* $(\perp)$ *symbols are for the longest time where dislocations are still observed. The diffusion rates for peak concentrations to decrease from the maximum non-annealed value to 1/2 of that value are also plotted,* $\tau_{1/2}^{-1}$, *as a function of 1/T (from ref. 5.41), (b) upper dashed region indicates defect free sample and the low dashed one a negligible dopant diffusion. Overlap of the two regions shows the optimum region to remove defects and to limit As diffusion. .*

with no oxygen precipitates from $2 \times 10^7 dyn/cm^2$ at 1100^0C to $10^9 dyn/cm^2$ at 800^0C. The Young modulus is $1.09 \times 10^{12} dyn/cm^2$ and the thermal expansion coefficient is 2.6×10^{-6} so that a difference of only 7^0C is enough to induce dislocations and slip lines at 1100^0C

Among the several methods used in these last years to overcome the disadvantages of furnace annealing, rapid thermal treatments by lamps with processing time of few seconds are suitable for production and will be discussed in the following. These treatments are indicated by RTA (Rapid Thermal Annealing) or by RTP (Rapid Thermal Processing). When first introduced commercially, RTP was intended for activation of shallow implants and other standard anneal processes, but in the past few years it has been used for contact alloying, silicide formation, thin dielectric growth and nitridation for diffusion barriers. A schematic view of an RTP system is shown in Fig. 5.23.

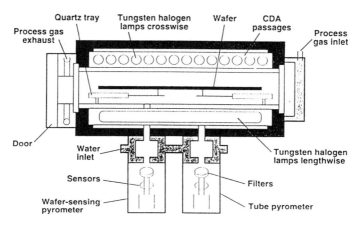

Figure 5.23 *Schematic of a one-light-source rapid thermal heating system. A silicon wafer is held on thin quartz pins in thermal isolation inside a cell, containing a controllable gas ambient. The tungsten halogen lamps heat the wafer through transparent windows aided by reflectors. The wafer temperature is monitored and controlled with an optical pyrometer that views the wafer back through a series of filters and windows (Source: AG Associates).*

A silicon wafer is held on thin quartz pins in thermal isolation inside a cell containing a controllable gas ambient. One or two light sources heat the wafer through transparent windows coupled with highly reflective mirrors. Tungsten-halogen lamps are commonly used as energy sources that can deliver up to 30,000 - 50,000 watts. The heat source must have a fast response time so that a feedback loop can be established between the wafer temperature and the source output. Temperature is measured usually by pyrometers that sens the thermal radiation from the back side of the wafer. The signal comes from a small spot 0.5 - 1 in diameter. The radiation is correlated to the temperature by Planck's law for black body radiation. Pyrometers do not give absolute temperature reading and need accurate calibration over the entire temperature range because of uncertaintes of the back side emissivity and on its dependence on temperature, wavelength, and surface condition.

The wafer is typically heated to about $\sim 1000^0 C$ for a time interval of 1-20 s. The thermal diffusion lenght $\sim \sqrt{2Dt}$ is about 0.5 cm for t=1 s and $D=0.1 cm^2/s$, so a wafer 0.05 cm thick is heated uniformely. Solid phase regrowth of an amorphous layer at these temperatures is completed in less than one millisecond: the time left may be used to remove defect clusters and dislocation loops in the a-c transition region. Impurity diffusion is limited to a depth typically $< 0.05 \mu m$.

The wafer holders must have low thermal mass and very low thermal conductivity so that the contact with the wafer does not produce a temperature gradient across the wafer. In many systems tungsten-halogen lamps are used. They are formed by a tungsten filament in a quartz tube with a small amount of one of the halogens (usually bromine).

When the tungsten filament is resistively heated, the tungsten evaporates and reacts with the bromine. When these molecules reach the filament, the tungsten will deposit on the filament and release the bromine for another cycle. This allows the filament to be operated at higher temperatures. Since each tungsten halogen lamp is rated for only a few kilowatts, an array of lamps is required to heat an entire wafer.

Upon irradiation the wafer becames hot at a rate that depends on the heat capacity, the optical absorption and reflection and the spectrum of the lamp. The heating rate and the temperature of lightly or heavily doped wafers are different. In Fig. 5.24a [5.43] the spectrum emitted by a tungsten-alogen source and the absorption coefficient of silicon wafers for several doping concentrations are shown.

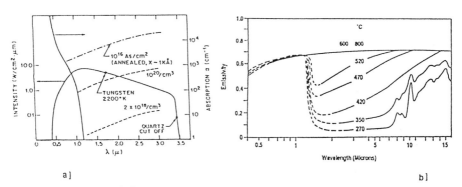

Figure 5.24 *(a) Normalized blackbody - like spectrum for tungsten lamps at $2200^0 C$ (leftordinate). Absorption coefficient for silicon (right ordinate) showing fundamental edge and free carrier components (from ref 5.43). (b) Emissivity of lightly doped silicon at different temperatures (from ref 5.44).*

Initially only the near infrared portion of the heater blackbody spectrum is absorbed, but as the wafer temperature rises, band gap narrowing and increased free carrier absorption results in a larger fraction of the spectrum to be utilized. The spectral emissivity of silicon is also a strong function of both temperature and wavelength as shown in Fig. 5.24(b) [5.44]. The rate of temperature rise for a uniformely doped wafer is given by

$$C\frac{dT}{dt} = P_a - P_e \qquad\qquad 5.6$$

where C is the heat capacity of the wafer, P_a the absorbed power and P_e the emitted power. A uniform temperature has been assumed throughtout the wafer volume. Pa is given by

$$P_a = \int (1 - R(\lambda))\mathrm{I}_\lambda d\lambda(1 - e^{-\alpha_\lambda d}) \qquad\qquad 5.7$$

where $\mathrm{R}(\lambda)$ is the wafer reflectivity at wavelength λ, I_λ the blackbody intensity of the lamp source, α_λ the absorption coefficient and d is the wafer thickness. The heat transport is dominated by the radiation over convection or conduction, expecially in vacuum systems. The emitted power is then given by the Stefan Boltzmann relationship: $\epsilon\sigma(T^4 - T_0^4)$ with ϵ the wafer emissivity and σ the Stephan-Boltzmann constant.

For a wafer as such, the emissivity is influenced by surface and by bulk properties. Thin dielectric layers, metal and composite layers can change drastically the emissivity of the wafer. The heating rate is given by

$$\rho C d\frac{dT}{dt} + 2\epsilon\sigma(T^4 - T_0^4) = \int_0^\infty (1 - R(\lambda)\mathrm{I}_\lambda d\lambda(1 - e^{\alpha_\lambda d}) \qquad 5.8a$$

and the cooling rate by

$$\rho C d\frac{dT}{dt} + 2\epsilon\sigma(T^4 - T_0^4) = 0 \qquad\qquad 5.8b$$

An example of a temperature-time profile together with the input power time dependence is shown in Fig. 5.25.

The power required to maintain a silicon slice at 1200^0C, assuming that the thermal radiation from the wafer is all radiated away is about $19W/cm^2$ or a total of 36KW for a 6 inch wafer. At high temperature the emissivity of a bare silicon wafer is $\simeq 0.7$ and

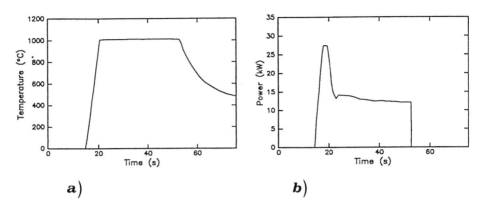

Figure 5.25   *Temperature-time profile of a silicon wafer (a) heated by the input-power time dependence (b).*

thus it radiates about 70% of the power emitted by a black body at the same temperature.

The previous treatment of thermal processing in lamp anneal-ers assumes an "isothermal" wafer. Temperature differences can develop across the slice due to a different balance between emis-sion and absorption in different regions. Among them one must consider the different geometrical and radiation situations at the edge as compared with the centre, the different energy balances in regions with different types, areas and shapes of patterned over-layer; the large temperature difference between adjacent regions because of the strong temperature dependence of both the absorp-tion coefficient and the thermal conductivity. A wafer heats more rapidly at the edge than at the center and cools more rapidly of the edge.

The slice temperature is almost universally monitored by se-lecting a band of the emitted radiation for measurement in a py-rometer. The reliability of the pyrometric measurement depends on the knowledge of the average emissivity of the body in the relevant wavelength band [5.45]. The intensivity increases espo-nentially with temperature:

$$I_\lambda = \frac{\epsilon C_1}{\lambda^5 (e^{C_2/\lambda T} - 1)} \qquad 5.9$$

$C_1 = 3.158 \times 10^8 W m^{-2} \mu m^{-1} \mu m^5$, and $C_2 = 14400 \mu m K$. The measurement error ΔT, as a function of the emissivity error Δe

is given by the following equation:

$$\Delta T = \frac{T^2 \lambda}{14400} \frac{\Delta \epsilon}{\epsilon} \qquad 5.10$$

It is seen that the error is increasing linearly with pyrometer wavelength and quadratic with temperature. Emissivity depends also on the wavelength so that there is no general way of measuring absolute wafer temperature by pyrometry alone. Thermocouple measurement is not reliable since it is never sure that thermocouple temperature is equal to the real temperature of the wafer. The thermocouple is never in contact with the wafer alone, and it is influenced by the radiation level in the surrounding. For that reason, a good pyrometry measurement on a known surface wafer can be more reliable than a thermocouple measurement.

In many cases the emissivity is not generally available and reliable indirect techniques of temperature measurement are required. A change in a material property that is sensitive to temperature can be the basis of such a technique. Suitable material changes in silicon processing are amorphous layer regrowth, silicidation, dopant activation after implant and oxidation. The same property can be chosen to characterize temperature uniformity. The adopted procedure, [5.46] say the dopant activation, must be determined as a function of temperature (see Fig. 5.26a).

A map of sheet resistance gives a map of the wafer temperature by assigning to each resistance value a corresponding temperature value. The implant should not fully actived because if the process is taken to completion, i.e. saturation, the sheet resistance no longer changes with temperature and a temperature map cannot be made. In this use the relationship between sheet resistance R_S and temperature follows the equation

$$R_S = \frac{A}{T} + B \qquad 5.11$$

where A and B are constants. Also

$$\frac{\Delta R_S}{R_S} = \left(\frac{A}{R_S T} \right) \frac{\Delta T}{T} \qquad 5.12$$

A similar relationship holds between standard deviation for sheet resistance and temperature

$$\frac{\sigma_R}{R_S} = \left(\frac{A}{R_S T} \right) \frac{\sigma_T}{T} \qquad 5.13$$

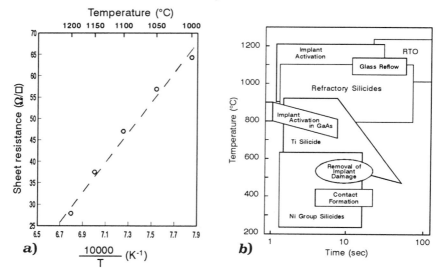

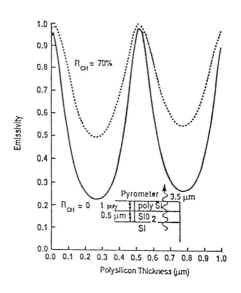

Figure 5.26   *(a)Calibration curve for relating sheet resistance to temperature* $(10^{16}/cm^2 - 50keV As$ *implanted 5 inch wafer, 10 sec anneal). (b) Time temperature space of some processes which can be used to evaluate uniformity (from ref. 5.46).*

Figure 5.27   *Variation in emissivity with polysilicon layer thickness for a planar structure of polycrystalline silicon on 0.5 μm oxide on the silicon substrate. The full line refers to a black body (R=0), and the dotted line to a body with R=0.7 (from ref. 5.47).*

Other material processes such as oxidation, silicidation, nitridation etc can be used to monitor the temperature of the wafer (see Fig. 5.26b). As already mentioned the procedure of overlayers like oxide, nitride, polycrystalline layers alters dramatically the temperature reading. Structure must be avoided, being extremely sensitive to emissivity. Consider for istance a pyrometer reading at $3.5\mu m$ and the structure is a two-layer, a polycrystalline silicon on oxide on the silicon substrate (Fig. 5.27) The oxide thickness is $0.5\mu m$ and acts a quarter-wave plate (i.e. there is a phase difference of 180^0 between radiation reflected off the top and the bottom surface of the oxide). The emissivity as shown in Fig. 5.27 [5.47] changes of an order of magnitude as the polycrystalline silicon thickness is varied. A variation of $\pm$ 10nm in the thickness of the polycrystalline silicon changes the emissivity can vary of 6% around a value of 50%, this would lead to a temperature uncertainty of $\pm48^0C$.

5.7 Impurity Diffusion During Annealing

Prolonged annealing at high temperature broads the implanted profile by diffusion. Assuming an initial gaussian shape for the implanted profile the post annealing distribution is given by

$$C(x,t) = \frac{N_\square}{\sqrt{2\pi(\Delta R_{p^2} + 2Dt)}} exp \left[\frac{-(x - R_p)^2}{2(\Delta R_{p^2} + 2Dt)} \right] \qquad 5.14$$

where D is the diffusion coefficient. Contributions from the defects created by the ions and loss from the surface are neglected. The first contribution is of relevance during RTA for shallow junction formation as we will detail in chapt. 7, the latter one for impurities that can escape from the surface. In some cases the loss can be reduced or prevented by capping the sample surface with an oxide or nitride layer. Such a boundary condition imposes the flux at the sample surface is zero, that is

$$\frac{\partial C}{\partial x}|_{x=0} = 0 \qquad 5.15$$

The surface "reflects" all the arriving impurities and as a first approximation the solution for this boundary condition can be constructed by adding two gaussian equations, one identical to eq. 5.14 and the other with R_p replaced by $-R_p$. This operation of adding two gaussian functions symmetrical to the plane of the sample surface (x=0) is equivalent to reflecting all out- diffusing atoms from the surface [5.47]

$$C(x,t) = \frac{N_{\square}}{\sqrt{2\pi(\Delta R_{p^2} + 2Dt)}}$$

$$\left[exp\frac{-(x - R_p)^2}{2(\Delta Rp^2 + 2Dt)} + exp\frac{-(x + R_p)^2}{2(\Delta R_{p^2} + 2Dt)} \right] \qquad 5.16a$$

The opposite case, a surface that is a perfect sink can also be described analytically with the boundary equation C(0,t)=0. Again the solution is constructed by subtracking the two gaussian distributions reported in eq. 516a, i.e.

$$C(x,t) = \frac{N_{\square}}{\sqrt{2\pi(\Delta R_{p^2} + 2Dt)}}$$

$$\left[exp\frac{-(x - R_p)^2}{2(\Delta Rp^2 + 2Dt)} - exp\frac{-(x + R_p)^2}{2(\Delta R_{p^2} + 2Dt)} \right] \qquad 5.16b$$

As an example of the previous case let us consider the diffusion of Al implanted in Si. Al is a p-type dopant, with a diffusivity near ten times that of boron and it is used to dope very thick layers ($\sim 10 - 100\mu m$) of silicon for power devices. Al is usually introduced by diffusion from a solid in a sealed quartz ampoule with silicon wafers. The ampoule is subsequently heated at high temperature for a long time. Several attempts have been made to replace this procedure with ion implantation. The main drawback is the low electrical activity that amounts to a few percent of the implanted dose. Among the several phenomena responsible for the poor electrical activity the escape from the surface playes a relevant role. No effective capping layers has been found so far [5.48].
The broadening of the implanted profile should be described then by eq. 5.16b. The experimental results [5.49] reported in Fig. 5.28(a) refer to $6.0 MeV - 1 \times 10^{14}/cm^2 Al$ implanted floating zone silicon and to hole distributions by spreading resistance after anneals at different temperatures and times. The dashed lines are calculated with eq. 5.16 and with the diffusion coefficient $D = 8.9 exp \left[\frac{-3.44}{kT(eV)} \right] cm^2/s$. The shift of the average penetration depth and the values of the residual doses are quite well reproduced by the out-diffusion mechanism, i.e. all the aluminum atoms that reach the surface escape immediately from the sample, or accumulate at the interface in the native oxide layer. The calculated and experimental profiles differ near the junction region,

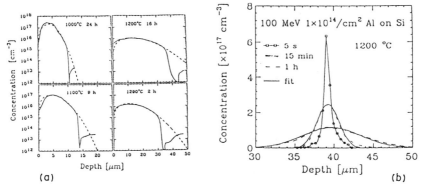

Figure 5.28 *(a) Experimental carrier profiles (full curves) and calculated chemical areas (broken curves) for $6.0MeV - 1 \times 10^{14}/cm^2$ Al implanted FZ silicon samples and diffused at different temperatures and times; (b) Experimental and calculated profiles for $100MeV - 1 \times 10^{14}/cm^2 Al$ - implanted FZ silicon samples annealed at 1200^0C for different times (from ref. 5.49).*

this is inherent to the electrical method adopted for profiling. The dashed lines represent the chemical profiles while the full lines the hole concentration that near the junction and in the depleted region is drastically reduced. The lost dose after an annealing at 1200^0C for 16 h is $7.6 \times 10^{13}/cm^2$, note the deep junction depth that one obtains.

If the species is implanted at very high energy far from the surface the broadening is limited to the bulk silicon and without on-diffusion the profile is governed by Eq. 5.14. The electrical profiles of $100MeV - 1 \times 10^{14}/cm^2 Al$ implanted FZ Si after annealing at 1200^0C for several times are shown in Fig. 5.28(b). The Al peak is located at about $40\mu m$ from surface and the broadening occurs well within the silicon substrate at least for the adopted the thermal treatment. The full lines represent fit of the profiles with the same diffusion coefficient as that used for the data of Fig. 5.28(a).

5.8 Interaction of Impurities with Ion Implanted Defects

Impurities present in the matrix can interact with the damage created by the implant and can change the amount and structure of secondary defects. Oxygen is the most popular impurity present in silicon with a content depending on the fabrication procedure.

In Czokralsky grown silicon the oxygen is in the $10^{17} - 10^{18}/cm^3$ range, an order of magnitude lower in floating zone material and even less in the epitaxial layer grown by chemical vapor deposition of silane at high temperature. CZ wafers are adopted in microelectronic industry overlayed, in some cases, with a few microns thick epitaxial layer. FZ wafers and thick epitaxial layers $(10 - 100\mu m)$ are instead used for power devices.

The oxygen present in CZ substrate is usually above solid solubility so that according to the particular thermal process it might precipitate either as isolated SiO_2 clusters or with damage if introduced by other means. These oxide clusters act as gettering of metallic impurities, like copper, iron, gold etc. The binding energy of the impurity increases in the stress field surrounding the defect and the free energy lowering is the driving force of the process. Gettering is a quite complex phenomenum and it has been detailed in literature [5.50]. In defective regions not only O atoms but also other common impurities such as N and C are gettered. High - temperature annealing induces an out diffusion of O and C in the CZ Si towards the surface; during this process oxygens and carbons interact with the defect bands. Nitrogen atoms are believed to be diffused from a N_2 atmosphere into the substrates during dry N_2 annealing. Usually the amount of O is two orders of magnitude higher than that of C and N.

The correlation among defect distribution and depth distribution of oxygen is shown clearly in Fig. 5.29a where the O SIMS profile is shown in CZ silicon substrate implanted with $6.0 MeV Si$ at a dose of $10^{14}/cm^2$ and annealed at $1200^0 C$ for $\frac{1}{2}$hr [5.51]. The formation of oxygen precipitates depends on the temperature of annealing. High temperature induces a long diffusion lenght of oxygen but increases also it solubility in Si. Both mechanisms trapping and dissolution of oxygen in defective regions, occur simultaneously at $T > 1100^0 C$. A sharp peak of O concentration occurs at almost the same depth where the defects exist. A similar implant in FZ produces no oxygen peak and the amount of damage is lower as shown by XTEM of Fig. 5.30a and 5.30b respectively. These secondary defects and small silicon oxide precipitates can act as a getter of metal impurities and can be used in the so called well engineering of CMOS.

The upper silicon region, above the defect peak, shows much better electrical behaviour if junctions are there present. The reduction of leakage current by a suitable implant is then another example of defect engineering. The use of silicon-ion as the implanted species avoids any dopant effect. Similar effects are shown by deep implants of common dopants in CZ substrate: forma-

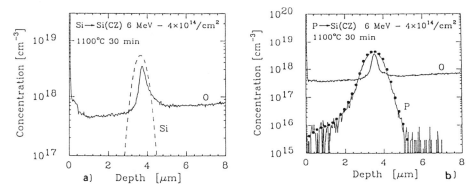

Figure 5.29 *(a) SIMS oxygen profile of a CZ-Si samples implanted with $6.0 MeV - 4 \times 10^{14}/cm^2 Si$ ions and annealed at $1100^{\circ}C$ for 30 min, the dashed line reports the calculated as-implanted Si profile; (b) SIMS oxygen and phosphorus profiles in a CZ-Si sample implanted with $6.0 MeV - 4 \times 10^{14}/cm^2 P^+$ ions and annealed at $1100^{\circ}C$ for 30 min, the full dots represent the measured electron concentration values (from ref. 5.51).*

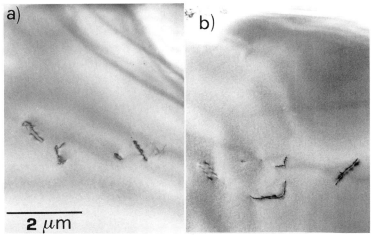

Figure 5.30 *Cross sectional TEM micrographs of CZ(a) and epitaxial (b) silicon samples implanted with $6.0 MeV - 4 \times 10^{14}/cm^2 Si$ ions and annealed at $1100^{\circ}C$ for 30 min (from ref. 5.51).*

tion of defect and silicon oxide precipitates as gettering centers of metallic impurities but no influence on the dopant distribution.

The dopant is electrically active in spite of the contiguous presence of defects, as shown in Fig. 5.29(b) where the electrical and chemical profiles of 6.0 MeV P- implanted CZ Si are reported together with the oxygen profile after the annealing.

Not all the dopants show this behaviour, as illustrated by the following analysis of Al implanted CZ and epitaxial Si substrates [5.52-5.53]. Implants of Al at high energy (6.0 MeV) locate the impurities at a large distance from the surface (few micrometers) $R_p \simeq 1.7 \mu m$. The electrical activation of the implanted Al species depends on the Si substrate, it is almost constant and equal to 80-100% in epitaxial and FZ layers, but 10% in CZ. As soon as the diffused Al atoms reach the surface, out-diffusion or segregation at the SiO_2/Si interface will take place. The SIMS profiles of O and Al are shown in Fig. 5.31a and Fig. 5.31b for CZ and FZ substrates, together with the electrical profiles by spreading resistance profilometry [5.52].

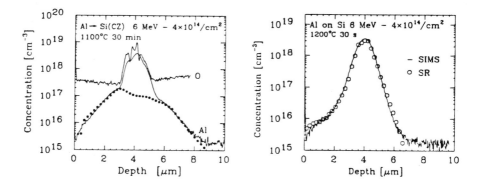

Figure 5.31 *SIMS Al and O depth profiles of* $6.0 MeV - 4 \times 10^{14}/cm^2 Al$ *ions implanted in CZ -Si (a) and in FZ-Si samples after* $1100^0 C$ - *30 min. The dots are the measured hole-concentration values (from ref. 5.51).*

A multipeak structure is evidenced around the depth of the projected Al range for both aluminum and oxygen atoms. This structure is more pronounced in the CZ sample, it is just detectable in FZ but absent in the epitaxial layers. The O concentration increases in the inactive region over the bulk level and a depletion, before and after this region, is observed.

The oxygen precipitates clearly act as gettering for Al and the precipitation phenomenon is enhanced by a kind of positive feed-back. Oxygen is collected by diffusion through the contiguous zones and its amounts increases with the square root of the product diffusivity× time at the annealing temperature. The resulting defect structure is quite complex: dislocation loops, coherent precipitates are found near the oxygen and aluminum peaks. This example illustrates the complexity of the annealing phenomenon where even the small amount of oxygen present in the substrate must be considered for its role on the damage evolution.

A TEM analysis is reported in Fig. 5.32.

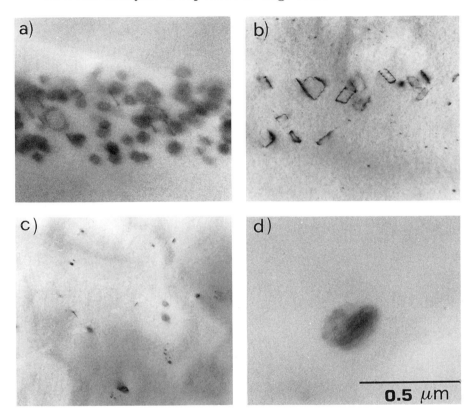

Figure 5.32   *Cross sectional TEM analysis of $6.0 MeV - 4 \times 10^{14}/cm^2 Al$ implanted CZ-Si samples and annealed at $1200^0 C$ for 30 s(a), 5 min (b), 4 h (c) and 120 h (d)(from ref. 5.51).*

After 30s at $1200^0 C$, only dislocation loops with sizes up to 500 nm are clearly observable. These structures are mainly located in a layer 700 nm thick at a depth corresponding to the Al projected ranges. After 5 min at $1200^0 C$ the extended defects tend to shrink while small coherent precipitates are formed. With increasing the

annealing time lattice extended defects are gradually annealed
and the precipitates grow in sizes and become incoherent with the
matrix.

5.9 Defect Engineering

The evolution of primary damage into secondary defects after
annealing is a quite complex phenomenon as seen in the previous
sections. It depends on several experimental parameters in a quite
subtle mode. As understanding of the defect formation mechanism
would be of relevance in the anneal handling. Recently the conditions of defect formation at least in the case of pre-amorphization
damage have been analyzed in details [5.54]. It has been found,
assuming as a criterion the presence or absence of extended defects
in the cross section of samples annealed at 900^0C-15 min, that it
is the total number of displaced silicon atoms to determine the occurrence of secondary defects. The reason why the total number
of displaced silicon atoms is the essential parameter for secondary
defect formation is that interstitials are highly mobile at the annealing temperatures and can diffuse over large distances before
being annihilated or trapped. Mobile Si interstitials throughout
the damaged region can contribute to extended defect formation.

Secondary defects form if and only if the integrated number
of displaced Si atoms exceeds a critical value. In the case of B
implants the critical number of displaced Si atoms for secondary
defect formation was $1-2 \times 10^{16}/cm^2$ by varying the beam energy
from 50 keV to 2.0 MeV and the dose over more than an order
of magnitude. The corresponding TRIM values for the number
of displaced Si atoms are higher $\sim 5 \times 10^{16}/cm^2$. The difference between experiments and simulation may be due to the fact
that the implants were performed at room temperature and TRIM
does not take into account dynamic annealing processes which occur during implantation. These results show that in the case of
boron the formation of secondary defects cannot be attributed to
exceeding a critical impurity dose or impurity peak concentration.

Fig. 5.33a shows schematically [5.55] the formation of secondary defects in medium and high energy implants. The point
defects are mobile enough to interact throughout the entire damage profile to form extended defects. These defects are in the form
of dislocation loops, rod-like defects, extended dislocations formed
along the $< 110 >$ directions, and according to the width of initial
damage concentration, i.e. the point defect concentration. Similar
results were found with Si and P implants, the critical number of

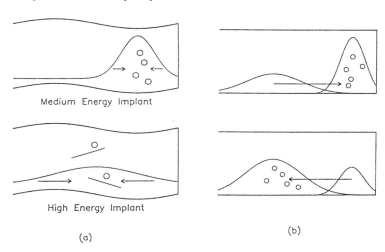

Medium Energy Implant

High Energy Implant

(a)

(b)

Figure 5.33 *(a) Criterion for secondary defect formation is the total number of mobile point defects generated by the implant, independent of the implant energy; the nature of defects (or loops or rods) changes as the width of the damage distribution increases. (b) Interplay between two spatially separated damage distributions: by varying the relative doses the position where secondary defects form can be altered (from ref. 5.55).*

displaced Si atoms for secondary defect formation lies in between 3 and $4 \times 10^{16}/cm^2$ independent of the implant energy and it is sligtly higher for P.

These critical values for secondary defect formation are higher for Si and P than for B implants. As discussed in chapter 4 low-dose B implants mainly results in the formation of isolated defects, whereas Si or P implants result in more complicated defect clusters or even small amorphous zones. Agglomeration of a sufficient number of mobile Si interstitials results in the formation of extended defects. These mobile interstitials cannot come from the small amorphous zones. The number of displaced Si atoms, as measured by RBS channeling, is then higher than the number of mobile interstitial Si atoms. The difference between the critical total number of displaced Si atoms between Si and P may be relate to the interaction of mobile point defects with impurities during the annealing, a similar case was considered previously for hot-implants of Ge into As-doped Si samples.

At high energies the number of isolated defects, which are produced per incident ion, increases. According to the combination projectile- ion-mass-energy secondary defect formation can therefore be obtained at high energy below amorphization threshold. At low energy a continuous amorphous layer is formed before to reach the critical value of the total number of displaced Si atoms. The way in which secondary defects nucleate from the mobile point defects and why the observed types of secondary form is still, however, unclear. It seems that when the damage level is sufficiently high dislocation loops will form instead of elongated dislocations. The total number of displaced Si atoms required for secondary defect formation does not depend only on the annealing temperature, but also on the specific annealing method. In the case of RTA the critical number of displaced Si atoms is higher than for furnace annealing. The difference may be due to the higher heating rate during RTA. In the heating an implanted silicon a competition takes place between annealing of implant damage and clustering of point defects. If the temperature is raised rapidly a larger number of point defects than during furnace annealing may annihilate before being captured by point-defect clusters. These clusters are responsible of secondary defect formation if their size is sufficiently large. Therefore RTA may result in a reduction of point defects available for secondary defect formation compared to furnace annealing.

Defects located at different depths may interact among themselves and according to the primary damage it is possible for istance to produce secondary defects either in the nearest or in the farest implant from the surface after annealing at high temperature. The interplay between two spatially separate damage distributions is shown schematically in Fig. 5.33b Experiments have been done on MeV B or As implants with an additional low-energy Si implant [5.56]. During annealing different processes can take place depending on the relative amount of disorder in the two regions. If the Si implant results in a total number of displaced Si atoms lower than the critical value for secondary defect formation the defect formation at larger depths is not influenced. The damage in the near surface region anneals out readily and no residual defect is found close to the surface. If the Si implant results instead in the production of a total number of displaced Si atoms above the critical value, residual defects will form during annealing in the near-surface region. These defects can then act as sinks for the mobile interstitial-type defects produced by the MeV As or B implant, reducing the number of available point defects for the formation of secondary defects. As the point-defect concentration at large depth increases the effect of sinks is reduced. With more point defects available at greated depths, dislocation-loop nucleation occurs fast enough relative to the surface so that the surface region is unable to getter the point defects from greater depths. The effects can be reversed by increasing the total number of defects at greater depth, i.e. the dose of high energy As or B implants. Dislocation loops form first at the greater depth and then act as sinks for point defects from near-surface region (see Fig. 5.33b).

This is an example of defect engineering, by a suitable combination of implants it is possible to form secondary defects either in the near surface region or in the deeper region. Dislocation formation is then engineered introducing alternative sinks for the interstitial type-defects [5.57]. An alternative method to avoid dislocation formation is to perform repetitive implant/anneal steps, where each step generates a number of displaced Si atoms below the threshold for secondary defect generation. This has been proved for several implants of B, P and As. As an example plan-view TEM micrographs of 1 and 4 step implants of 80 keV B implants to a total dose of $3.2 \times 10^{14}/cm^2$ are shown in Fig. 5.34 [5.58].

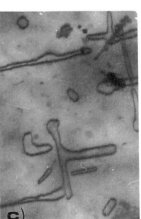

Figure 5.34 *Plan-view TEM micrographs after 900^0C annealing of $3.2 \times 10^{14}/cm^2 80keV/B$ implanted samples (a) single implant and 20 min anneal; (b) four implants of $0.8 \times 10^{14}/cm^2$ and in between them 900^0C - 20 min, (c) single implant and 80 min anneal (from ref. 5.58).*

The first implant gives rise to a dislocation density of $2 \times 10^9/cm^2$ after 20 minutes at 900^0C. Only one dislocation is observed in the TEM micrograph of the 4 step implant. The critical dose at 80 keV B RT implant is then $\sim 8 \times 10^{13}/cm^2$. The concentration profiles of electrically active B for multiple step implanted samples as obtained by spreading resistance measurements are shown in Fig. 5.35. No large differences are observed for the two profiles.

Figure 5.35 *Hole concentration profiles for Si implanted with* $3.2 \times 10^{14}/cm^2$ *80 keV B in 1 and 4 steps (from ref. 5.58).*

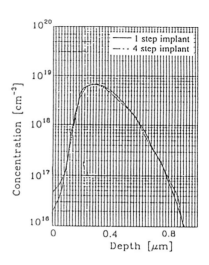

The diffusion of B after the 4 anneal steps is not significantly larger than in a single step, also the activation is not much affected by the presence of dislocations. A prolonged anneal at $900^0 C$ of the single step implant does not modify substantially the amount and type of dislocation.

Carbon is a peculiar species for what concerns the formation of secondary defects. After C implants it was found that dislocations do not form [5.59]. It was suggested that the formation of secondary defects is suppressed because the small C atom acts as a sink or trap for the Si self-interstitials, either generated by the C or by other implanted species overlapping the C profiles [5.60-5.61]. It has been found that the trapping of C is maximum when the carbon profile overlaps the damage region by the other implanted ions. A distance of 0.1 - $0.2\mu m$ between the C and the damage profile already degraded the influence of the C implant for high implant doses. When the implanted dose is low the interaction takes place over a larger distance $\sim 1\mu m$.

CHAPTER 6

ANALYTICAL TECHNIQUES

6.1 Introduction

The development of semiconductor technology has been extended also to analytical instrumentation. The trend is quite similar for the ion implantation process. All the available techniques, either destructive or nondestructive, have been used to characterize the implanted layers mainly for these two aspects: depth profile of the implanted species and of the damage. The dopant profile measurements require techniques able to reach the low concentration range of $10^{14} - 10^{15}/cm^3$, to determine the lattice location e.g. interstitial or substitutional, of the implanted species and the electrical activity. Methods that provide two-dimensional profiles are also necessary for the device fabrication.

Elemental determination and chemical state identification represent a large fraction of the analysis requirements but electrical measurements are among the most sensitive means to measure parameters that ultimately affect device performance. The other relevant characteristic of the implanted layer is the residual damage in terms of defect structure, amount, localization and electrical behaviour. Here too several analytical tools are available, starting from an atomistic determination of the defect structure by high resolution transmission electron microscopy to the quantitative and non destructive analysis by $MeV He^+$ backscattering in combination with channeling effect technique [6.1]. Point defects and small clusters have been analyzed by electron paramagnetic resonance, positron annihilation, optical response, atomic force microscopy etc. In principle any physical property that depends on the ordered arrangement of lattice atoms can be used to evaluate the defects, their nature and their evolution. Each technique has its own limitations, advantages and disadvantages.

The detailed evaluation of implanted layers can only be achieved by applying complementary analytical tools. Such an evaluation procedure can be long. It is then important to select the most appropriate techniques for a given system and to follow a logical sequence of analyses. In chapter 2 some analytical techniques have been briefly discussed such as four point probe and thermal wave absorption to evaluate implant uniformity and dose. Experimental data on depth profiles of impurities and on damage have been also presented and these data are based on analytical methods such as SIMS, spreading resistance profilometry and transmission electron microscopy.

It is impossible to cover even briefly all the analytical tools employed in the evaluation of the implanted layers, so that in the following the most common techniques will be discussed. The choice depends of course on the knowledge and on the previous work of the author that usually considers the techniques to him familiar as the most important. In addition to SIMS, spreading resistance, van der Pauw and channeling, new two dimensional delineation techniques will be presented.

6.2 Secondary Ion Mass Spectrometry

In secondary ion mass spectrometry (SIMS) [6.2 - 6.3] the sample maintained in a vacuum chamber is bombarded by a monoenergetic beam of primary ions ($Ar^+, O^+,$) at keV energies. Sputtered species are emitted as neutrals in various excited states, as ions both positive and negative, simply and multiply charged and as a clusters of particles. The positive or negative ions are then analyzed by a mass spectrometer and separated according to the mass-to-charge, m/q, ratio. The mass-separated secondary ions are detected by a suitable detector and the signal fed to a recorder or to a computer. The sputtered particles are originated from the first monolayers of the sample, so that the mass spectrum reveals the elements present at the surface. The bulk composition is obtained by eroding the sample via continuous sputtering and by recordering simultaneously the mass spectrum at various depths. The principle and the apparatus of a SIMS are shown skematically in Fig. 6.1a and Fig. 6.1.b respectively [6.4].

Fixed a particular, m/q, ratio, i.e. a selected peak and recording the intensity as a function of bombardment time, a depth profile is obtained of the chosen species.

The most probable energy of the secondary ions is lower than

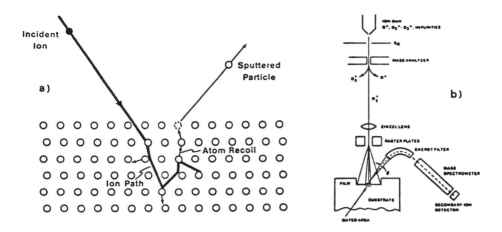

Figure 6.1 *(a) Secondary ion mass spectroscopy (SIMS) process, (b) schematic diagram of a spectrometer (from ref. 6.4).*

10 eV. They are emitted as a result of momentum transfer from the several target atoms in motion in the collision cascade to the surface atoms. (Fig. 6.1a) In SIMS only the sputtered ions are monitored. A mass spectrum resulting from Ar^+ bombardment of an aluminum target is shown in Fig. 6.2a for positive secondary ions and in Fig. 6.2b for negative ions respectively [6.5].
Electropositive elements appear with greater intensity in the positive spectrum while electronegative elements favor negative ion formation. Matrix effects are clearly expected because the ionization cross section depends on the environment.

The sputtering rate of the bombarded sample [6.6] is given by

$$\dot{x} = \frac{J_p \cdot S}{N} \qquad 6.1$$

where J_p is the primary ion flux density (ions/$cm^2 \cdot s$), S the sample sputtering yield and N the atomic density.

The flux density is determined by the rastered area Ar that is usually much larger than the area A from which secondary ions are accepted from the mass analizer. In the case of depth profile, i.e. dynamic SIMS the ion current density is over $1\mu A/cm^2$ and a sputter rate of a few microns per hour is obtained. The sputter rate cannot be too large since the depth sampled per data point is $\dot{x}\tau$, where τ is the counting time. The sputtering yields usually range between 0.2 and 20.

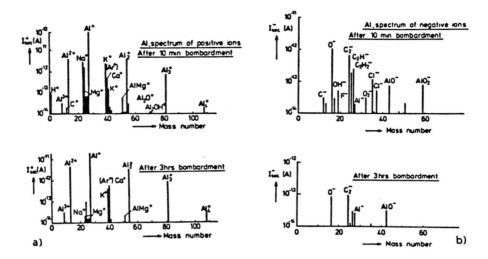

Figure 6.2 *Mass spectrum of a magnesium -doped aluminum sample bombarded with Ar^+ ions. (a) positive secondary ions, (b) negative secondary ions (from ref. 6.5).*

The degree of ionization $I^\pm$, is defined as the ratio of positive M^+ or negative M^- emitted ions to the total amount of the isotope M sputtered over the same amount of time. The secondary ion yield is given by the product $I^\pm S$ and it represents the number of secondary ions formed per incident atom. More useful is the practical ion yield β i.e. the number of ions detected per atom present in the sputtered volume; finally the instrumental detection efficiency η is given by the ratio of detected to emitted ions from the analyzed area A.

The ionization degree ranges between 10^{-5} and 1: elements of low ionization potential forms M^+ ions easily while elements of high ionization potential forms instead M^- ions. The ionization process depends on the chemical composition of the upper surface layers [6.7]. The presence of electronegative elements like oxygen enhances positive secondary ion emission, viceversa electropositive elements enhance the emission of negative secondary ions [6.8].

For this reason instead of sputtering with an inert gas like Ar when an high sensitivity is required oxygen O_2^+ primary beam is used for positive SIMS and cesium C_s^+ primary beam for negative SIMS. The secondary ion yield changes during the first instant of analysis until a steady state situation is reached with an equilibrium surface composition. To a first approximation the equilibrium surface concentration of implanted atoms is equal to

$1/S$, with S the steady state sputtering yield of sample atoms. A further enhancement in the yield is achieved if the sample when bombarded with Ar^+, O_2^+ or C_S^+ primary beam is maintained in an oxygen ambient or simultaneously exposed to a beam of evaporated cesium atoms in order to increase the content of the reactive species on the surface.

The secondary ion signal will depend upon the element, sample, bombardment conditions, species selected, instrument and sample consumption. The count rate of the positive M^+ or negative M^- ions is given by

$$I_{M\pm} = I_p \cdot SC_M \cdot \beta_{M\pm} \eta_{M\pm} \qquad 6.2a$$

where C_M is the fractional atomic concentration of the isotope M^* and I_p the primary beam current (ions/s), or in terms of the analyzed area

$$I_{M\pm} = A \cdot \dot{x} \cdot \beta_{M\pm} \eta_{M\pm} \cdot C_M \; N \qquad 6.2b$$

Count rates in the range $10^4 - 10^8$ are usually encountered in normal analysis.

The elements are identified by their characteristic peaks in the mass spectrum. The measurement of secondary ion intensities is quite straightforward and accurate. The conversion of intensity and time scales to concentration and depth can be done by means of eq. 6.1, 6.2a or 6.2b but due to the large variation of β and η with the particular experimental conditions and on the matrix composition it is necessary to use standards of close chemical composition to the sample for calibration. Accurate quantitative measurements are performed in homogeneous matrix with low concentration impurities. Ion implantation provides a convenient means to prepare suitable standards. The dose ϕ is known with an accuracy of $\sim 1\%$, so that the integral I of the count rate of the implanted isotope over the analyzed depth provides the product $A\beta_{M\pm}\eta_{M\pm}$ that can be used for other measurements.

$$I = \int I_{M\pm}(t)dt = A\beta\eta \cdot N \int C_M dx = A\beta\eta \cdot \phi \qquad 6.3a$$

and then

$$A\beta\eta = \frac{I}{\phi} \qquad 6.3b$$

The sputtering rate is determined by measuring the sputtered depth of the crater in the elapsed time with a mechanical stylus profilometer or an interface microscope. Errors introduced by a different sputter rate between the sample and the standard are removed by taking the ratio of the SIMS signals to that of a matrix element.

Depth profiling for ion implanted samples is performed limiting the area from which emitted ions reach the mass spectrometer either mechanically or electronically. The scale is actived electronically only when the primary beam strikes the sample within a small central part of the sputtered area. The detection limit for several common dopants in silicon is about $10^{15}/cm^3$. A state of the art profile spanning six decades is shown in Fig. 6.3 for 70 keV $^{11}B \cdot 1 \times 10^{16}/cm^2$ implant in Si [6.9].

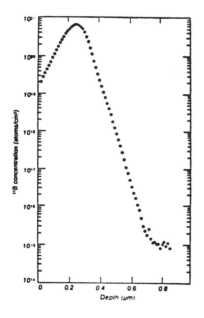

Figure 6.3 *Depth profile of 70 keV$^{11}B - 1 \times 10^{16}/cm^2$ implanted in Si (from ref. 6.9).*

This is a nice case because no interference peaks are present at the ^{11}B mass, Arsenic cannot be profiled when monitoring low energy secondary ions due to the coincidence of ^{75}As with $^{29}Si^{30}Si^{16}O$. The analysis can be performed by an energy filter that allows to monitor secondary ions of kinetic energy higher than 50 eV. This can be accomplished easily by biasing the target to - 50 V to suppress low energy positive secondary ion. Arsenic can also be detected at concentration of $10^{15}/cm^3$ by monitoring $^{75}As^-$ under C_S^+ bombardment in ultrahigh vacuum with a relatively small potential biased target. High mass resolution is required to profile phosphorus in silicon to separate $^{31}P^{\pm}$ from $^{30}SiH^{\pm}$ and in any case the improvement of the vacuum system to minimize residual gas and the $^{30}SiH^{\pm}$ formation is a good practice [6.5].

Sample contamination can fix the background level when depth profiling the gaseous elements H, C, O and N in silicon. Background signals for $H^{\pm}, ^{12}C^{-}$ and $^{16}O^{-}$ arise from the elements themselves on account of residual gas adsorption onto the sample. The background can be reduced to at the best the $10^{16}/cm^{3}$ concentration range by using a very fast sputter rate under ultra high vacuum conditions. Detection limits can be improved by monitoring the minor isotopes $^{2}H, ^{13}C$ and ^{18}O whose concentration in the residual gas will be much less.

Problems arise also when the monitored species is distributed in two contiguous and different material, e.g. SiO_2/Si or poly-crystalline silicon / silicon substrate. The change in secondary ion intensities at such interfaces does not merely reflect compositional changes but also a change in the sputter rate and changes in the degree of ionization due to the matrix effect.

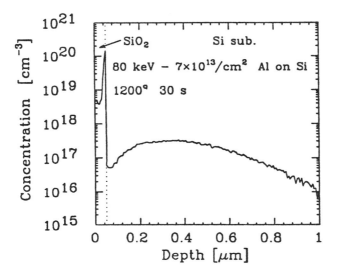

Figure 6.4 *SIMS profile of* $7 \times 10^{13}/cm^2 \cdot 80 keV\,Al$ *implanted in Si through a 50 nm thick SiO_2 after* $1200^0C \cdot 5s$ *(from ref. 6.10).*

But if they are properly taken into account the technique becomes unique in answering different questions. Fig. 6.4 reports the Al profile of a $7 \times 10^{13}/cm^2 - 80 keV\,Al$ implanted Si through 50 nm

thick SiO_2 after annealing at 1200^0C - 5sec. The Al atoms diffuse
from the substrate and accumulate at the SiO_2/Si interface. The
experimental profile can be fitted assuming a diffusion process of
Al and a repartition coefficient of 700 between SiO_2 and Si. The
detection limit is also below $10^{15}/cm^3$ [6.10].

Depth resolution is limited by the sputtering process and is
given as the exponential decay length of the trailing edge of the
profile. It is difficult in some cases to separate sputter-induced
broadening from intrinsic range straggling of shallow implants.
Comparison between profiles under different processes is instead
meaningfull. The profiles of $10keV B$ implanted in preamorphized
Si layers, along the $< 100 >$ axis and in a random direction are
shown in Fig. 6.5 vs the sputtering time.

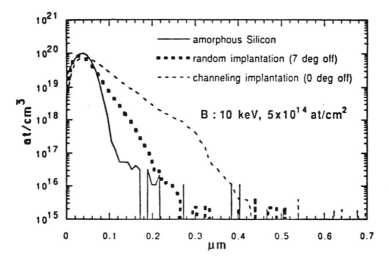

Figure 6.5 *SIMS profiles of 10 keV $^{11}B-1\times10^{14}/cm^2$ implanted
in amorphous silicon, along the $< 100 >$ axis of a Si single crystal
and in a random direction.*

Clearly the tails differs substantially and shallow junctions require
the use of preamorphized silicon layers with self-ion or with Ge
ions.

The implants must be activated by thermal annealing. Even
the use of rapid thermal processes causes the broadening of the
profile as shown in Fig. 6.6 for BF_2 implants in silicon.
This phenomenon named transient enhanced diffusion and due
to the release of point defects at the end of range damage was
discovered by SIMS analysis and it is of relevance in the shallow
junction formation as discussed in chap.7.

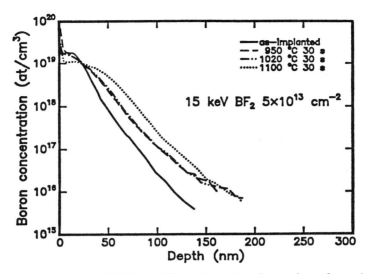

Figure 6.6 *Boron SIMS profiles of as-implanted and rapid thermal annealed 15 keV $BF_2 \cdot 5 \times 10^{13}/cm^2$ implanted silicon wafers. Annealing performed in N_2 ambient.*

The inherent depth resolution is ≤ 1 nm and is limited by the secondary ion escape depth and by the random nature of the sputtering process on an atomic scale. In practice several factor limit the depth resolution, as inhomogenity in the primary ion current density with a non-flat creater bottom, bombardment - induced roughness, atomic mixing caused by the collision cascade, radiation-enhanced diffusion and chemical effects. The relative depth resolution $\Delta X/X$ is constant and usually equal to a few %. The beginning of the profile is affected by large uncertains and variations due to removal of the contamination layer, to the build-up of implanted oxygen within the surface and to a possible change of the composition in a compound target due to the "preferential sputtering" of one component until a steady state is reached in which the components are sputtered at a rate proportional to their bulk concentrations. A useful trick is to deposit, if possible, a thin layer of the substrate material onto the specimen so to establish the steady state conditions before the actual sample is profiled.

The use of SIMS is not limited to analized depths of few microns, it is possible to extend the range up to several tens of microns as illustrated by the profile reported in Fig. 6.7b for 50 MeV $^{11}B \cdot 2 \times 10^{16}/cm^2$ implanted in Si [6.11]. Clearly the usual SIMS analysis cannot be done and it was perfomed on the beveled samples shown in Fig. 6.7a

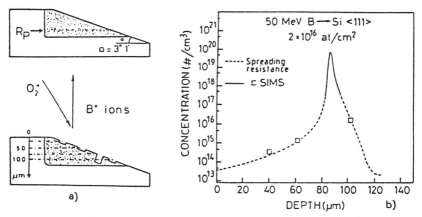

Figure 6.7 *(a) Geometry adopted for SIMS depth profile in beveled samples, (b) SIMS and spreading resistance profile of 50 MeV $^{11}B \cdot 2 \times 10^{16}/cm^2$ implanted into $< 111 > Si$ (from ref. 6.11).*

The other achievment of SIMS is the possibility to profile simultaneously different species in spite of their electrical interaction. Implants of the same type dopant such as P and As can be resolved in the two species profiles while from the electrical point of view only the overall electron concentration is obtained. Similarly SIMS allows also to establish space correlation and interaction between different impurities. The gettering of Al by oxygen precipitates in implanted CZ silicon substrates and described in chapter 5. sect. 7 illustrates clearly the powerful of this technique that is now considered as routine in the microelectronic industry.

Instead of ionized sputtered species, neutral sputtered species can be used for mass analysis. The sputtered neutral species are ionized and subsequently mass analyzed in a quadrupole. The detection limit is routinely in the 1-10pmm range. The sputtered neutral mass spectrometry SNMS is a technique for elemental and isotopic analysis with similar sensitivity for all the elements and without substantial matrix effects. The flux of particles sputtered has the same composition as the substrate [6.12].

6.3 Spreading Resistance Profilometry: One and Two Dimensional Analyses

The knowledge of the chemical profile is not enough to the device fabrication by ion implantation, it is necessary to measure also the electrical profile, after a suitable annealing.

The electrical counterpart of the SIMS is the spreading resistance profilometry [6.13]. This technique is used to determine
carrier depth profiles in bevelled samples. Two probe tipes, usually of osmium or tungsten, are positioned very close together,
at a distance of about 20 μm on the sample. A small difference
of potential $\sim 5mV$ is applied between the two tips, the current
flows and spreads out from a small contact on the surface of a
conducting sample giving rise to a resistance. If the contact area
of the probes is made small enough the largest component of the
measured resistance $\sim 80\%$ arises from a region within a distance
of five times the probe radius. Since metal probes can be made
a few micrometers in diameter, the effective sampling volume of
the probe is very small.

The experimental set up is shown in Fig. 6.8. The sample
is beveled by lapping the silicon wafer at a specified angle, that
ranges from 16" to 5'.

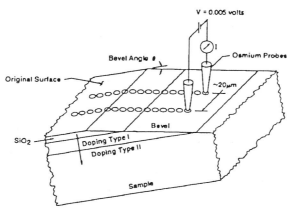

Figure 6.8 *Configuration for making a spreading resistance measurement on a beveled sample. (Source: Solid State Measurement
brochure).*

The two probes are then automatically stepped along the bevel
surface and at each step the current is measured. The probe tip
material is harder than the silicon and consequently fractures the
silicon leaving "probe marks". The relation between the lapping
angle, θ, the distance along the bevel y and the normal to the surface x is $y = \frac{x}{tg}$ so that a shift of $2.5\mu m$ with $\theta = 10'$ corresponds
to a vertical depth of 7nm which is quite sufficient for most VLSI
application today. For two point probes of circular contact of radius a with a semi-infinite semiconductor the sheet resistance R_S

is given by [6.14]

$$R_S = \frac{\rho}{2a} \qquad\qquad 6.4a$$

where ρ is the resistivity of the semiconductor. Due to the small dimension of a the measured spreading resistance may be a factor 10^3 larger than the numerical value of the resistivity. The relationship of eq. 6.4a is an oversimplification. The contact areas of the probes are never truly circular or of the same size. If there is an appreciable resistivity gradient through the sampling volume, then correction need to be made for this.

The first step is then the conversion from the spreading resistance to the resistivity ρ. This is accomplished usually by a calibration procedure using several wafers of different resistivity and homogeneous type and orientation. In general the semiconductor is not of uniform resistivity as an implanted one, and is not of semiinfinite extent. In this case equation 6.4a is modified into

$$R_{MEAS} = \frac{\rho}{2a} \cdot CF(\rho, a, s) \qquad\qquad 6.4b$$

s is the probe separation. CF is the so called "sampling volume correction factor" which accounts for structure variations layer boundaries and resistivity gradients [6.15]. The correction factor has been calculated for depth variations in resistivity. This is a complex problem as CF is itself a function of the resistivity and implicit equations must be solved [6.16]. Algorithms are now available that can performe these corrections in an efficient manner. The in-depth distribution is modelled as a series of distinct sublayers of uniform resistivity, each sublayer being represented by a single measurement point. The Laplace equation is assumed to be valid and is solved in cylindrical coordinates, with suitable boundary conditions. The procedure requires in addition an iterative process to ensure that the solution is physically correct. Among the different algorithms available that of Berkowitz and Lux is the most popular and is incorporated in the software of spreading resistance apparatus on the market [6.17].

For homogeneous bulk samples, for which the correction factor is one since all the layers have the same resistivity, the basic equation 6.4a implies a linear relation between resistance and resistivity. Actual measurements, however, do not often show this trend. The basic contact model is then sometimes unsatisfactory and several modifications have been introduced to improve its correspondence with the real calibration data [6.18]: either a resistivity dependent barrier resistance term is added, or a resistivity dependent variable radius can be considered, or a combination of both can be applied. Modern spreading resistance apparatus gives the choice between a barrier resistance contact model or a variable radius one. The first model is used when the calibration curve in a log-log scale has a slope larger than one. This is the case for n-type silicon when measuring with well conditioned probe tips. The variable radius model is used for p-type silicon where the barrier resistance model is often inapplicable because the calibration curve for p-type has usually a slope smaller than one.

The correction factors act as an amplifier of noise on raw spreading resistance data, since a small change in measured spreading resistance between two closely sublayers implies a large change in resistivity. Data smoothing is therefore essential before correction factors are applied. But the smoothing procedure requires a particular attention to give reasonable hints to the problem. Following this complex procedure performed now by computer the raw data of spreading resistance are converted into a resistivity profile and then into carrier concentration profile by the use of published conversion date. Several assumptions are made in this conversion procedure, mobility values are taken from literature data for bulk and perfect substrates. Disorder can be still present after the annealing of the implant so that the carrier concentration may be understimated. In addition the published data are for conversion between resistivity and dopant concentrations, SR measures carrier concentrations and not dopant concentrations. It is assumed that the two concentrations are equal.

In spite of all these assumptions and uncertains SR is a quite powerful analytical tool. A wide concentration range $(10^{13} - 10^{21}/cm^3)$ may be profiled with a spatial resolution of 20-30 nm. In favourable systems depth resolution of a few nanometers has been achieved.

Some examples are shown in Fig. 6.9. In samples beveled for spreading resistance measurements, the "on bevel" carrier distribution may be distorted from the corresponding dopant atomic distribution. The origin of this effect is related to a phenomenon inherent in SR called "carrier spilling" [6.19]. This is caused by the redistribution of mobile carriers in samples consisting of thin sublayers with different doping levels. The final effect of this "carrier spilling" is the shift of the junction depth toward the surface. The calculation of dopant profiles starting from spreading resistance measurements is based as previously outlined on the solution of multilayer Laplace equation. Several authors have shown that the assumption of charge neutrality in each layer is incorrect due to the carrier distribution in non uniformly doped silicon. Current research is focused at this problem solving the one-dimensional

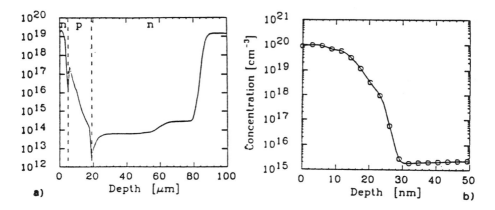

Figure 6.9 *Spreading resistance profiles of (a) n-p-n power transistor and of (b) very shallow junction. (Courtesy of G. Galvagno)*

Poisson's equation to predict the on bevel carrier profile [6.20]. These complex calculations will not be treated here, but an example that demonstrates the entity of this phenomenon will be given. In Fig. 6.10a a SR measurement (symbols) is compared with a SIMS profile (continuous line) for 1 MeV - $2 \times 10^{13}/cm^2 P$ ions implanted into p-typesilicon.

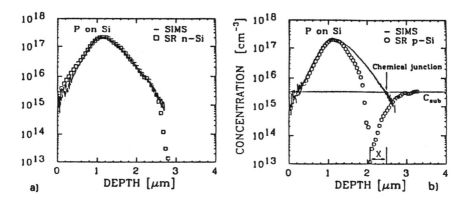

Figure 6.10 *Spreading resistance and SIMS profile of $2 \times 10^{13}/cm^2 \cdot 1 MeV P$ ions implanted into p-type (a) and n-type (b) silicon substrate. (Courtesy of V. Privitera and G. Galvagno)*

The chemical junction indicates the point where the dopant profile concentration becomes equal to the substrate dopant level. The difference between these two profiles is evident. The on bevel junction shift X is the distance over which the electrical junction of the SR profile is shifted with respect to the metallurgical junction due to the carrier spilling. The shift depends on the shape of the dopant distribution and is more evident in low shape profiles. No significant difference is however present when dopants are introduced into the the same type materials (see Fig. 6.10b).

A special sample preparation and measurement procedure has been developed recently to measure two-dimensional carrier profiles in silicon by SRP with 50 nm resolution: by this technique iso-concentration contour lines at the substrate level are obtained and the lateral spread of impurities under a masking layer, due to implantation and/or diffusion, can be measured [6.21]. The method is based on a magnification of the lateral and the depth scales using a double bevel with very small angles. The normal SRP measurement configuration is changed to provide the lateral resolution and to be capable of determining iso-concentration lines at a level fixed by the used substrate. In order to apply the 2D-SRP technique a mask is needed which forms a regular stripe pattern on the wafers (alternatively mask and window). Oxide stripes, normally used as a mask, are obtained using dry etching in order to provide nearly perpendicular mask edges. The oxide layer has to be thick enough such that no ion penetrates through the oxide.

Let us consider as an example a structure consisting of stripes of implanted and unimplanted regions. A complete determination of the doped region includes the one dimensional vertical profile as well as the determination of the lateral profile under te oxide masks. In a normal SRP measurement the bevel is made parallel to the stripe direction to expand the depth scale. To determine the lateral distribution the sample is instead rotated such that the bevel edge forms a very small angle (β) with respect to the stripe direction as shown in Fig. 6.11 [6.22].

The latter implies that when moving across the beveled surface in a direction parallel with the bevel edge one obtains a magnification of the lateral direction of $1/sin\beta$. The distance perpendicular to the bevel edge provides an expanded depth scale. The actual SRP measurement is performed by measuring the resistance between the two probe points which are placed at the same depth (same distance from the bevel edge) with an interspacing of about $100\mu m$. The probes are stepping across the beveled surface parallel to the bevel edge (i.e. the lateral direction).

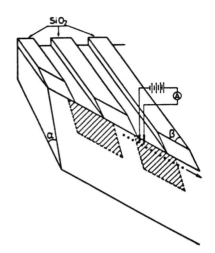

Figure 6.11   *View of a double bevelled sample composed of oxide stripes with implanted region in between. The hatched areas represents the implanted regions on the bevel plane. The probes are stepping parallel to the bevel edge (from ref. 6.22).*

A typical result of such a scan is shown in Fig. 6.12. This sample was implanted with 1 MeV - $1 \times 10^{14}/cm^2 P$ in a stripe structure and the annealing process was performed at $1100°C$, 10 s. The substrate is n-type with a concentration of $10^{15}/cm^3$. In the high resistance region (1-2) one or both probes are completely outside the implanted region and the spreading resistance corresponds to the substrate value; in the low resistance region the second probe reaches the boundary of the implanted region whereas the first probe is still in the highly doped region. The lowest value corresponds to the spreading resistance as it would be measured in a normal one dimensional experiment.

From the above it is clear that the width of the low resistance region is commensured with the width of the doped region. If this measurement is repeated at different distances from the bevel edge, i.e. at different depths, the lateral straggling as a function of depth in the sample is obtained.

In order to understand the result of Fig. 6.12, it is necessary to perform a coordinate transformation from the bevel plane (x",y") to the sample plane (y,z), as shown in Fig. 6.13 [6.22].

One can observe, by the transformation

$$z = cos\alpha cos\beta \cdot x" + sin\beta \cdot y" + sin\alpha cos\beta \cdot z" \qquad 6.5$$

that $|z| \simeq |x"|$ because α and β, respectively the depth and the lateral bevel angles, are quite small. The projections of the approximately circular probe marks, from the bevel surface (x",y") into the (x,z) plane will look approximately like very flattened ellipses with the long side equal to the size of the probe in the x"

Figure 6.12 *SRP pro-file measured in a n-type silicon substrate implanted with* $10^{14}/cm^2 - 1 MeV P$ *ions and annealed at* $1100^0C \cdot 10s$. *The line scan is done* $5\mu m$ *below the bevel edge, corresponding for a* $1^0,9'$ *bevel to a depth of* $0.098\ \mu m$ *(from ref. 6.22).*

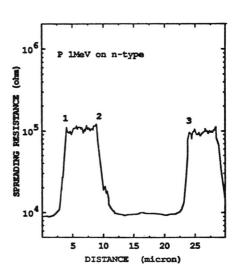

Figure 6.13 *Schematic representation of the probe contacts projection from the bevel plane* (x'',y'') *to the sample surface plane* (y,z). *The separation between the centers of the probe contacts, parallel and orthogonal to the bevel edge, is indicated.*

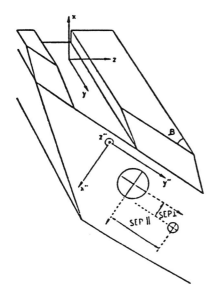

direction. In order to avoid making contact with the second probe in the already damaged silicon (as a result from the imprint of the first point) the probes can be separated in a direction orthogonal to the bevel edge over a small distance $(3 - 5\mu m)$. *Orthsep* is defined as the value of this separation. *Paralsep* refers to the distance between the two probes in a direction parallel with the bevel edge. Eq. 6.6 describes the projection of the probe positions from the beveled surface onto the original cross section (x,z) of the sample. The separation between the centers of the probes in the (x,z) plane will be

$$Sep = OrthSep \cdot cos\alpha cos\beta + ParalSep \cdot sin\beta \simeq$$

$$\simeq OrthSep + ParalSep/Mag \qquad 6.6$$

where the reciprocal value of $sin\beta$ is the lateral magnification. If $\Delta Meas$ is defined as the width of the low resistance region then this corresponds to the part where the second point enters the doped region until the first one leaves the doped region. Finally the formula giving the width of the doped region can be written:

$$\Delta z = \Delta MEAS + ParalSep/Mag + OrthSep - (P1 + P2)/2 \qquad 6.7$$

where Δz is the width of the stripe plus twice the lateral diffusion, P1 and P2 are respectively the lengths of the first and the second probe and $\Delta MEAS$ is the width of the low resistance region in the spreading resistance profiles. Eq. 6.7 relates the distances observed on the scan of Fig. 6.12 with the width of the doped region and the probe geometry [6.23].

From eq. 6.7 one can obtain Δz (lateral width) provided the probe separation (orthogonal + parallel) and the probe sizes (P1,P2) are known. In principle the separation and probe sizes can be measured by a suitable optical microscope. Giving the magnification obtained from the small angles ($\alpha = \beta = 40' - 1°$) a depth and lateral resolution of 25-50 nm can be obtained with a stepsize of $2.5\mu m$.

Several parameters are needed to determine the absolute value of Δz, such as the correct determination of the magnifications, i.e. of the α and β angles. α is determined using a surface profilometer and β by an optical microscope using a repetitive stripe structure, for instance $10\mu m$ *oxide*/$10\mu m$ *doped region*, such that always more than one stripe can be measured at the same time. Since in the 2D scan a repetitive structure is analyzed the magnification can be obtained, quite accurately, by equating the measured distance between two periodic peaks to the mask plus window dimension (in the case of Fig. 6.12 was $20\mu m$).

The evaluation of the probe size and separation is also of great importance. These two parameters can be determined by high magnification optical microscope, but the uncertainty on both of them is still at least 300 nm. Therefore, if this technique has to be applied to shallow junction characterization, a calibration method for better determining probe size and separation is necessary.

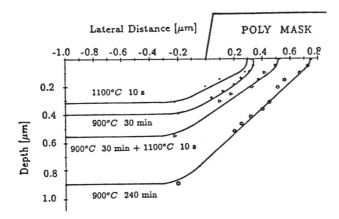

Figure 6.14 *Comparison between contour lines of a B implanted sample after several diffusion processes at* $900^\circ C$ *and at* $1100^\circ C$ *respectively. A polysilicon masking layer is used (from ref. 6.24).*

Two-dimensional contour lines, relative to a 20 keV, 2 × $10^{15}/cm^2$ B implant, after several diffusion processes, are shown in Fig. 6.14. A polycrystalline silicon mask was used [6.24]. These iso-concentration curves are positioned at a level of 1 × $10^{15}/cm^3$ fixed by the used substrate. By the maximum lateral spread and depth of the 2D contour lines, it is possible to calculate the diffusion coefficients in the lateral and vertical direction.

This technique has been also applied to the measurements of the lateral spreading of implants performed in channeling geometry. The data reported in Fig. 3.30 were obtained by this 2D spreading resistance procedure and revealed the large reduction in the lateral spread of the channeled ions. As another example let us consider the diffusion of metallic impurities, such as Pt and Zn, implanted in Si through a mask. The two-dimensional diffusivity can be monitored by 2D-spreading resistance. Some results are shown in Fig. 6.15a for Pt and 6.15b for Zn respectively [6.25]. The data indicate that the diffusion of Zn is slower than that of Pt, but that in both cases the lateral spread under the mask is higher than the vertical penetration. The difference between the vertical and the horizontal diffusivity can be attributed to the peculiar atomic mechanism that governs the transport of transition metals in silicon: i.e. the kick-out process. In this case the atomic transport is provided by interstitial metal atoms (A_i) which become substitutional (A_s) by replacing a Si atom which becomes interstitial. The substitutional impurity is immobile and then the effective diffusion coefficient is determined by the in flux of interstitial metal atoms and by the outflux of silicon self interstitials produced by the kick-out reaction. The surface is a quite good sink of Si-interstitials so that the lateral diffusion is faster than the vertical being no-limited by the outflux of Si-interstitials. These impurities are used for the lifetime control in power devices and the method will be described in sec. 7.6.

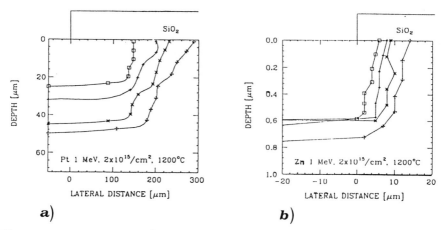

Figure 6.15 *Two dimensional contour line of the region containing Pt (a) and Zn (b) after diffusion at $1200^0 C$ for ($\square$) 10, (+)20, ($\times$) 30, (+) 40 s respectively (from ref. 6.25).*

6.4 Carrier and Mobility Profiles

The conversion from spreading resistance to carrier and to dopant concentration relies on the following main assumptions: all the dopant is electrically active and the mobility is due only to the scattering of carriers from ionized impurities and from lattice vibrations.

In spite of a complete activation residual damage reduces the carrier mobility so that the conversion made by the SRP is affected by errors. It is important to measure in an independent way the carrier and the mobility profiles.

By using a combination of Hall effect and sheet resistance measurements it is possible to determine the carrier concentration as well as the mobility of a doped sample. This method has the advantage to be absolute and not comparative (as SRP) so that calibration samples are not required. The measurements determine the amount of carrier concentration by the Hall coefficient $R_H \propto \frac{1}{n}$ and the mobility by considering the sheet resistance value $\mu = \frac{1}{ne}$. The advantage to measure the mobility permits the method to be applied also if mobility values are altered by some reasons as the presence of defects or precipitates in heavily implanted samples. If the measurements are coupled with a layer by layer removing procedure the carrier concentration profile can be determined. Special structures are required so that test patterns have to be included in the masks for wafer processing. Other limits of the technique include the necessity to isolate electrically by junctions all the layers and the difficulties in determining the carrier profile beyond the junction depth.

Among the several methods used to measure the Hall effect that based on the van der Pauw pattern [6.26] has been recognized as the most advantageous. van der Pauw had demonstrated that for an arbitrary sample shape its resistivity and Hall coefficient can be determined using the formulas:

$$R_S = \frac{\pi}{ln2} \frac{V_{34}}{I_{12}} F\left(\frac{V}{I}\right) \qquad\qquad 6.8$$

$$R_{HS} = \frac{1}{2B} \frac{\Delta V_{H,13}}{I_{24}} \qquad\qquad 6.9$$

If four arbitrary point contacts 1, 2, 3, 4 on the border of a simply connected surface are realized. The asymmetries are considered in the correction function $F\left(\frac{V}{I}\right)$ of eq. 6.8.

To eliminate structural asymmetries and interfering magneto electrical effects (Nerst, Ettinghausen, Righi-Leduc, etc.) the current electrodes as well as the voltage electrodes are periodically exchanged and the measurements are averaged on all the possible configuration.

In practice the structure used is illustrated in Fig. 6.16. Implanted contacts (1, 2, 3, 4) are adopted in this case to maintain the low contact resistance independently of the resistivity of the central region (6) that increases during the measurements cycles. In this way carrier concentration can be determined down to $1 \times 10^{15}/cm^3$. Symmetrical samples are used for practical reasons.

Using this method the main sources of errors are the geometry and the scattering factor. The scattering factor r is defined as the ratio of the Hall to the drift mobility:

$$r = \frac{\mu_H}{\mu_d} = \frac{<\tau_R^2>}{<\tau_R>^2} \qquad\qquad 6.10$$

where τ_R is the relaxation time of the carriers. Theoretical considerations [6.27] result in $r = 1$ for degenerate semiconductors

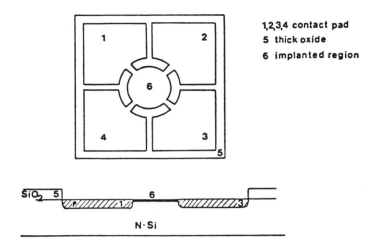

Figure 6.16 *Plan view and cross section of a Van der Pauw geometry adopted for Hall effect and resistivity measurements. 1, 2, 3 and 4 are the contact regions, 6 is the central region where the analysis must be performed, 5 is the SiO_2 layer on top of the Si substrate.*

and $3\pi/8$ for non-degenerate ones, assuming that lattice scattering is the dominant scattering process. The experimental values are between 1 and 1.3 for n-type silicon and between 0.7 and 1 for p-type silicon. Values of 1 for n-type silicon and 0.8 for p-type silicon are normally used.

Measurements can be performed also in polycristalline materials. In this case both the carriers trapped in the grain boundaries or the carriers in the grains contribute to determine the total mobility of the sample.

Depth profiles can be determined if the measurements are sequentially repeated after etching a layer from the sample. (see Fig. 6.17) [6.28].
In this case the mobility and the carrier concentration in the layer (shaded) are given by the following equations [6.29]:

$$\mu_i = \frac{R_{HS,i}\sigma_{S,i}^2 - R_{HS,i+1}\sigma_{S,i+1}^2}{\sigma_{S,i} - \sigma_{S,i+1}} = \frac{\Delta(R_{HS}\sigma_s^2)}{\Delta\sigma_s} \qquad 6.11$$

$$n_i = \frac{\sigma_{S,i} - \sigma_{S,i+1}}{q(x_{i+1} - x_i)\mu_i} = \frac{1\Delta\sigma_s}{q\mu_i\Delta x} \qquad 6.12$$

The mobility and concentration values are located at the middle of the layers $x_i + \frac{\Delta x}{2}$. The thickness of the removed layer has to

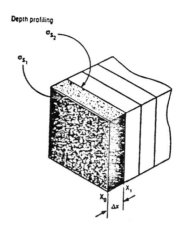

Figure 6.17 *Schematic of the layer removal procedure. The measurements are performed after removing a layer of thickness* ΔX.

be estimated from case to case in order to keep its influence on the calculation of the values of a single layer negligible. If a very thin layer is chosen the differences in eq.6.11 and eq.6.12 become small with respect to the absolute value increasing the errors in determining μ_i and n_i. If a large step is used instead the evaluation of the shape of the profile can be a problem when an abrupt fall is present. In this case the measurements can be performed removing a thin layer and the differences can be calculated considering values a distance of 2 or 3 times the etched layer. Both a high definition and a low absolute error are maintained in all the part of the profile.

Anodic oxidation is the most exact and the simplest method to remove thin layers if the oxide is eliminated by means of a suitable etchant. In silicon the oxidation rate is dependent on the surface carrier concentration. Anodic oxidation is performed illuminating the sample to increase the carrier concentration and to make the oxidation rate independed of the sample doping. (see Fig. 6.18a). Futhermore the oxidation is performed by using a limited constant current until a presetted voltage value is reached. The final voltage is proportional to the thickness of the silicon layer [6.30] oxidized.(see Fig. 6.18b). Being the final voltage value dependent on different parameters (such as the cell geometry, the oxidized area, the used electrochemical solution) a calibration of the oxide thickness as a function of the final voltage is required. Both the oxidation and the etching rates using a 10% HF solution are uniform and well reproducible.

The errors in the thickness of the removed layer are principally due to the calibration curve determination. Using different technique to measure the oxide layer grown on silicon it is possible to

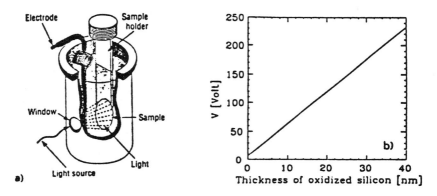

Figure 6.18 *(a) Schematic of the apparatus adopted for anodic oxidation of a silicon sample. The process is usually performed illuminating the sample. (b) Relationship between the anodic potential and the thickness of the oxidized silicon. (Courtesy of V.Raineri)*

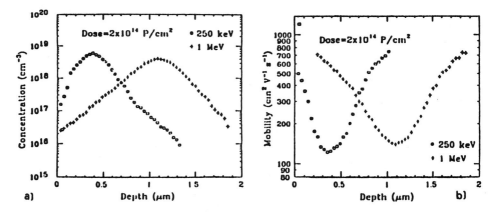

Figure 6.19 *Carrier (a) and mobility (b) profiles for $2 \times 10^{14}/cm^2 P$ ions implanted into Si at 250 keV (o) or at 1 MeV (+) energy respectively. The samples were annealed at $1000^0 C$ for 10 s (from ref. 6.31).*

remove in a controlled way silicon layers down to 5 nm of thickness. Automated system are realized where the oxidation step is one of the step of a sequence of operations.

Carrier and mobility profiles are shown in Fig. 6.19 for $2 \times 10^{14}/cm^2 P$ implanted into Si at 250 keV and 1.0 MeV en-

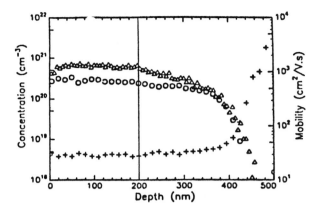

Figure 6.20 *Carrier (0) and mobility (+) profiles for a 200 nm thick polycrystalline silicon layer on top of a Si single crystal implanted with $2 \times 10^{16}/cm^2 As$ ions. The sample was annealed at 975^0C for 30 s. The triangles report the chemical concentration obtained by Rutherford backscattering (from 6.32).*

ergy respectively [6.31]. As another example the carrier and the mobility profiles are reported in Fig. 6.20 for a polycrystalline silicon layer 200 nm implanted with $2 \times 10^{16} As/cm^2$ and annealed at 975^0C for 300 s [6.32]. The profiling technique is then not limited to single crystal substrate.

6.5 Rutherford Backscattering and Channeling Effect

Rutherford backscattering spectrometry (RBS) is a high-energy ion beam technique primarily applied to measure elemental combination or impurity concentration as a function of depth in the near surface region of the solid [6.33]. The combination with channeling effect [6.34] allows the determination of the lattice location of impurities and the depth profile of lattice damage if the analysis is performed in a single crystal substrate. It is a quantitative and for many purposes nondestructive method, with 1 to 5% accuracy, does not require standardization and is matrix independent. It is fast and usually simple to interpret. Channeling experiments are instead more complex and require a deeper date analysis. The analysis is typically performed with $2MeV He^+$ beam produced by accelerators such as Van de Graaff or Tandem machines. RBS has practically no lateral resolution, several attempts have been made to make available micrometer beam size but their use has been limited to a few applications.

In ion implantation this technique has been adopted to study the lattice location of implanted species heavier than the host atoms and the damage associated to the process. The solid epitaxial regrowth of amorphous layer has been studied by channeling technique measurements. In the following the basic mechanism of RBS and channeling analysis will be described together with some examples of interest to ion implantation.

Large angle scattering of high energy light particles carriers information on the different masses present in a sample and on their distribution in depth. Projectiles such as $2MeV\,He^+$ impinge on the sample and those backscattered are detected by a solid state detector that monitors their energy after the collision events. (See Fig. 6.21).

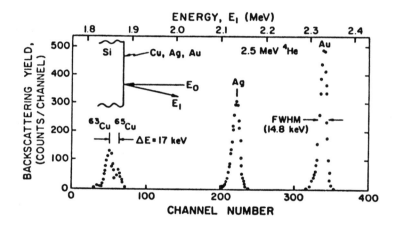

Figure 6.21 *Backscattering spectrum ($\theta = 170^0$) for 2.5 MeV He ions incident on a target with approximately one monolayer coverage of Au, Ag and Cu (from ref. 6.35).*

The energy transfer from the mono-energetic projectile to a target atom can be calculated from the conservation of energy and momentum. Therefore by measuring the energy of the scattered ions, one obtains a direct measurement of the mass of the target atom. The ratio of the projectile energy after the collision E_1 to the incident one E_0 is called the kinematic factor k_M and is given by

$$k_M = \frac{E_1}{E_0} = \left[\frac{(M_2^2 - M_1^2 sin^2\theta)^{\frac{1}{2}} + M_1 cos\theta}{(M_2 + M_1)} \right]^2 \qquad 6.13$$

with θ the scattering angle and M_2 the target atom. For a head-on collision the projectile is scattered with the minimum energy.

If on the sample surface equal quantities of atoms with different masses are present the number of backscattered projectiles at the several energies is determined by the cross section or the probability of the event. In formula the relationship between the number i.e. the yield Y, of detected projectiles and the number N_S of target atoms/cm^2 is given by

$$Y = \sigma(\theta) \cdot Q \cdot \Omega \cdot N_S \qquad 6.14$$

where Q is the total number of incident beam particles, Ω the solid angle subtended by the detector and $\sigma(\theta)$ the cross section that for a pure coulomb potential becomes the Rutherford cross section

$$\sigma(\theta) = \left(\frac{Z_1 Z_2 e^2}{4E} \right)^2 \frac{1}{sin^4 \theta/2} \qquad 6.15$$

Relationships 6.13, 6.14 and 6.15 allow the composition ratio at the surface to be determined. RBS is also sensitive to depth analysis. Only a small fraction of the impinging ions suffers large angle scattering ($\sim 10^{-5} - 10^{-6}$), all the remaining ones continue their path in the target and are slowed down by anelastic collisions with electrons. If a backscattering event occurs at a depth x from the surface the projectile is detected at a lower energy due to the loss during the incoming and outcoming path (see Fig. 6.22). The combination and interplay of kinematic factor, scattering cross section and stopping power allows RBS analysis to perceive masses and mass ratios (compositions) as a function of depth.

As a first example consider the analysis of a Si simple implanted with $2 \times 10^{15} As/cm^2$ at 250 keV. Fig. 6.23 report the energy spectrum of 2.0 MeV He^+ backscattered particles. The shift in energy between the energy signal, of As, atoms on the sample surface and the actual peak signal ΔE_{As} indicates that the As atoms are located below the surface of the Si. The yield of As signal if referred to the height of the Si signal in the same spectrum gives the implanted dose. Detector angle and efficiency as well as integrated counts are not required in the evaluation. Only the energy width on channel is needed. RBS has been adopted infact to calibrate with accuracy ($\sim 1\%$) the implanted dose. The signal of elements lighter than the substrate is superimposed to the substrate yield and then difficult to separate if the impurities are present in a small amount.

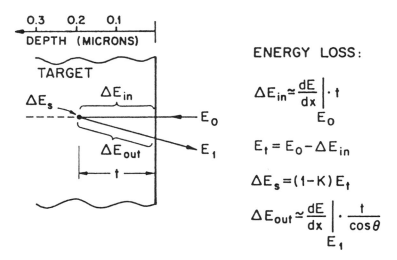

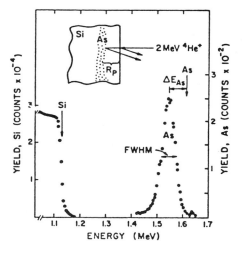

Figure 6.22 *Depth perception of Rutherford backscattering due to the sequence of energy losses: energy lost in electronic stopping on in ward path, ΔE_{in}: energy lost in the elastic scattering event ΔE_s and energy lost to electronic stopping on the outward path, ΔE_{out}. $E_1 = E_0 - \Delta E_{in} - \Delta E_s - \Delta E_{out}$ (from ref. 6.35).*

Figure 6.23 *Energy spectrum of 2.0 MeV ^4He ions backscattered from a silicon crystal implanted with a $1.2 \times 10^{15}/cm^2$ - 250 keV As ions (from ref. 6.35).*

Depth perception can be enhanced by detecting the backscattered projectiles at a glancing angle, a small change in depth measured along the normal corresponds to a large outcoming path and then to an increased energy loss [6.36]. The silicon signal is affected by the large amount of As dissolved in its near surface region that causes a decrease in the yield.

The minimum amount of surface impurities that can be detected by 2 MeV He^+ backscattering is roughly given in terms of impurity atoms $/cm^2$ by $(Z_{substrate} /Z_{impurity})^2 \times 10^{14}$, i.e. $\sim 10^{-2}$ of a monolayer for heavy impurities on a light substrate.

As an example of the procedure the analysis of a Si sample implanted with 40 keV - $2 \times 10^{14}/cm^2 Pb$ ions and annealed at $650^0 C \cdot 30$ min is shown in Fig. 6.24 [6.37].

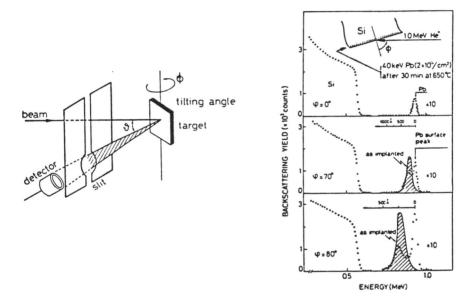

Figure 6.24 *Energy spectra of* $1.0 MeV^4 He^+$ *ions backscattered from a Si crystal implanted with* $2 \times 10^{14}/cm^2 \cdot 40keV \cdot Pb$ *ions. The spectra are recorded for several tilting angles (from ref. 6.37).*

For tilting angles larger than 65^0 the Pb signal exibits a double peak: one located at the implantation range and another at the target surface. The Pb higher - energy signal remains unaffected by tilting whereas the lower signal shifts to lower energy values. The splitting occurs only in the annealed sample and is associated to the reordering of an amorphous layer containing an impurity at concentration above the solid solubility.

As discussed in some details in chapter 5 ions impinging along a low index axis or plane can be steered by a series of gentle collisions with the target atoms and become channeled. When positively - charged particles travel along channeled trajectories in a solid all nuclear and atomic phenomena for which a close encounter is necessary (such as nuclear reaction, large angle scattering, X-ray emission from particle-induced inner shell atomic excitation) are strongly suppressed. Virtually all Rutherford backscattering setups are equipped for channeling experiments by reduction of the beam divergence and by the addition of a goniometer on which the target is fixed. The yield of backscattered particles is shown in Fig. 6.25 for 2 MeV He^+ impinging in a random direction and along the $< 100 >$ axis of a Si target. In the aligned spectrum the scattering yield from the bulk of the sample is reduced by almost two orders of magnitude.

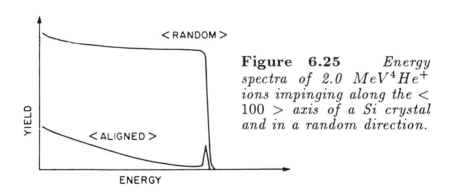

Figure 6.25 *Energy spectra of 2.0 MeV $^4He^+$ ions impinging along the $< 100 >$ axis of a Si crystal and in a random direction.*

The surface peak corresponds to the scattering from the silicon atoms present in the native oxide layer and the surface. The minimum yield is due to the yield of ions that impinge within a small distance from the atomic rows.

Channeled particles interact with atoms on nonsubstitutional lattice sites: interstitials in regular or random positions, vacancy-associated atoms displaced from their regular sites, oversized substitutional atoms [6.38].

Most information comes from a comparison of the backscattering spectra from the host with those from the impurity atoms. Strong yield attenuation of the channeled beam particles with the foreign atoms prove that they lie within the shadow of the aligned set of atomic rows. Successive measurements along two or more

Figure 6.26 *Atomic configuration in a $< 110 >$ plane of a Si lattice of several impurity-atom locations. The shaded areas represent the respective forbidden regions into which a channeled beam cannot penetrate (from ref. 6.38).*

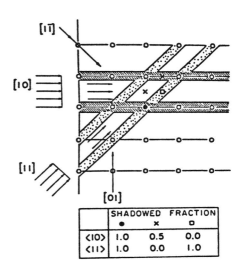

	SHADOWED FRACTION		
	●	×	□
$\langle 10 \rangle$	1.0	0.5	0.0
$\langle 11 \rangle$	1.0	0.0	1.0

different directions can be used to locate the exact position of the impurity in the lattice.

Fig. 6.26 schematically illustrates, for a simplified two dimensional lattice how the channeling effect can be used to determine in a diamond lattice (like Si) the lattice site of impurities for three different locations (● substitutional site; X tetrahedral interstitial site; □ a non substitutional position). The substitutional impurity ● would give a channeling behaviour similar to the lattice atoms for channeling along both the $< 111 >$ and the $< 110 >$ directions. The tetrahedral interstitial impurity X would exhibit the same attenuation as the substitutional impurity along the $< 111 >$ direction but a consistent yield along the $< 110 >$ direction. The other nonsubstitutional position (□) would have a yield along the $< 110 >$ direction rather similar to that observed for tetrahedral interstitial impurity (X). Measurements along different low-index directions are, in general, required to assign the lattice location. In some cases a detailed knowledge of the spatial distribution of channeled particles is required to compare yields for different axes and for nonsubstitutional impurities. For well behaved impurities, the substitutional fraction is given by the ratio

$$\% = \frac{\chi_i^{(r)} - \chi_i^{<110>}}{1 - \chi_{Subst}^{min}} \qquad 6.16$$

where $\chi_i^{(r)}$ and $\chi_i^{<110>}$ are the impurities yields for random and aligned incidence while χ_{sub}^{min} is the minimum yield of the substrate for aligned incidence.

Fig. 6.27 reports as an example the lattice location of 40 keV
Pb implanted at two doses $6 \times 10^{13}/cm^2$ and $6 \times 10^{15}/cm^2$ into Si
and annealed at $650^0 C$ - 30 min. At low dose the thermal process
anneals out the damage, as seen by the decrease of the aligned
yield for the Si signal, and locates in substitutional lattice sites
the Pb atoms as shown by the same reduction in the Pb signal for
both the $< 111 >$ and the $< 110 >$ direction. The same annealing
is not enough to reorder the Si sample implanted at $5 \times 10^{15}/cm^2$,
as indicated by the aligned spectra [6.39].

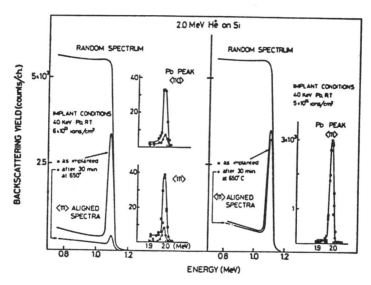

Figure 6.27 *RBS spectra of $2.0 MeV$ ^4He impinging along the
$< 111 >$ and the $< 110 >$ axis of a Si sample implanted with 40
keV $6 \times 10^{13} Pb/cm^2$ (left hand side) and $5 \times 10^{15}/cm^2$ (right hand
side). The analysis has been performed on the as-implanted and
on the annealed samples at $650^0 C \cdot 30$ min [6.39].*

The "channeling ability" of single crystals reflects their
monocrystalline quality. The simplest and most extreme case of
a defective crystal is that of an amorphous layer, for which the
aligned yield remains equal to the random yield. The analysis
reported in Fig. 6.28a refers to a sample implanted at low temperature 77K with multi-energies of Si to a measured amorphous
thickness of 450 nm [6.40].
    The aligned yield follows the random yield for a certain en-
ergy width and then decreases. The decrease occurs at the interface between the amorphous and the single crystal regions where

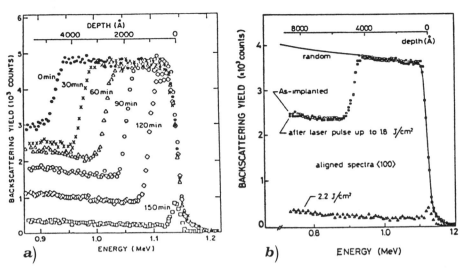

Figure 6.28 *(a) Aligned spectra for 2.0 MeV $^4He^+$ ions incident along a < 100 > channeling direction of Si (100) samples, implanted at liquid nitrogen temperature, and annealed at 550°C for the indicated times (from ref. 6.40), (b) channeling analysis of a same as-implanted sample before and after irradiation with a laser pulse of 1.8 and 2.2 J/cm² energy density (from ref. 6.41).*

the particles can be steered by the crystal rows and and planes. The yield is greater than that from a perfect single crystal because a fraction of the particles in the beam after traversing the amorphous layer emerges with an angle greater than the critical angle for channeling. Change of the energy width, i.e. of the amorphous layer thickness with thermal processes can be analyzed to provide information on solid phase epitaxial regrowth. (see chapter 5) Several spectra are also reported in Fig. 6.28a and they correspond to samples analyzed after annealing at 550°C for the time indicated in the figure.

Channeling can also monitor the phase transition induced by a laser pulse of nanosecond duration as shown by the spectra reported in Fig. 6.28b. The transition to single crystal of the 4000 thick amorphous silicon layer occurs at a well defined energy density of the laser pulse. In the case shown in figure the amorphous layer becomes single crystal at 2.2 J/cm² while it is still disordered at 1.8 J/cm² as indicated by the height of the aligned yield. The threshold character of the transition implies a different mechanism, liquid phase epitaxial instead of solid phase epitaxial growth. If the energy density of the laser pulse is enough to melt all the amorphous layer the subsequent solidification of the liquid occurs on a single crystal seed and a crystalline layer results. Below threshold only part of the layer is molten and the solidification occurs on a disordered substrate. Usually a polycrystalline structure is obtained. Channeling is not able to distinguish between an amorphous and a fine grain polycrystalline material.

Each type of defect present in a single crystal is expected to increase the backscattering yield above the aligned yield of a virgin crystal. This will be visible in the signal from the defective region of the single crystal as well as in the region below it, even if the latter is in perfect condition. Defects interact with the channeled beam and cause a change in the particles trajectories. Some particles are dechanneled and see the crystal as a random medium. The dechanneling depends on the type of defects in the disordered layer [6.42]. Point defects, stacking fault [6.43], dislocations [6.44] have each been found to give rise to a different energy-dependence of the dechanneling factors. (see Fig. 6.29) Differing energy-dependence can be adopted to enhance the sensitivity to the type of defect studied. If more than one type of defect is present a quantitative separation of the defects cannot be obtained generally [6.34].

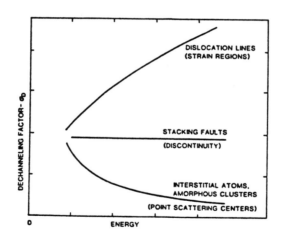

Figure 6.29 *Schematic of the energy dependence of the dechanneling factor of various types of defects (from ref. 6.34).*

Several problems limit the accuracy of the analysis: the stopping power of channeled and non channeled particles differ substantially, the backscattering of channeled particles from displaced atoms is hard to separate from the backscattering of dechanneled particles from lattice atoms, displaced atoms generally do not occupy random positions in an otherwise perfect lattice, and the channeling flux is not constant across the channel area. In spite of all these cautions channeling analysis provides a simple method to determine the depth distribution of damage, expecially for as-implanted samples. The damage distributions reported in Fig. 4.9 for 80 keV - $2 \times 10^{14}/cm^2 B$ ions impinging at 0^0 and at 7^0 were obtained by the aligned and random spectra shown in Fig. 6.30.

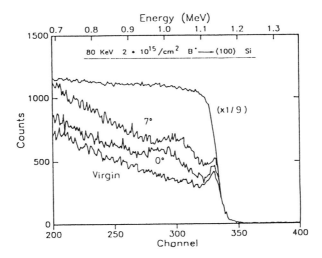

Figure 6.30 *Random and* $< 100 >$ *aligned yields of 2.0 $MeV\,^4He^+$ ions backscattered from silicon samples implanted at room temperature with $2 \times 10^{15}/cm^2$ - 80 keV B ions with a tilt angle of 0^0 and 7^0, respectively. The yield of a virgin sample is also shown for comparison (from ref. 6.45).*

The experimental higher yield at depths larger than the implanted region is in agreement with the previous consideration on the dechanneling by defects. The dechanneled ions can be backscattered by all of the atoms of the crystal and so they contribute to the aligned yield. The problem consists now in the separation of the two contributions: backscattering of the channeled component from defects and backscattering from the dechanneled component from all the crystal atoms. Just behind the surface peak the contribution of dechanneling is minimal and so the disorder concentration in terms of displaced silicon atoms is given by

$$N_0(0) \; = \; N \frac{\chi(0) - \chi_v(0)}{1 - \chi_v(0)} \qquad\qquad 6.17a$$

χ is the ratio of aligned to random yields at the energy (or depth) where the disorder is evaluated $\chi_v(0)$ is equal to χ_{min}. In the above equation $(1 - \chi_v)$ represents the channeled component and the term $\chi(0) - \chi_v(0) = [(1 - \chi_v(0)) - (1 - \chi(0))]$ represents the difference between the channeled fraction in the virgin and in the damaged crystal. The analysis at a certain depth x can be performed on the basis of eq. 6.17a if the dechanneled component at the same depth x is known:

$$N_0(x) = N\frac{\chi(x) - \chi_d(x)}{1 - \chi_d(x)} \qquad 6.17b$$

Several procedures have been discussed in the literature to extract the random component $\chi_d(x)$ and its variation with defects. A simple method approximates the shape of the random component with a straight line (see the dashed line in Fig. 6.30). The depth distribution of displaced Si atoms can then easily extracted and the result for the samples analyzed in Fig. 6.30 are reported in Fig. 4.9. The profiles can be compared with TRIM simulation that yields the number of vacancies vs depth.

6.6 Transmission Electron Microscopy

The electron microscopy is one of the most powerful technique for material analysis. Electron micrographies give often unique information about the morphology and the atomic structure of wide range of materials. Mainly two different kinds of electron microscopes are employied: scanning electron microscope (SEM) and transmission electron microscope (TEM) [6.46]. In the former case a focussed electron beam hits the sample with a spot of 30-100 nm in diameter. The beam is scanned over an area whose lateral extension ranges between 1 and 100 μm. For each spot position the sample produces secondary electrons which are collected by a photomultiplier detector. The signal coming from the detector is used to modulate an image in a CRT display. Those regions of the sample which are morphologically shadowed to the detector will appear dark. Thus the contrast in the image is basically due to the surface topography of the analyzed sample. The lateral resolution of a SEM is directly given by the diameter of the beam spot.

In transmission electron microscopy a parallel electron beam cross through the sample whose thickness has been reduced at 50-300 nm. The diffracted beams produced at the outside of the sample. are focussed in the focal plan of a magnetic objective lens which produces a first magnification of the illuminated sample area. The final image is obtained by using subsequent magnetic lenses (projectors) and a total maximum magnification of 1-3 millions is obtained. The lateral resolution of a TEM mainly depends on the spherical aberration of the objective lens: typical values are of 0.1 - 0.3 nm. Since the image contrast is mainly due to the interference between the diffracted beams, crystallografic structure and defects can be resolved.

Sample preparation is an important step for TEM observation: a sample area with large lateral extension (typical values 10 - 100 μm) is required to be less than 50 - 300 nm thick, as shown in Fig. 6.31a. If the sample has to be observed in cross section the thinning procedure starts by the realization of a sandwhich structure as shown in Fig. 6.31b [6.35].

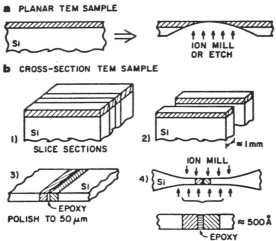

Figure 6.31 *Schematic of sample preparation for planar (a) and cross section (b) TEM analysis. The sequence in (b) involves: (1) and (2) slicing sections about 1 mm thick, epoxy to hold two sections (3) together and then polishing to about $50\mu m$ and (4) ion milling to reduce the thickness to about $0.1\mu m$ (from ref. 6.35).*

The surfaces of two different samples (500-$600\mu m$) are glued together by means of an epoxy resin. The structure is completed by glueing two silicon pieces (1 mn1 thick) on the backside of each sample. The sandwich is now ready to be cut in order to obtain slices of a thickness of 500 μm. Each slice is polished on both sides and a circular hole (3 mm in diameter) is produced on one side by using an ultrasonic disc cutter (Fig. 6.31c). The depth of the hole is about 20-$40\mu m$. Furthermore the opposite side of the slice is mechanical lapped and polished up to the trace of the scavo appears. A copper ring (100 μm thick) is applied over the thin disc to allow an easy handling of it (Fig. 6.31d). Final milling is obtained by ion sputtering: two Ar^+ beams with energies of 3-7 keV are focussed on the disc at incident angles of $12-18^0$. In order to ensure an uniform thinning the disc is continuosly rotated.

Once the sample has been prepared the conditions of electron illumination depend on the features to be evidenced during TEM analysis. In the case of device observation the sample is usually tilted in such a way that the electron beam crosses through the sample parallel to the substrate crystallografic axis along which the device is built (the (110) axis is frequently used). This allows the observation of well defined interfaces between the different layers present in the device structure as shown by the micrograph of Fig. 6.32a that refers to the structure of 1 Mbit Flash Eprom cell. The $\sim$ 10nm thick gate oxide between the silicon substrate and the polycrystalline silicon floating gate is well resolved. If the sample is not correctly oriented the observed interface appears more broad than in reality because of the finite thickness of the sample (Fig. 6.32b).

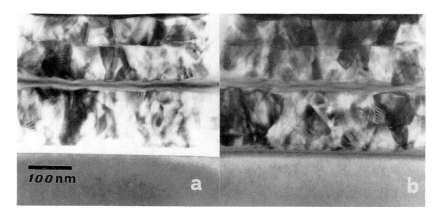

Figure 6.32 *Transmission electron cross section of 1Mbit Flash Eprom memory cell. (a) Electron beam parallel to the (100) substrate orientation, (b) sample tilted of 15^0. (Courtesy of C.Spinella).*

In order to obtain an image contrast which essentially depends on the different electron scattering by the layers present in the structure, a relative large number of diffracted beams ($\sim$ 6) is used to form the image. Fig. 6.33 shows the TEM micrograph of the emitter region in a high frequency bipolar transistor. The analysis was performed by transmission electron microscope with an acceleration energy of 200 keV.

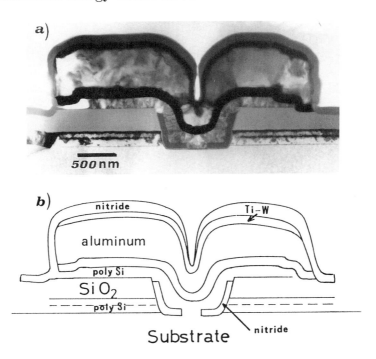

Figure 6.33 *(a) Cross section of the polycrystalline emitter of a high frequency bipolar transistor, (b) Schematic illustration with the labels of the different layers). (Courtesy of C.Spinella)*

The emitter contact of Fig. 6.33 was realized by an *As* doped polycrystalline silicon layer 200 nm thick. The micrograph also resolves the other layers of the structure: the contact window is defined by lateral nitride spacers whilst the metallization consists of a layer of aluminum ($1 \mu m$ thick) between two thin films ($\sim 100 nm$ thick) of titanium which are evidenced as the regions with darkest contrast in the image.

Other important information can be obtained by high resolution analysis. In order to reach the limit resolution of the microscope the aberration effects on the image must be reduced as much as possible. The cromatic aberration pratically vanishes if the sample is very thin (less than 30 - 40 nm) so that the electron energy loss can be neglected. The spherical aberration due to the geometry of the objective lens is reduced by using a finite aperture in the focal plane of the objective lens. In this way a limited number of diffracted beam at lower scattering angles partecipates to the image formation and the resolution ρ_s will be given by:

$$\rho_s = C_s \theta^3 \qquad\qquad 6.18$$

where C_s is the spherical aberration coefficient and θ is the diffraction angle. C_s is 1 mm. If the objective aperture is too small the resolution can worse due to the Abbe limit:

$$\rho_a = 0.61 \frac{\lambda}{\theta_a} \qquad\qquad 6.19$$

where λ is the electron wavelenght and θ_a is the maximum scattering angle which is defined by the diameter of the objective aperture.

The total spatial resolution ρ will be:

$$\rho = \sqrt{\rho_s^2 + \rho_a^2} \qquad\qquad 6.20$$

Fig. 6.34 shows a high resolution micrograph of the interface between the single crystal silicon substrate and the polycrystalline silicon layer at the emitter contact region of the sample analyzed at lower magnification in Fig. 6.32. The analysis was performed in the as-deposited sample and after annealing at $1000^0 C$ - 10s. The contrast at the interface in the as-deposited sample indicates the presence of a native oxide layer ~ 0.1 nm thick. In the annealed samples brigth spots appear, and the order extends all over the analyzed cross section [6.47].

The periodic structure of the lattice planes in Bragg conditions are evidenced by the alignement of the bright points in the image. This periodicity is locally interrupted at the interface because of the presence of amorphous silicon oxide bubbles. These bubbles are the result of the evolution of the continuous native oxide film present at the interface after the thermal treatments to diffuse the *As* dopant contained in the polycrystalline silicon layer. The micrograph was taken by using 23 diffracted beam from the (110) silicon pole.

So far we have discussed the most common applications of TEM to the ion implantation field: namely defects either in plan or in cross section. Recently the experimental procedure previously developped for delineation of bidimensional dopant concentration profiles by scanning electron microscopy [6.48] has been extended to TEM analysis gaining in resolution [6.49].

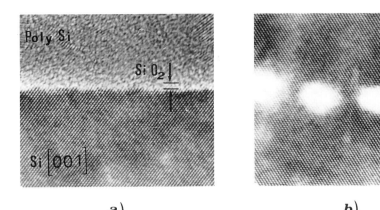

a)

b)

Figure 6.34 *High resolution transmission electron micrograph of the interface between the single crystal substrate and the polycrystalline deposited silicon layer. (a) as-deposited, (b) after $1000°C \cdot 10s$. The white regions represent the silicon oxide bubbles formed after annealing (from ref. 6.47).*

Junction delineation is also done by the junction-staining procedure. The method is based on the staining of a p-n junction after exposure to ultra violet light in presence of a suitable chemical solution. The staining solution consists of 5% concentrated HNO_3 in 95% HF (48%). The p and n regions are then distinguishable by the difference in colouring or by the fine line seen at the transition. The sample is usually lapped at a small angle as for the spreading resistance measurement. The analysis is performed in a optical microscope with a resolution of $\sim 0.5 \mu m$ [6.50]. In the case of electron microscopy the samples are thinned in cross-sectional configuration and subsequently etched in a $HF : HNO_3 : CH_3COOH(1 : 3 : 8)$ chemical solution for times of 1-10s. This solution etchs more easily those regions of silicon doped with As and in presence of UV illumination also those doped with B. The difference in etched thickness is then correlated to the dopant concentration. A calibration with spreading resistance profiles allows a quantitative evaluation of the dopant concentration. As an example the XTEM image of a masked sample implanted with $80keV B - 10^{15}/cm^2$ at a tilt angle of 7^0, annealed at $1200^0C - 10$" and selectively chemical etched is shown in Fig. 6.35a [6.51].

The equicontrast line indicates the equiconcentration profile and the asymmetry around the mask edges related to the angle of incidence is clearly shown. Fig. 6.35(b) reports instead the similar

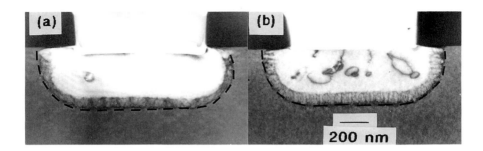

Figure 6.35 *Cross section TEM analysis after chemical etch of $10^{15}/cm^2 \cdot 80 keV B$ ions implanted at 7^0 (a) and at 0^0 (b) tilt angle and annealed at 1200^0C - 10". The dashed lines are the equicontour profiles at a concentration of $10^{17}/cm^3$, calculated according SUPREM-4 program.*

analysis of a masked Si sample implanted at 0^0 with $10^{15}/cm^2$ — $40 keV B$ and annealed at 1200^0C - 10". The dashed line is the equicontour profile at a concentration of $10^{17}/cm^3$ calculated by the SUPREM-4 program.

CHAPTER 7

SILICON BASED DEVICES

7.1 Introduction

Ion implantation is now routinely adopted for the doping of almost all-silicon-based devices. The key advantage of ion implantation is the precise control in the number of introduced ions. The most complex memory and microprocessor chips require during their fabrication from 10 to 15 implant steps. With the development of new structures other implants are added to improve the electrical behavior of devices. Implants are used to form source and drain regions, but also to control the threshold voltage of the MOSFET, to improve the latch-up immunity and to decrease the hot carrier production. New designs and new integrated circuits require also an accurate control of the dopant concentration gradients and profiles. The precise control of the dose is not any more enough and the other degrees of freedom offered by ion implantation, such as the independent control of the beam energy and of the incident angle, are employed with success. In the well of MOSFET the dopant profile is not that determined by the diffusion law with a maximum at the surface and a steadily decreasing concentration with depth. Several implants at different doses and energies are performed and the resulting profile looks like that shown in Fig. 7.1 [7.1]. Each peak has a specific aim in the optimization of the device electrical behaviour.

The shape takes advantage of the high energy implants now available by the existence of industrial implanters [7.2]. A section will be devoted to the present and future applications of high energy implants in microelectronic device fabrication [7.3].

The first large scale application of ion implantation was the

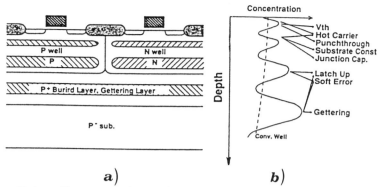

Figure 7.1 *Cross section of a retrograde CMOS well (a) and schematic profile of the dopant in a multiple implanted well of a CMOS and relationship with the device electrical characteristics (b) (from ref. 7.1).*

adjustment of the MOSFET threshold voltage, by the precise and unique dose control in the low $10^{12}/cm^2$ range [7.4]. This application will be described in view of its routinely relevance in device fabrication. Shallow junctions are required in the VLSI and ULSI technology, ion implantation is also a key process in this issue and several attempts have been made [7.5]: low energy implants; molecular ion, such as BF_2^+, implants to decrease the B energy and to suppress the channeling tail by the increased lattice damage; implants in pre-amorphized silicon layer; use of implanted silicide layers as diffusion sources. Some of these procedures will be considered and described later.

Bipolar devices relies also on implantation. Emitter, base and collector are formed by ion implantation and several improvements are indeed associated to the exploitation of the different degrees of freedom offered by the implant technology [7.6]. The formation of polycrystalline emitters and of buried collectors will be detailed for high speed bipolar. The use of ion implantation is not limited to the common dopants, such as B, P or As, in some cases other impurities are introduced. As an example the control of minority carrier lifetime by Au or Pt implants in power devices is also discussed, just to give an idea of the very broad range of applications offered by ion implantation. The content of this chapter is not at all exaustive for the use and for the impact of ion implantation in the silicon device fabrication, it is just a guideline for further analysis.

7.2 Threshold Voltage Control in MOSFET

The metal-oxide-field-effect-transistor MOSFET is of great commercial importance, expecially in digital circuits. It is the main component of the DRAM and is the dominant device used in VLSI circuits because of its semplicity and because it can be scaled to smaller dimensions than other types of devices. The working principle of the MOSFET was already described in Chapter 1, but the basic elements of an n-channel MOSFET are again reported in Fig. 7.2, for convenience [7.7].

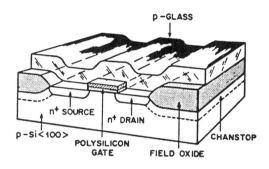

Figure 7.2 *Basic elements of a n-channel MOSFET (from ref. 7.7).*

Two regions, drain and source respectively n-type doped are embedded in a p type material. In between them a thin oxide layer is overlayed by a metal gate. The two n-p junctions are electrically disconnected unless a n-type inversion layer at the surface provides a conducting channel between them. When the surface is inverted and a voltage is applied between source and drain electrons can pass through the channel. The FET is in the "on" state and the current is transported by majority carriers.

The n-channel MOSFET is called NMOS; if all the n-type and p-type layers are changed to p-type and n-type respectively the p-channel MOSFET or PMOS is fabricated. By a suitable choice of the doping it is possible to fabricate on the same chip, one near the other a p- and a n-type MOSFET, the so called complementary MOS or CMOS structure.

The conductivity of the on channel surface layer is determined by the difference in gate and silicon work functions, by the presence of charges in the oxide and at the SiO_2/Si interface and by the external applied voltage, V_G, to the gate. If for $V_G = 0$

there is no inversion layer a gate voltage must be applied to induce a channel: MOSFET of this type are called enhancement-mode devices. If instead for $V_G = 0$ surface potential resulting from the work difference and from the charge is enough to create the inversion layer the MOSFET is called a depletion-mode device. The gate voltage control is then usually employed to reduce the conductance of the built-in channel.

The MOSFETS are the basic elements of the memories, i.e. of devices which can store digital information in terms of bits (binary digits). For large memories such as RAM (Random Access Memory) the memory cells are ordered in a matrix structure, and they can be accessed in random order, independent of the exact location, to store (write) or to retrieve (read) the data [7.8]. The simplest cell is realized by a single MOSFET with a storage capacitor. The transistor serves as switch and some electrons can be storage in the capacitor providing the bit of information. The terms bit line and word line refer to the row and column organization of the memory.

The circuit diagram of a one-transistor DRAM cell is shown in Fig. 7.3.a [7.9]. If at time t_0 the voltage on the column

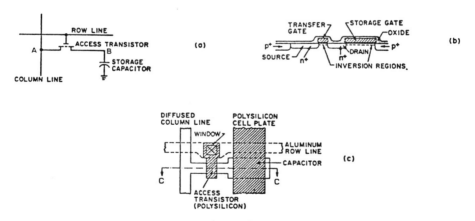

Figure 7.3 *Dynamic random access memory cell (a) circuit diagram schematic, (b) cross section through A-A, (c) lay out of the cell (from ref. 7.9).*

line (or bit line) is V_0 and if the gate potential (row line) is above threshold to induce a conductive channel, the MOS acts as a transfer-gate or pass transistor and the capacitor is charged so that the potential in B becomes equal to that of A. If the potential of the gate is reduced below threshold the potential in B remains fixed to the previous value. The potential in A can be changed but not that in B. The capacitor acts then as a memory element because it can save the information associated to the value of the potential in B. (Fig. 7.3b) The storage capacitor uses the channel region as one plate, the polysilicon gate as the other plate and the gate oxide as the dielectric layer. The internal drain serves as a conductive link between the two inversion layers under the storage gate and the transfer gate.

The memory is called dynamic RAM or DRAM. The cell layout is reported in Fig. 7.3c. Because of its small cell area and low power consumption the DRAM has the highest component density per chip. To day the 256 Mbit DRAM is in the development stage. This astonishing accomplishment relies on the shrinkage trend of the FET cell. The minimum feature size is reaching $0.25\mu m$ with a cell area of $0.5\mu m^2$.

Decreasing the size it is necessary to increase the storage capacity of the capacitor and several approaches are now investigated, among them the use of materials with high dielectric constant such as ferroelectrics, the extension of the capacitor to three dimensions (i.e. stacked or trench capacitors) or the increase of the surface area [7.10]. This can be achieved by burying the capacitor under the silicon layer in a silicon-on-insulator (SOI) substrate. The trench and the stacked cell are shown in Fig. 7.4. In the trench cell the storage capacitor is folded into a trench ($\sim$ few microns deep and less than $1\mu m$ wide) etched into the silicon substrate. The two electrodes of the capacitor are formed by doped polycrystalline silicon deposited in the trench and the silicon substrate separated by a thin oxides. The doping of the trench wall has been considered in sect. 3.8. The charge is stored in the inversion region surrounding the trench. In the stacked cell the capacitor is implemented on top of the transfer transistor and field oxide isolation. Both electrodes are formed by polycrystalline silicon.

The capacitance of the storage capacitor is in the 10 fF range and the storage charge in the 30 fC range assuming a voltage of 3 V. The charge in the capacitor cannot remain indefinitely due to the leakage current of the reverse biased $n^+ - p$ of the transfer gate. The time interval during which the capacitor is discharged is of about $10^{-2}s$, so that these cells must be refreshed every $10^{-3}s$. The time between successive refresh cycles is called refresh delay time. An increase in the leakage current, as that due to the presence of secondary defects in the depletion layer, causes an increase of the refreshing frequency and a decrease of the refresh delay time. As an example the relationship between

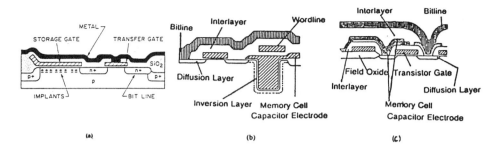

Figure 7.4 *Schematic cross sections of (a) adjacent planar MOS capacitor (b) trench and (c) stacked DRAM cell.*

the cumulative number of bits in a 512 k block of a 4Mbit DRAM that fail a refresh test as a function of the refresh delay time is shown in Fig. 7.5. The analysis of the failed cells has shown the presence of dislocations in the trenches, and their number agrees quite well with the failed cells [7.11]. The DRAM are then volatile memories which lose their information if the power supply is switched off.

The threshold-voltage adjustment of MOSFET was the first industrial application of ion implantation. In enhancement mode n-channel MOSFET the adjustment is done by low dose ($\sim 10^{12}/cm^2$), low energy (30 - 40 keV) boron implant. This implant makes possible the use of less heavily doped substrates which improve MOS circuit performance.

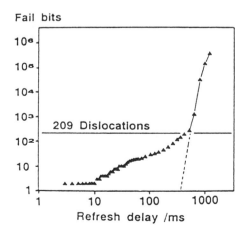

Figure 7.5 *Relationship between the cumulative number of bits in a 512 block of a 4M DRAM that fail a refresh test as a function of the refresh delay time. The number of dislocations detected by etching in the trench capacitors of the memory is also shown (from ref. 7.11).*

The threshold voltage, V_T, is given for an n-channel MOSFET by

$$V_T = \Phi_{MS} - \frac{Q}{C_0} + 2 \mid \phi_p \mid + \frac{\sqrt{4\epsilon_s q N_A \phi_p}}{C_0} \qquad 7.1$$

where Φ_{MS} is the difference between the gate and the Si substrate work function, Q the fixed charge at the oxide-silicon interface, ϕ_p the voltage difference between the intrinsic level E_i and the Fermi level, E_{F_p}, C_0 the capacitance of the polysilicon-oxide-silicon capacitor, and N_A the acceptor concentration of the substrate [7.12].

The previous relation does not take into account the contribution of trapped and mobile ionic charges distributed in the oxide. The interface-trapped charges are due to the $Si - SiO_2$ interface properties and their amount depends on the chemical composition of this interface. The density of these states is orientation dependent, in < 100 > wafers is about one order of magnitude lower than in < 111 >. The density in terms of electronic charge can be as low as $10^{10} cm^{-2}$. The fixed-oxide charge is located within 3nm of the Si/SiO_2 interface, and it is positive. It amounts also to $10^{10}/cm^2$ in well behaved oxide grown over < 100 > substrates.

The low density of the interface-oxide-charge was the main reason to adopt < 100 > instead of < 111 > oriented substrate. The trapped charge are associated to defects in the oxide while the mobile charges to contaminants such as sodium or other alkali ions.

At the threshold voltage V_T, the Si surface becomes of n-type and the difference between the intrinsic and the Fermi level E_F is equal and opposite to that in the inside p-type material. Above V_T, in the so called strong inversion condition, a mobile electron charge is present near the interface. The electron density increases exponentially with the surface potential. The threshold voltage values of n-channel devices, V_{T_n}, and p-channel devices, V_{T_p}, are shown in Fig. 7.6, as a function of the dopant substrate concentration and for three oxide thicknesses 10nm, 25nm and 65nm respectively. The gate was assumed to be n^+ polysilicon and the surface charge density equal to $10^{10}/cm^2$. For n-channel devices V_{T_n} is negative for low dopings and thin oxide layers but becomes positive at high dopings and thick oxide layers. For p-channel devices V_{T_p} is always negative.

In the working conditions it is desirable to have positive values for V_{T_n} and negative for V_{T_p}. In addition being V_T quite sensitive to the exact thickness of the oxide and to the cleaning procedure a method that allows the adjustment of V_T after the gate oxide formation is quite relevant.

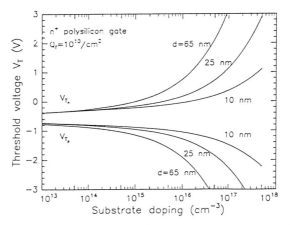

Figure 7.6 *Calculated threshold voltage of n-channel (V_{T_n}) and of p-channel (V_{T_p}) MOSFETs as a function of dopant concentration, assuming a n^+-polysilicon gate and a surface charge density of $10^{10}/cm^2$, and for several gate oxide thicknesses.*

Boron implantation through the oxide layer, fulfills these requirements. The energy of the implant is chosen so that the peak concentration occurs at the SiO_2/Si interface or just below it. (See Fig. 7.1 and 7.7).

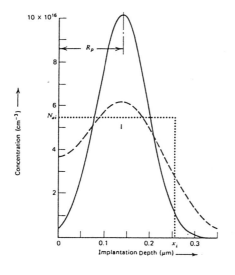

Figure 7.7 *Depth distribution of the implanted boron ions before (solid line) and after (dashed line) the thermal treatment. The dotted "box" curve shows the distribution used for calculation (from ref. 7.12).*

In the simplest treatment the ion distribution is approximated by a sheet of negative inprobile B^+ ions located at the interface. This charge reduces the effects of both the positive depletion charge and the positive fixed charge. By this assumption the change in threshold voltage, ΔV_T, is given by

$$\Delta V_T = \frac{qN_\square}{C_0} \qquad\qquad 7.2$$

with N the implanted dose and C_0 the MOS capacitance. The threshold voltage of an n-channel MOSFET increases of 1 Volt by a boron implant of $2.1 \times 10^{12}/cm^2$ or of $2.1 \times 10^{11}/cm^2$ for a 10 nm or 100nm thick gate oxide respectively.

If the implantation is performed at higher energy or if the subsequent thermal budget is enough to spread out the profile, the effects of the distributed implanted boron charge must be considered. The assumption of a sheet charge is not valid and the treatment is more complicated. Numerical solutions are available and the change of the threshold voltage with implantation dose is usually obtained empirically.

To pursue further the subject let us consider for simplicity a box distribution for the final profile. The implanted dopant is assumed to have a constant concentration N_0 from the surface to a depth R_0, so that $N_0 = \frac{N_\square}{R_0}$. The dopant concentration in the region $0 - R_0$ is given by $N_0 + N_a$ and for depth $> R_0$ by N_a, being N_a the substrate acceptor concentration. The implanted profile is assumed to be contained within the surface space - charge region, as it is usually. (see Fig. 7.7). By solving the Poisson's equation for this charge distribution the depletion-layer width x_d is given by

$$x_d = \sqrt{\frac{2\epsilon_0 \epsilon_r}{qN_a}(\phi_s - \mid \phi_p \mid) - R_0^2 \frac{N_o}{N_a}} \qquad\qquad 7.3$$

At inversion the surface potential ϕ_s becomes $\phi*_p = (kT/q)ln[(N_0 + N_a)/n_i]$. The charge density, Q_d, for the maximum width, x_{dmax}, of the depleted region at the inversion becomes

$$Q_d = qN_0 R_0 - qN_a x_{dmax} =$$

$$= -qN_0 R_0 - \sqrt{2qN_a \epsilon_s(\mid \phi_p^* \mid + \mid \phi_p \mid) - q^2 R_0^2 N_a N_0} \qquad\qquad 7.4a$$

or in terms of $N_\square$,

$$Q_d = -qN_\square - qN_a x_{dmax} =$$

$$= -qN_\square - \sqrt{2qN_a\epsilon_s(|\phi_p^*| + |\phi_p|) - q^2 R_0 N_\square N_a} \qquad 7.4b$$

The threshold voltage becomes

$$V_T = \phi_{MS} + \frac{Q'}{C_0} + |\phi_p^*| + |\phi_p| + \frac{qN_\square}{C_0} +$$

$$+ \frac{1}{C_0}\sqrt{2qN_a\epsilon_s(|\phi_{*p}| + |\phi_p|) - q^2 R_0 N_\square N_a} \qquad 7.5$$

The implant affects the threshold voltage in three ways:
1) the voltage drop, at the inversion, across the surface depleted region is $|\phi_p^*| + |\phi_p|$ instead of $2|\phi_p|$ (dopant concentration increases);
2) the implanted dose adds a new linear term to the voltage;
3) the square root expression for the depletion-charge term is altered essentially by the added term $-q^2 R_0 N_a N_\square$.

The first term depends logaritmically on the implanted dose, the second linearly and the third mainly on the square root of the dose and of the implantation depth. The main effect is then that associated to the added charge, i.e. the term considered in eq. 7.2 [7.13].

In the more general case the diffusion of the dopant with the appropriate boundary conditions at the SiO_2/Si interface must be taken into account [7.14]. The capping oxide is for almost all the common dopants a near perfect reflecting boundary but for boron the segregation coefficient is lower than one and boron has a tendency to be retained in the oxide. The behaviour of boron during field oxidation will be considered at page 283.

For p-channel MOSFET the boron implant decreases the voltage threshold. It compensates dopant atoms in the inversion region. It is important that such an implanted layer be sufficiently near the surface for complete depletion when the MOS is in the "off" state in order to avoid source-drain leakage. It is desiderable also to form the channel at a certain distance from the surface, to improve the mobility of the carriers. The surface mobility due to the increased number of scattering centers is about 0.5 the bulk mobility. An increase in mobility improves the electrical characteristics of the device, as the transconductance, i.e. the change of drain current with the gate voltage at constant source-drain voltage.

7.3 Short Channel Effects

The main trend in the DRAM technology, is the reduction of the cell size and then of the channel lenght. Several effects, named short -channel effects, complicate the simple electrical behavior of the MOSFET. Some of them will briefly considered in the following, to show the use of new implant steps to alleviate in part these effects. Source and drain depletion regions can extend into the channel even without bias, as these junctions are brought closer together in short - channel devices. Drain and substrate forms a n^+p junction reverse biased. As the drain depletion region continues to increase with bias it can actually interact with the source-to-channel junction and lower the barrier potential (see Fig. 7.8a and Fig. 7.8b). This effect is called drain-induced-barrier-lowering (DIBL). When the source junction barrier is reduced electrons are easily injected into the channel and the gate voltage no longer controls the drain current [7.12].

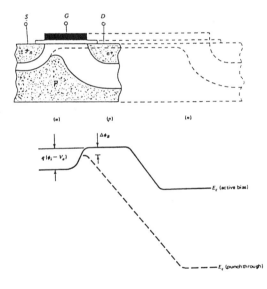

Figure 7.8 *(a) Schematic of the intrusion of the source and drain depletion regions into the channel as the device dimensions are reduced from the long-channel case, (b) conduction-band energy along the $n^+ - p - n^+$ horizonthal structure for long and short-channel cases. The reduction in barrier height can result in loss of gate control (from ref. 7.12).*

Source and drain regions under extreme conditions can meet and this punch-through effect results in a continuous depleted region from source to drain. The current becomes exponentially dependent on the drain voltage instead of being independent of it. Another effect of relevance is the large value of the subthreshold current due to the presence of some electrons in the channel before strong inversion is established. In short channel these electrons can diffuse easily from source to drain and make impossible to turn off the device below threshold.

Scaling the dimension of the device is not compensated by the simultaneous reduction of the applied voltage, so that electric fields tend to be increased at small geometries. Carriers acquire energies from the electric field and become hot, i.e. they have kinetic energy in excess of that corresponding to the ambient temperature. The field in the reverse biased substrate-drain junction can lead to impact ionization and carrier moltiplication. The resulting holes contribute to the substrate current and some may reach the source lowering the barrier and injecting electrons from the source to the p region. A lateral n-p-n transistor action can take place and it prevents the gate control of the current.

Another effect of the hot electrons is their transfer by tunneling to the oxide layer where they become trapped. The threshold voltage as well as the I-V characteristics of the MOSFET are changed. Some effects are schematically represented in Fig. 7.9 [7.15].

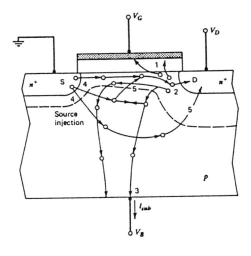

Figure 7.9 *Cross section of an n-channel MOSFET showing hot-carrier processes: energetic carriers give rise to avalanche electron-hole pair production (2). Holes are collected by the substrate (3), and hot electrons (1) may enter the oxide layer and produces a gate current. Electron (4) can be injected into the substrate from the source due to the forward local bias created by the holes for the avalanche effect and are collected (5) at the drain (from ref. 7.15 -* [(c)]*1982 IEEE).*

Hot electron effects can be reduced by decreasing the junction electric field and then the doping of the source and drain. Low doped regions present, however, expecially in small geometries, some other disadvantages as the high contact resistance so that a particular design has been adopted.

It consists of two doping levels, with heavy doping over most of the source and drain areas and with low doping in the region adjacent to the channel. This design is called lightly doped drain (LDD) [7.16]. The structure decreases the electric field between the drain and channel regions and separates the maximum current path in the channel from the maximum electric field location. The injection of electrons into the oxide, the amount of the impact ionization and other hot electron effects are reduced.

The simplest LDD structure consists in a double implant (DDD or double diffused drain) of low dose $P^+(\sim 10^{13}/cm^2)$ and high dose $As^+(\sim 10^{14}/cm^2)$ through the same gate mask and in the subsequent thermal diffusion. Phosphorus has an higher diffusion coefficient than arsenic so that phosphorus penetrates behind the gate edge and produces a less steeper profile. (Fig. 7.10) The implants are performed usually at 0^0 tilt to avoid shadowings effects [7.17].

The high temperature anneal has a deleterious effect on the junction depth so that the LDD (light doped drain) structure consists usually of a low dose implant aligned to the gate (n or p implant) followed by a high dose implant through the gate with an oxide sidewall spacer formed between the two implants (Fig. 7.10b) [7.18].
Even this structure presents some disadvantages related to the degradations of device performance.

Several other approaches have been utilized to optimize the channel and drain profiles. Among them particular importance has been given to a new implant process characterized by programmable large tilt angle and target wafer rotational repositioning during implantation. The process is named "Large Angle Tilt Implanted Drain" (LATID). The adopted sequence of implants for the process is shown in Fig. 7.11a and the achieved lateral profile in Fig. 7.11b [7.19]. At first, after the gate fabrication, phosphorus ions at a dose of $4 \times 10^{13}/cm^2$ and energy in the 60-80 keV range are implanted at an angle of 45^0 to the surface normal, a second implant is made after a rotation of the wafer through 180^0. A symmetrical penetration under the gate is obtained. This implant is followed by a 0^0 tilt high dose As^+ implant ($\sim 10^{15}/cm^2$) self-aligned to the gate. Lateral oxide spacers are not necessary. By this process it is possible to adjust independently the length of the gate to the n-region overlap, the n-dose under the gate and the n^+ - to gate region overlap. The process does not require

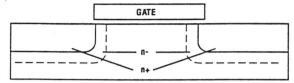

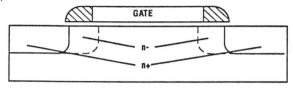

Figure 7.10 *(a) Double-diffused drain obtained by low - dose P$^+$ and high-dose As$^+$ implants through the gate oxide and by a subsequent thermal diffusion, (from ref. 7.17 - $^{(c)}$1985 IEEE) (b) conventional lightly doped drain realized by low dose implant aligned to the gate and by high dose implant through oxide sidewall (from ref. 7.18 - $^{(c)}$1982 IEEE).*

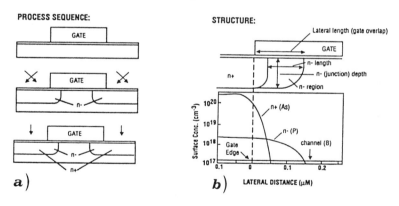

Figure 7.11 *(a) Large-angle-tilt-implanted-drain process sequence: $4 \times 10^{13}/cm^2 - 80keV$ P implant at 45^0 tilt with 180^0 repositioning followed by a $10^{15}/cm^2 - 80keV$ As$^+$ implant at 0^0; (b) structure of LATID and surface concentration vs the lateral distance from the gate edge (from ref. 7.19 - $^{(c)}$1988 IEEE).*

diffusion and it is compatible with rapid thermal annealing.

A similar process is also utilized as punch-through-stopper in buried channel p-MOSFET. These devices suffer of undesirable device performance , i.e. high subthreshold current, due to the drain-induced barrier lowering. The electric field from the drain can "punch through" the depleted counter-doped buried layer and reach the source region. A simple solution to the problem consists in a deep implanted n-type region adjacent to the p^+ source/drain regions. The sequence of adopted process is shown in Fig. 7.12 [7.20].

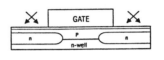

Figure 7.12 *Process for the formation of large-angle-tilt-implanted-punch through stopper (LATIPS) (from ref. 7.20 - (c)1988 IEEE).*

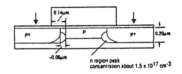

The large angle tilt implant of phosphorus at two different positions, 180^0 rotated, locates these ions deeper laterally than the p^+ source/drain implants. The counter-doped p-region was realized by $1.7 \times 10^{12}/cm^2 - BF_2 - 50keV$ implant, the n-region by $2 \times 10^{13}/cm^2 - 90keV P^+$ at 25^0 tilt in two steps, with the target wafer rotated of 180^0. The source and drain implants were done with $BF_2 - 3 \times 10^{15}/cm^2$ 40 keV. These devices show great improvements in punchthrough characteristics. The subthreshold current in the off-state was many orders of magnitude lower than in the conventional p-MOSFET.

These examples show clearly the powerful of the implant process when the three degrees of freedom $\div$ energy, dose and beam direction are employed. The tilt angle controls the lateral penetration of the ions underneath the mask, the energy the depth and the dose the concentration respectively. Complex channel structures in both p and n-type MOSFET can be created by this new technology [7.21].

7.4 Shallow Junctions

The common trend to the development of solid state devices,

either memory or microprocessor is the increase in packing densities, device operating speed and low power consumption. In the scaling of the device, the channel length and gate dimensions are reduced together with the gate oxide thickness and the source/drain junction depth. To maintain a threshold voltage for the FET independent of the channel length it is necessary to scale the source/drain junction depth and then the lateral penetration under the gate.

MOS devices with submicron dimensions require junction depth of $0.1 - 0.2\mu m$. Several methods have been investigated for the formation of shallow junction and some of them will be considered in the following, such as implants of the desired species in silicon or diffusion from implanted silicides and polycrystalline silicon layers.

Shallow n-type source/drain can be formed relatively easy due to the high mass of the arsenic ions. The high dose implant results in a fully amorphized layers. The main problems are the end of range damage and the secondary defects formed after the solid phase epitaxial growth in other than $< 100 >$ directions (see sect. 5.5).

Shallow p-type junctions are more difficult to form because the low mass of boron does not form an amorphous layer and channeling tails are often present. Anneal temperatures $> 900^0C$ are required to get full electrical activity and considerable diffusion occurs either enhanced by the residual defects or by the equilibrium thermal point defects. To overcome these problems boron implant is commonly carried out by first amorphizing the silicon substrate to a depth beyond the intended junction position or by implanting molecular ions such as BF_2^+, BCl_2^+ in order to obtain a sufficiently shallow implant range.

The profiles of 10 keV - $5 \times 10^{14}/cm^2 B^+$ implanted in preamorphized silicon and in silicon single crystal at 7^0 tilt or at 0^0 were reported in Fig. 6.5. The SIMS analysis clearly show the influence of the targe structure on the dopant depth distribution. The preamorphization is usually performed by Si or by Ge implants. In the case of BF_2^+ implant, the dissociation of the molecular ion following its first atomic scattering event produces a lower-energy boron. The energy of the boron atom is $(M_B/M_{BF_2})E_0 = (10/49)E_0$, being E_0 the incident energy of the BF_2 molecular ion. If the BF_2^+ implant is performed in a preamorphized wafer even shallower junctions are formed. In all of these approaches rapid thermal techniques of annealing are adopted to reduce the diffusion and to improve the quality of the layer. In some cases the RTA is subsequent to a previous furnace anneal at 550^0C to regrow the amorphous layer. The post-implantation anneal results in deeper profiles than one may anticipate on the basis of the B intrinsic diffusion even in the pre-amorphized sample.

Implants of boron in bare silicon are characterized by a channeling tail that extends deeper after the annealing. The anomalous diffusion is transient in nature, after 10s at 900^0C it stops. The investigation so far performed has evidenced that transient B tail diffusion results from high temperature anneals for short times (10-20s). The behaviour is similar for doses between 10^{13} and $5 \times 10^{15}/cm^2$. The anomalous diffusion persists at larger times if high B doses are implanted [7.22].

The transient diffusion is probably associated to the interaction of mobile Si interstitials with B. The diffusivity of B is enhanced by the presence of interstitials in excess of the equilibrium concentration. The transient diffusion can be then associated to the interaction of interstitial Si atoms with B atoms through the kick-out of B atoms that as interstitial species diffuse very fast. In the samples implanted at high B dose the evolution of secondary defects at prolonged annealing time result in a subsequent release of interstitials responsible of the B prolonged diffusion. Transient B tail diffusion can be significally reduced if released interstitials are driven toward other damaged regions as those produced by high energy Si irradiations, to form extended defects at larger depths [7.23].

If the samples are annealed for longer times or at higher temperatures, however, extended defects evaporate and further anomalous B diffusion is observed. Extended defect formation at the peak of the B damage distribution does not affect the transient diffusion. Si interstitials from the tail region have to cross the B tail distribution before being trapped by the extended defects at the peak of the distribution. The anomalous B diffusion stops once the interstitials are trapped by the extended defects or annihilated [7.24].

Implants in a pre-amorphized surface layer eliminate channeling tails, but although reduced, anomalous diffusion is still present. Again Si interstitials at the end of range of the amorphizing species are responsible of the enhanced diffusion. The SIMS profiles reported in Fig. 7.13 refer to 10 keV B - $1 \times 10^{15}/cm^2$ implants in layers preamorphized by $10^{15}/cm^2 Ge$ ions at different energies $(A - 125keV, B - 80keV, C - 50keV, X$-no implant$)$. The profiles were measured in the as-implanted samples (Fig. 7.13a) and after an anneal at 550^0C-30min followed by 1050^0C-10sec

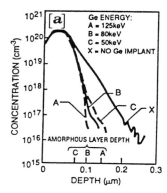 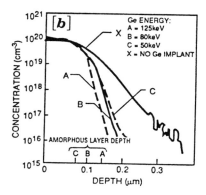

Figure 7.13 *(a) SIMS profiles of 10 keV·B^+ · 1 × $10^{15}/cm^2$ implants in Si preamorphized with 1 × $10^{15}/cm^2$ − Ge ions at energy of 125 keV (A), 80 keV (B) and 50 keV (C), the X profile refers to a single crystal. (b) same as in (a) but after $550^0 C$·30min and $1050^0 C$ - 10 sec. anneal (from ref. 7.25 -* [(c)] *1988 IEEE).*

RTA (Fig. 7.13b). Clearly junction depths below $0.2\mu m$ are obtained [7.25].

The diffusion depends on the location of the B profile with respect the amorphous layer depth. If the B profile cross the amorphous layer (as in B and C) during annealing a considerable diffusion takes place. The amorphous to single crystal transition region is reach of displaced knock-on Si atoms. These Si interstitials agglomerate during annealing in the end-of-range dislocation loops. If these Si interstitials during their life interact with the B atoms cause an enhanced transport as before considered for the implants in crystalline silicon.

It is then necessary in the Ge preamorphized B-implanted junction that the amorphous layer extends deeper than the profile. The near-surface concentration of interstitials must be reduced by lowering if possible the implantation energy of the amorphizing ion and the implant dose. The anneal should be performed in a non-oxidizing or slightly reducing surface reaction to avoid the injection of silicon-self interstitials. Not only the final junction depth but also the leakage current is very sensitive to the amount of post-anneal residual damage and to its location relative to the junction. As a secondary effect the Ge atoms present in the implanted layer interact with B and retards its diffusion during high temperature annealing [7.26].

Shallow junctions are also formed in combination with silicidation of source and drain to form contacts with a reduced para-

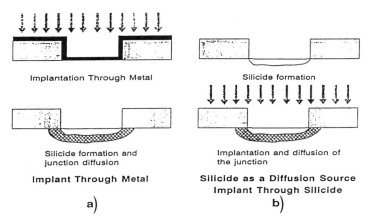

Figure 7.14 *Schematic of shallow junction formation by dopant implants in a metal (a) or in a silicide (b) thin layer followed by thermal diffusion-in the silicon substrate. In both cases the silicides are self-aligned.*

sitic resistance. The series resistance of junctions increases as the device scaling factor, while the contact resistance increases as the square of the scaling factor, being proportional to the area [7.27 - 7.28].

The implants are performed through a deposited metal layer (Co, or Ti mainly) before silicide formation or in the already formed silicide that acts then as a diffusion source. The two cases are illustrated in Fig. 7.14. In the first case a silicide layer is formed after annealing with a steep implant profiles and a shallow junction. $TiSi_2$ and $CoSi_2$ are used as metal contacts for their lower resistivities, $\sim 15\mu\Omega \cdot cm$, and superior thermal stabilities. The process is not of easy control. Silicon is consumed during the silicidation reaction and for these two silicides as much silicon is consumed as the amount of formed silicide. In many cases the silicon consumption during silicidation is considerably greater than would be predicted on the base of density from the simple ratios of silicide thickness/silicon consumed.

In the case of $TiSi_2$ some silicon is lost in the $TiON$ layer formed during annealing in nitrogen ambient [7.29]. If Argon is used in the annealing atmosphere, considerable silicon is lost laterally into titanium. Dopant depletion by the silicide can reduce the dopant concentration at the silicide silicon interface by a factor of ten and must be carefully controlled by proper annealing to avoid unacceptably large increase in contact resistance. Furthermore the presence of a polycristalline silicide over heavily doped junctions provides a rapid pathway for the evaporation of excessive quantities (over 50 %) of dopant. It is then necessary that the dopant distribution, silicon consumption and silicide metallurgical phase be precisely known and reproducible.

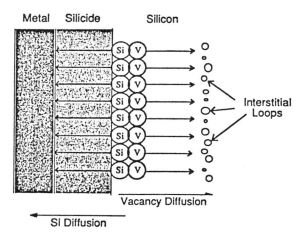

Figure 7.15 *Formation of Si vacant sites during silicidation with Ti (from ref. 7.29).*

Moreover silicidation reaction perturbs the point defect concentration (both interstitial and vacancy) in the substrate. The interfacial point defect generation depends on the details of the interdiffusion mechanism between Si and the metal films. For instance the $TiSi_2$ formation occurs by the predominant diffusion of Si atoms through the growing silicide. As silicon break away from the substrate, a vacant lattice site is left behind as shown schematically in Fig. 7.15 [7.29].

A beneficial effect can be the shrinkage of dislocation loops by recombination of vacancy with self-interstitial present in the loops. The silicon/silicide interface must be free of contaminants otherwise it may act as a sink of the generated point defects.

With the diffusion of dopants from a silicide source the dopant ions are introduced directly into the silicide thus, the implantation damage is confined to the silicide and away from the silicon. Since diffusion within the polycristalline silicide is very fast, the junction depth in the silicon is controlled as in the classical constant-source diffusion process. Furthermore the use of silicides as a diffusion source has been shown [7.30] to be a clever way to eliminate the edge leakage.

The use of silicides as diffusion source (SADS) requires low energy ion implantation and it is very vulnerable to excessive dopant evaporation from the silicide surface during diffusion because of the shallow starting profile. The dopant diffusion through the silicide must be rapid and the silicide must be stable against the thermal cycles required to drive the dopant into silicon.

The two main advantages of a polycristalline source are, as already said, the confinement of the ion implantation damage in the poly layer rather than in the silicon near the p-n junction and the achievment of a uniformly, heavily doped source from which a classical predeposition diffusion can be obtained. The heavy doping provides a high interfacial doping concentration and thus a low contact resistance. Actually the silicide-silicon interface is not at all smooth on the scale of 10 - 100 nm. Since the junction is seen to follow the roughness of the silicide [7.31], the junction is therefore rough also. Even if the junction roughness follows that of the silicide the impact of such rough junctions on junction leakage or other device parameters has not been clearly evaluated.

Furthermore in many systems, as As in $CoSi_2$, the bulk diffusivity is very small compared to the grain boundary diffusion. Thus the dopant is not distribuited uniformely within the silicide. If the diffusivity at the silicide-silicon interface is lower than along the silicide grain boundary, at the intersection of one silicide grain boundary with the silicide/silicon interface a higher concentration of dopant is present with respect to the rest of the interface. Then close to this high concentration sources the junction depth is deeper than between two adjacent grain boundary.

This effect has been observed by XTEM on samples where a chemical etch has been performed to delineate the junction, as described in sect. 6.6. Fig. 7.16(a) shows the morphology of 40 keV *As* ions implanted into 170 nm thick $CoSi_2$ layer to a dose of $2.5 \times 10^{16}/cm^3$ and annealed at 900^0C for 5 min. The Si substrate is etched only in well defined sites where a grain boundary of the silicide film is in contact with the $CoSi_2 - Si$ interface. Instead when the diffusion coefficient in the bulk silicide is comparable with that along the grain boundary, as in the case of the B diffusion in $CoSi_2$, the junction is less rough with respect the previous case as shown by the XTEM reported in Fig. 7.16(b). In this last case 30 keV B ions were implanted into 170 nm thick $CoSi_2$ layer and annealed at 900^0C for 5 min. The etched region in the Si substrate extends $\sim$ 50nm below the $CoSi_2 - Si$ interface. The lateral shape of the diffusion profile is quite uniform and follows the degree of roughness of the interface. The formation of very shallow junction is self adjusting to the silicide-silicon interface only in the case of boron. The effects of high grain boundary diffusion are further complicated when additional grain growth accompanies the diffusion process. This grain growth changes the dopant distribution between the grains and the grain boundaries [7.32].

Figure 7.16 *Cross section transmission electron micrographs of $CoSi_2$ layers implanted with As (a) and B (b) ions. The samples were annealed at 900^0C for 5 minutes and chemical etched before TEM analysis (from ref. 7.32).*

In addition to the roughness of the interface and to the different diffusivity of dopant in bulk and in the grain boundaries other problems have been found in the formation of shallow junctions using a silicide layer as a diffusion source: the formation of metal-dopant compounds and the stress induced precipitation of dopants. The metal-dopant compound formation has been observed for B and As implanted into $TiSi_2$ and $TaSi_2$ [7.33], but not after the diffusion from $CoSi_2$ and WSi_2 [7.34]. Instead the stress induced precipitation of dopants has been observed for the diffusion of B, As and P from both $TiSi_2$ and $CoSi_2$.

The TEM cross-section of an arsenic implanted $CoSi_2$ layer diffused at 900 ^oC for 1 hour is shown in Fig. 7.17a [7.34]. Three arsenic precipitates are visible in the silicon substrate at the interface with the silicide layer. They lye in the (111) planes and extend down to a depth of 30-60 nm. The contrast behaviour is typical of precipitates coherent with the silicon matrix. The same kind of particles have also been observed at the titanium silicide/silicon interface (see Fig. 7.17b) after the same diffusion process. Precipitates of different shape have been observed at the $TiSi_2$/silicon interface after phosphorus diffusion at 900 ^oC for 1 hour (Fig. 7.17c) and close to the $CoSi_2$/Si interface after boron diffusion (Fig. 7.17d) [7.34].

The precipitation is probably due to the change of the dopant

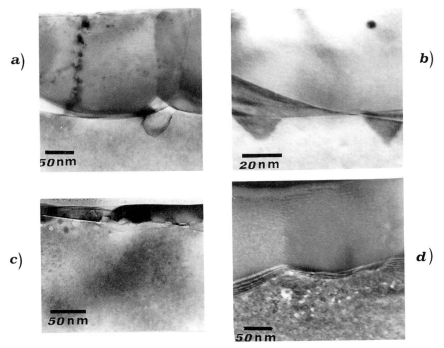

Figure 7.17 *(a) Bright-field micrograph of a 170 nm thick $CoSi_2$ implanted with 40 keV As − $10^{15}/cm^2$ and diffused at 900^0C - 1 hr; (b) same process as in (a) but for a 66 nm thick $TiSi_2$; (c) bright-field micrograph of a 48 nm thick $TiSi_2$ implanted with 20 keV P − $10^{15}/cm^2$ and diffused at 900^0C - 1 hr; (d) dark-field micrograph of a 170 nm thick $CoSi_2$ implanted with 30 keV B · $10^{15}/cm^2$ and diffused at 950^0C · 30 min. All the micrographs refer to cross - sections (from ref. 7.34).*

solid solubility introduced by the different thermal expansion co-efficient of the silicide layer and of the silicon substrate.

7.5 Complementary MOS Devices and Technology

A particularly useful device in digital circuits is a combination of n-channel and p-channel MOSFETs on adjacent regions of the chip. This structure is named complementary MOS transistor circuit or simply CMOS circuits, and it has a low power dissipation in the steady state operation. A diagram of the circuit and a schematic cross section of the device are shown in Fig. 7.18a and 7.18c respectively [7.35]. It serves as an inverter, i.e. a logic circuit whose binary output is the inverse of the input (Fig. 7.18d). A suitable arrangement of inverters allows the fabrication of arbitratly complex logic circuits. In the inverter the devices of the two transistors are connected together and form the output, while the input terminal is the common connection to the transistor gates.

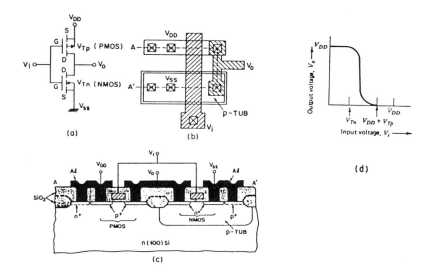

Figure 7.18 *CMOS inverter: (a) circuit diagram, (b) circuit layout (c) cross-section along dotted A-A' line (b), (d) voltage-transfer characteristics (from ref. 7.35).*

Both devices are enhancement mode MOSFETs with threshold voltage V_{T_p} less than zero for the PMOSFET and V_{T_n} greater than zero for the NMOSFET (typically V_{T_p} ranges between -1.0 and -0.5 V, and V_{T_n} between 0.5 and 1.0 V). If the input voltage is lower than the threshold voltage for the n-channel but higher than that for the p-channel MOSFET, e.g. $V_i = 0$ (or ground), the PMOS is in the 'on' state and it provides a conductive path to the power supply V_D. The output voltage, V_D at the drain of the PMOSFET is in the "high" state V_{DD} (logic 1). Alternatively a positive value of V_i turns the n-channel "on" and the p-channel "off". The output voltage reaches now the "low" state (logic 0) at ground potential. Thus the circuit operates as an inverter with a binary "1" at the input the output is in the "0" state and viceversa "0" at the input produces "1" at the output. The voltage-transfer characteristics are reported in Fig. 7.18c, the transition is well defined and quite abrupt between the two logic states. In each logic state one or the other MOSFET is always "off" and there is no direct current path to carry current from the power supply, except for the small leakage current or during switching from one state to the other. The device has also a good immunity to noises for the low impedance between power supply and ground voltage. Thus the average power dissipation is in the order of nanowatts. The low power consumption is one of the most attractive feature of the CMOS circuit.

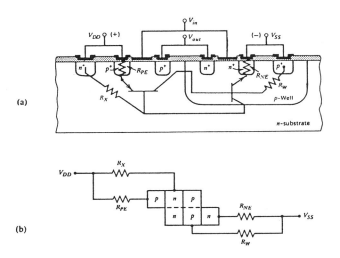

Figure 7.19 *(a) Parasitic p-n-p and n-p-n bipolar transistors responsible of latch-up and associated resistors in a CMOS structure. (b) Circuit and schematic representation of the cross-coupled parasitic npn and pnp transistors (from ref. 7.12).*

The main drawback of the device is the presence of an effect called "latchup", a condition in which regenerative parasitic bipolar-transistor action induces a low resistivity path between the power supply and the ground [7.36]. Parasitic n-p-n and p-n-p bipolar transistors are present in a CMOS structure as indicated in Fig. 7.19 [7.12]. In the first one the n^+ source or drain acts as emitter, the p-tube as base and the adjacent n-tube as collector. Similarly the second transistor p-n-p is formed by the p^+ source or drain as emitter, the n-tube base and the adjcent p-tube as collector. The base-collector of the p-n-p transistor is in common to the collector-base of the n-p-n transistor. Under active bias conditions the collector of the p-n-p transistor delivers current to the base of the n-p-n transistor and viceversa the n-p-n collector delivers current to the p-n-p base. Even for small current gains, β , this interconnection can lead both devices to saturation. The power supply is then connected to ground via low resistance path and two small voltages drop across the saturated base-collector and base-emitter junctions respectively.

The behaviour is similar to that encountered in silicon controlled rectifier (SCR) devices that switch large amount of power. Normally the emitter-base junctions of both parasitic transistors are reverse-biased so that latch-up does not occur. Minority carriers can be however generated by several processes: forward biased junctions during transient phenomena, impact ionization of hot carriers, charging and discharging of capacitors. Any generation of carriers can be amplified by the run-way process of the structure. Several possibilities have been adopted to avoid or to reduce the latch-up. The current gains of the bipolar transistors must be reduced and the well resistors should be as small as possible. The resistivity can be lowered by a high doping of selective regions.

Another and quite effective method adopt a trench isolation between the two wells. The trench is formed by anisotropic reactive-sputter etching. An oxide layer is grown in the bottom and in the walls of the trench, which is then refilled by deposited polysilicon or silicon dioxide. The two regions are typically isolated by the refilled trenchs and the latchup is avoided. This last technological process is quite complex and alternative procedures are preferred [7.37].

Ion implantation offers unique solutions and advantages to the CMOS technology [7.38]. Conventional CMOS was realized by the fabrication of the n-channel transistor in a p-well formed by diffusing boron atoms into an n-type substrate. The p-channel device was built in the n-type substrate. The p-type well was realized by boron implants followed by high- temperature diffusion. Subsequently a similar structure with the n-well replacing the p-well was also fabricated. The n-channel devices formed by this CMOS process are equivalent to the FETs produced by NMOS technology. The n-well is formed by implanting phosphorus ions instead of boron in a lightly doped p-substrate. P-well versus n-well has some disadvantages and advantages. At design rules of $\sim 1\mu m$ n-channel devices provide about two times more drain current because electrons move faster than holes, but a substrate current about few orders of magnitude higher due to the higher impact ionization coefficient of electrons. Another limit is that the doping concentration in the well must be higher than in the starting substrate, thus resulting in higher junction capacitance.

A better approach expecially at $0.5\mu m$ design rule or lower, is the twin- tube i.e. the formation of two separate wells for n and p channel transistors in a lightly doped substrate or in an epitaxial layer.

A quite complete device structure is shown in Fig. 7.20 [7.39], all the implants are indicated. The p-well is realized by boron implants at energies in the 100-200 keV range and in the $2 \times 10^{12}/cm^2 - 8 \times 10^{12}/cm^2$ dose range. The n-well is realized by phosphorus implants with the same energy range and dose. The implants are followed by diffusion at high temperature, typically several hours ($\sim 10 - 20$ hours) at $1100^0 C$, to get a deep junction and to avoid the punchthrough from the transistors at the well surface to the substrate under the well. The twintube CMOS process allows independent adjustment and optimization of both the n-channel and p-channel transistors over a wide range of parameters. The substrate can be either n or p-type. The doping concentration near the surface of the well, the gate material and the thickness of the gate oxide layer set the voltage threshold at which the device channel will conduct. The near-surface concentration of the well also affects the performance of the device, as the carrier mobility, the source-drain capacitance and the inter-devices leakage currents. As in the MOS drain and channel optimization, several implants are adopted to control the threshold voltage, the hot carrier formation, the punchthrough and the field isolation.

The thick oxide layer between the two regions is formed by the local oxidation of silicon (LOCOS process) and the well doping below the oxide is increased by a second implant to prevent surface inversion of the well surface by stray voltages from active interconnect lines running over the field oxide. The threshold voltage of such parasitic transistors must be kept higher than any possible operating voltage so that spurious channels will not be inadvertently formed between devices. The V_T of the field region needs to be 3-4 volt above the supply voltage. The increase of V_T can be obtained by increasing the field-oxide thickness and raising the doping beneath the field oxide (see eq. 7.1). Due to planarity and coverage problems reduced field-oxide thickness are preferred and to achieve a sufficiently large field threshold voltage the doping under the field-oxide must be increased. This is usually done by ion implant and the step is called a field isolation implant or "channel stop" or "guard ring". It is performed by B^+ or P^+ ions in the energy range 50-150 keV and in the dose range $5 \times 10^{12} - 6 \times 10^{13}/cm^2$.

During field-oxidation several phenomena influence the boron distribution. The oxide has a tendency to take up boron, the seg-

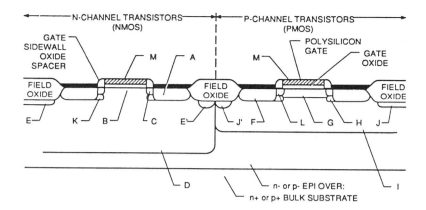

Figure 7.20 *CMOS structure with all the implants (from ref. 7.39).*

Implanted Regions and their Functions:

A = NMOS source/drain; basic transistor structure.

B = NMOS channel threshold voltage adjust; sets n-channel V_t (or V_p).

C = NMOS LDD; hot carrier suppression.

D = p-well ("tub") structure; contains NMOS transistors.

E = p-type "channel stop" for p-well; intra-well (E) and inter-well (E') field isolation.

F = PMOS source/drain; basic transistor structure.

G = PMOS buried-channel threshold voltage adjust; sets p-channel V_t (or V_p).

H = PMOS "punchthrough" suppression.

I = n-well ("tub") structure; contains PMOS transistors.

J = n-type "channel stop" for n-well; intra-well (J) and inter-well (J') field isolation.

K = NMOS "punchthrough" suppression.

L = PMOS LDD; hot carrier suppression.

M = polysilicon gate doping (typically n+); improves conductance.

regation coefficient, i.e. the ratio between the equilibrium concentration of B in Si and of B in SiO_2, is lower than 1. Impurity may have a tendency to escape through the oxide layer. If the diffusivity of the impurity in the oxide is small this factor will be of no relevance, otherwise the impurity distribution is substantially affected. The extent of the redistribution depends on the relative rate of oxide growth as compared to the diffusion rate. This rate is also enhanced by the Si interstitials released during oxidation. Thus relatively high boron doses are needed ($\sim 10^{13}/cm^2$) in order to achieve acceptable field threshold voltages. This also implies that the peak concentration of the boron implant must be deep enough that it is not included by the growing field-oxide interface. If the channel-stop doping is too heavy, it will cause high source/drain - to - substrate capacitances and will reduce the source/ drain to substrate p-n junction breakdown voltages.

A layer of patterned nitride masks the phosphorus implant for n-well. Subsequent to the growth of a masking oxide, the nitride layer is removed and the oxide is used as a mask during boron implant for the p-well. The two wells are separated by oxide thick layers grown by LOCOS process. This technology requires four micrometers or more p^+ to n^+ lateral isolation separation. A thin gate oxide ($\sim$ 10nm) is grown on the active areas as the gate oxide. A layer of polycrystalline silicon is deposited and n-type doped by As or P implantation. The poly-layer is patterned and reactive-ion-etched to form MOS gates for both n and p-channel FETs. Source and drain regions are formed by boron implantation into the p-channel active areas. The energy of the boron ions is in the 5-40 keV and the dose in the $2 \times 10^{15}/cm^2 - 6 \times 10^{15}/cm^2$ range. In some cases BF_2^+ is used as species with the same dose and energy. A similar process is applied to form n^+ source and drain for n-channel FETs. As is implanted at an energy in the 40-80 keV range and $2 \times 10^{15}/cm^2$ dose. The process requires several other implants to improve the electrical characteristics of the devices. New possibilities have been offered by increasing the energy of the ions and some applications of high energy implants will be described in sect. 7.7.

7.6 Lifetime Engineering in Power Devices

In the last ten years the field of power electronics has experienced a considerable evolution for the advances in fabrication technology and a significant expansion in the market due to the increasing requests of power components for variable speed motor drives, serve drives and power supplied in general [7.40]. Ion implantation became a key step process also in the fabrication of power devices being able to tailor both dynamics and static electrical characteristics. In the following, as an application, the use of ion implantation as control of minority carrier lifetime will be described.

The requirement of minority carrier lifetime reduction or control in a semiconductor device goes under the name of lifetime engineering. The motivations for a fine control of lifetime are strongly dependent on the specific power device structure and its circuital applications. For example a precise control of lifetime is required in a power bipolar transistor for the fine tuning of the storage time, while a reduction of lifetime is necessary in order to decrease the turn-off time of a power insulated gate bipolar transistor (IGBT). Very often the achievement of this result is obtained at the expense of a degradation of other electrical properties. In power devices the introduction of deep levels in the silicon band gap, which is necessary for lifetime reduction, causes electrical compensation in the material and hence an increase in the on-state resistance and in the power losses. Therefore a compromise between on-state losses and faster turn-off is the main target of the different lifetime control techniques, which embody metal doping and irradiation with high energy particles such as electrons, neutrons, protons or alpha particles.

The main argument against Au or Pt doping by a film deposition as lifetime killers has been the fact that, due to the diffusion mechanism of these elements [7.41], the resulting concentration profiles are U-shaped with a maximum at the front and at the back side of the wafer and greatly sensitive to the details of the thermal process. Reliability is therefore a serious issue when this method is adopted. However, it has been demonstrated [7.42] that, using low fluence ion implantation as prediffusion source this problem is solved and a uniform concentration all over the wafer thickness is obtain. Solid solubility of Au in $c - Si$ is $\sim 5 \times 10^{15} at/cm^3$ at 970^0C. Therefore only $2.5 \times 10^{14} at/cm^2$ are necessary to saturate a $500 \mu m$ thick wafer at the solid solubility value. If higher fluences are implanted the thermal diffusion results again in U-shaped profiles, while at lower fluences and longer diffusion times the Au atoms redistribute until a uniform profile is obtained. Since this redistribution occurs at concentration lower than solid solubility, the process is faster as the effective diffusion coefficient is larger at lower concentrations [7.43]. In fact, as shown in Fig. 7.21, only 5 hrs at 970^0C are necessary to get a uniform profile.

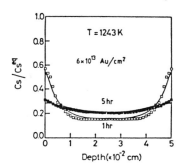

Figure 7.21 *Gold concentration profiles after diffusion at 1243 K for 1 and 5 hr. Implant fluence 6 × $10^{13}/cm^2$ (from ref. 7.42).*

The final concentration is given by $C = N/d$ where N is the dose and d the wafer thickness. Therefore this value is not determined by the thermal process. Both the reduction in annealing time and the achievment of a uniform profile guarantee a higher reliability and feasibility of the process.

The possibility to achieve uniform concentration profiles allows also an improvement of the tradeoff curves between static and dynamic characteristics as will be shown for power metal-oxide-semiconductor field-effect transistor.

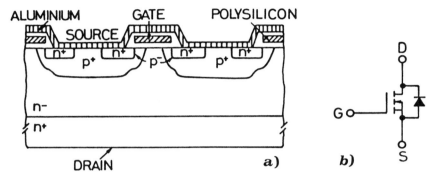

Figure 7.22 *(a) Power MOS structure, (b) equivalent circuit (from ref. 7.44 - [c]1992 IEEE).*

Power semiconductor devices are more complicated in structure and operational characteristics than their low-power counterpart. The power MOS device consists of multi-MOS basic cells interconnected in parallel on a single die; each one of these cells consists of a vertical diffused MOS (VD-MOS) silicon gate structure where the current flows vertically from the drain to the source. The device consists of a two-level structure (Fig. 7.22a) where the lower level is the polycristalline silicon gate and the upper is the source metallization. The fabrication sequence involves several processing steps and it will be briefly detailed in the following. As as example the skeme of a VD-MOS with a breakdown voltage $\geq 500V$ is considered [7.44]. The devices, are fabricated on a lightly n-doped (n^-) epitaxial silicon layer with a resistivity of $20\Omega \cdot cm$ and a thickness of $50\mu m$. This epitaxial layer is grown on a heavily n-type doped (n^+)CZ substrate, $< 100 >$ oriented, with a resistivity of $10^{-2}\Omega \cdot cm$ and a thickness of $600\mu m$; the device area is $15.7mm^2$. Inside each single cell a heavily p-type doped (p^+) region is formed in order to limit the gain of the parasitic bipolar structure $n^+/p^+/n^-$. The polysilicon gate is then deposited onto a thin oxide layer (85 nm) and is insulated from the source metallization by an intermediate layer of phosphorus-doped vapor-deposited oxide (P-vapox). The polysilicon gate is used as a mask against the B implantation to realize the lightly p-doped p^- body and the As implantation for the n^+ source contact. The MOS channel regions are obtained by difference in lateral diffusion of the two impurities distributions (p^-) body and n^+ source. After the formation of the channel Au or Pt is implanted or deposited. Subsequent thermal processes to diffuse these impurities are performed before the metallization step. Surface metallization consists of a 3-μm-thick Al layer passivated with P-vapox. Then the fabrication of the device is completed by reduction to 300 μm of the wafer thickness and subsequent full backside metallization.

Due to the particular structure of the device, a parasitic diode is present between the body (p^+ region) and the drain. In some circuital applications of the MOSFET, such as half or full bridge converters for motor speed control, reverse current flow through the device is required. The possibility of utilizing the parasitic diode inherent to the structure is attractive because it eliminates the complexity and the cost of an external diode. However, in order to keep power dissipation low, this diode must exhibit good reverse recovery characteristics. In particular, it is necessary to perform metal doping of the diode in order to reduce the value of the reverse recovery time t_{rr} (turn-off time during which minority carriers escape from the n^- region) and of reverse recovery charges Q_{rr} (integral of reverse current I_R which flows inside the diode during t_{rr}). However, the deep levels introduced by Au or Pt in the gap will partially compensate the material causing an increase in resistivity particularly in the lightly doped n^- region. This compensation produces an increase in the on-resistance R_{on} of the device, which is the total resistance between source and drain in the *on*-state. As the power dissipation in the *on*-state is given by $P = I_D^2 R_{on}$, the increase in R_{on} causes a reduction in the current carrying capability of the device.

Trade-off curves between Q_{rr} and R_{on} for the different prediffusion sources are shown in Fig. 7.23 [7.44], in order to determine the conditions which produce the lowest increase in R_{on} for a given value of Q_{rr}. Both R_{on} and Q_{rr} values are normalized to those measured in undoped devices. The three curves refer to wafers with a prediffusion source realized by gold deposition, gold ion-implantation or platinum ion-implantation, respectively. In gold plated devices the different values of Q_{rr} and R_{on} are obtained by changing the diffusion temperature (from 850 to $880^0 C$) of the isochronous annealing (1 h). In ion implanted devices Au

or Pt are introduced at 40 keV with fluences between 10^{12} and $6 \times 10^{13} ions/cm^2$ and annealed at 970^0C for 5 h. As shown in Fig. 7.23 a reduction of Q_{rr} always corresponds to an increase of R_{on}. The best compromise is obtained for platinum implantation. In platinum implanted samples, in fact, it is possible to reduce Q_{rr} by a factor of 10 with a limited increase ($\sim 10\%$) in R_{on}.

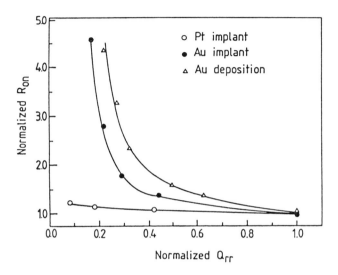

Figure 7.23 *The resistance in the on-state, R_{on}, versus the reverse recovery charge, Q_{rr} values for gold deposition (Δ), platinum (o) implant and gold ($\bullet$) implant in P.MOS devices (from ref. 7.44 - [c] 1992 IEEE).*

As Q_{rr} mainly depends on the concentration of recombination centers in the n^- epitaxial and R_{on} on the resulting resistivity profile in this layer, the best compromise between Q_{rr} and R_{on} is achieved by the method which produces the lowest increase in the resistivity for a given density of recombination centers.

The impurity distributions in the epitaxial layer of these devices are reported in Fig. 7.24a. For Au plated sample the concentration profile is strongly depth dependent and high Au concentration values are measured close to the wafer surface. This is due to the peculiar mechanism of Au diffusion and to the fact that, for an infinite source, the surface concentration is fixed by the Au solid solubility C_s^{eq} at the diffusion temperature. Due to this boundary condition the diffusion temperature must be kept below 900^0C to avoid an excessive compensation of the n^- region.

At these temperatures any attempt to obtain a complete redistribution of the impurity is hampered by the low value of diffusion
coefficient.

Figure 7.24 *Impurity (a) and resistivity profiles (b) in the epitaxial layer of the P.MOS device after the drive-in diffusion: gold deposition (Δ) followed by a thermal process at 880^0C for 1 hr; ion implantation of Au (●) or Pt (o) with a fluence of 7.5 × $10^{12}ions/cm^2$ followed by a thermal process at 970^0C for 5 h. The dashed line in (b) represents the resistivity of the epitaxial layer of undoped wafer (from ref. 7.44 - (c) 1992 IEEE).*

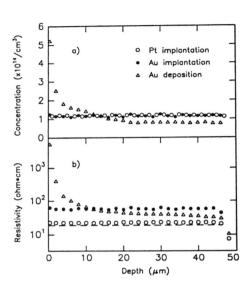

For ion-implanted samples, however, the boundary conditions
are changed and a thermal process at 970^0C for 5 hrs produces a
uniform concentration profiles at a value of 1.2×10^{14} atoms/cm^3.
Despite the completely different depth profiles, the total amount
of Au in the epitaxial layer is roughly the same for both Au plated
and Au implanted samples. This is consistent with the observation that the devices are characterized by the same value of
Q_{rr}. However, the advantages of a uniform distribution are evident when the resulting resistivity profiles are considered (Fig.
7.24b). For Au plated samples, the high concentration close to
the wafer surface causes a very large increase in resistivity. In
fact, as the gold acceptor level is very close to midgap the compensating effect is quite strong. In particular the silicon resistivity,
which remains almost constant for a gold concentration lower than
the shallow dopant concentration (2×10^{14}atoms/cm^3), increases
quite abruptly at higher concentration: it increases by about two
orders of magnitude when the gold concentration increases from
1.5×10^{14} to 5×10^{14} atoms/cm^3 [7.45]. Thus the high concentration values near the surface contributes noticeably to the overall
increase in R_{on} and determines the difference between Au plated
and Au implanted samples. As shown in Fig. 7.23 the best trade-

off between Q_{rr} and R_{on} is obtained for platinum. This can be understood by observing (figs. 7.24a and 7.24b) that it is possible to introduce in the material the same concentration of Pt with an increase in resistivity which is a factor of 3 lower than for Au. In fact, the compensation of the n^- layer is less effective for Pt than for Au due to the large difference in the acceptor level energies, $E_c - 0.21eV$ and $E_c - 0.54eV$ for platinum and gold respectively [7.46].

7.7 High Energy Implant Applications

In several sections of this and of previous chapters the use of high energy implants for device fabrication has been briefly considered. The range of high energy is quite broad and starts already at about 200 keV. The next generation of equipments for microelectronics includes high energy implanters being forecasted an increase of applications for high energy ion implants [7.47]. For this reason the present section will treat in a more details the subject. The main capability of the high energy implant is the location of dopant atoms at a certain depth from the wafer surface. It is possible to form a buried doped layer with peak concentrations as high as $10^{18}/cm^3$ but with the surface at a concentration in the $10^{15} \cdot 10^{16}/cm^3$ range. Several profiles are shown in Fig. 7.25 for boron [7.44], and phosphorus ions [7.49].

With increasing the energy a thicker surface layer remains undoped and in this region devices can be built. The high energy profiles are quite asymmetrical and are simulated by TRIM or MARLOWE code. In many cases a simple evaluation of the profile is necessary and the following polynomial expression provides the first four moments of the distribution, i.e. the projected range R_p, the straggle ΔR_p, the skewness γ and the kurtosis β in terms of beam energy [7.50]:

$$(R_p, \Delta R_p, \gamma, \beta) = a_0 + a_1 E + a_2 e^2 + a_3 E^3 (E \ in \ MeV) \qquad 7.6$$

The values of a_0, a_1, a_2 and a_3 are reported in table 7.1 for B, P and As

Among the several applications [7.49] let us mention the change of the threshold voltage of some MOSFET in programmable SRAM at the end of the process, the fabrication of retrograde wells for CMOS, the formation of buried grids for soft error prevention, the elimination of the epitaxial layer and the creation of heavily doped buried layer for gettering. The first large

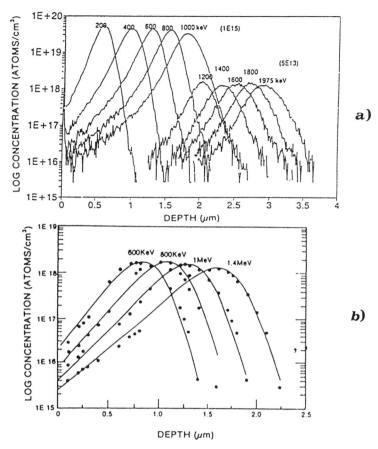

Figure 7.25 *SIMS profiles of high energy B(a) (from ref. 7.48) and P(b) ions (from ref. 7.49).*

scale application of high energy implant was instead the doping of charge coupled devices (CCD) [7.51].

In conventional technology the p and n-well of CMOS are fabricated by low-energy low-dose implant of the chosen dopant followed by a drive-in process at elevated temperature ($> 1100^0 C$) for several hours. The diffused region extends typically to several microns. The thermal process is costly, time consuming and may induce slip dislocations or wafer warpage expecially on large wafer sizes, metal contamination, deformation of the tubes and boats. Diffusion non uniformity is also associated with differential temperature gradient when 150 or more large size wafers are placed in side a diffusion furnace. With boron or phosphorus a temperature

Table **7.1** *Values of coefficients* a_0, a_1, a_2 *and* a_3.

	a_0	a_1	a_2	a_3
		B		
$R_p(\mu m)$		2.19	-0.49	0.069
$\Delta R_p(\mu m)$	0.16	0.023	——	——
γ	-0.9	-0.52	——	——
β	3.9	7.5	-0.78	——
		P		
$R_p(\mu m)$	——	1.2	-0.16	0.011
$\Delta R_p(\mu m)$	0.26	0.012	——	——
γ	0.88	-0.58	0.03	——
β	5.3	0.16	0.16	——
		As		
$R_p(\mu m)$	——	0.78	-0.056	0.002
$\Delta R_p(\mu m)$	0.34	0.014	——	——
γ	-0.007	0.11	——	——
β	1.67	0.56	——	——

variation of $15^0 C$ at $1100^0 C$ leads to a 113% variation of dopant diffusivity. During the process impurity atoms diffuse laterally as well as vertically. The lateral diffusion and the width of the depleted region at the p-n junction limits the packaging density, i.e. the number of devices allocated per unit area. The structure is illustrated schematically in Fig. 7.26a.

The profile refers to $3 \times 10^{12}/cm^2 - 40keV\, B^+$ ions implanted through a mask made by a SiO_2, LOCOS processed, and diffused at $1150^0 C$ for 6 hours and 30 min. The surface concentration is $2 \times 10^{16}/cm^3$ and the junction with the $20\Omega \cdot cm$ n-type substrate is located at 4.0 μm. The width of the depletion region (see Eq. 1.2) extends for $0.5\mu m$ in the p-well and for 0.7 μm in the n-substrate. The contiguous n-well for the complementary MOS should be at such a distance to prevent punch-through between the two depletion layers. The width of the field oxide should be μm at least. For the same reason the well should extend in depth up to several micrometers to prevent vertical punch-through. The well

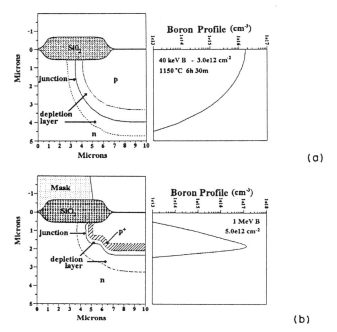

(a)

(b)

Figure 7.26 *(a) diffused well structure for* $2 \times 10^{13}/cm^2$ — $40keVB^+$ *implant after thermal annealing at* 1150^0C *for 6 hours and 30 m. The dashed area represents the depletion width of the p-n junction. Substrate concentration* $5 \times 10^{15}/cm^3$. *(b) retrograde well formation by high-energy boron implant,* $1.0MeVB \cdot 5 \times 10^{12}/cm^2$. *The dashed area represents the depletion width. On the left part the dopant profile is shown. (Courtesy of M. Saggio)*

concentration cannot be increased over $10^{16}/cm^3$ otherwise a high threshold voltage is required to invert the channel conductance in the n-MOS.

The formation of a retrograde well by high energy implant overcomes some of the previous limitations as shown in Fig. 7.26(b). Unlike a diffused profile in which the peak concentration is always at the Si wafer surface, the peak of the implanted profile is buried at a certain depth and the impurity concentration decreases as it approaches the surface. This type of profile is named retrograde profile and the implanted well is often referred as retrograde well. The data in Fig. 7.26(b) simulate the profiles of $5 \times 10^{12}/cm^2$ — $1.0MeVB$ with a peak concentration of $10^{17}/cm^3$ at a depth of $2\mu m$. The width of the depletion layer in the p-well is now of $0.2\mu m$ and shallower depths can also be used with this profile. As a result smaller field oxide spacing are needed and the packaging density of devices increases. The thermal budget is considerably reduced, it is only necessary the activation of the and the annealing of the damage, no impurity transport is required.

The reduction in size of the well implies also a thinner epitaxial layer and in some cases it can be replaced by a buried doped layer. To prevent latch-up doped p^- layers are grown epitaxially on p^+ substrates. The CMOS device is built in the epilayer and the heavily doped substrate improves latch-up immunity. The high conductivity substrate can be replaced by a blanket boron implant in a ligtly doped p-type substrate to create a p^+ buried layer below a deep n-well. The epitaxial layer is not anymore needed. This represents one of the more interesting applications of high energy implant for manifacturing cost reduction.

The buried layer with a peaked dopant profile has a built-in electric field created by the diffusion of majority carriers. This electric field repels minority carriers and this feature can be useful to suppress substrate generated minority carriers due to thermal excitations, alpha-particle, cosmic rays, etc. Soft errors, i.e. the spontaneous memory bit state changes [7.52] by injected charges are then drastically reduced. Energetic incident particles such as cosmic rays or alpha particle from a package of the chip can strike a critical part of a device, create electron-hole pairs below active memory elements such as the well or the storage capacitors as shown in Fig. 7.27(a). Dependent on the device bias the well is inverted or depleted. The negative charges created by the energetic particle move toward the positive gate and the positive charges move away from the gate. If the well is already filled with electrons, it will remain filled and no change of state (i.e. bit flip) will occur. However, initially empty wells may now become partially filled with electrons and if enough electrons are collected the state changes and a bit flip occurs.

A critical device parameter is the critical charge or the number of electrons associated with the difference between the "0" and "1" state. Depending on the used technology and design rules, typical pockets are around $10^5 \div 10^7$ electrons. A 5 MeV α-particle generates about $1 : 5 \times 10^6$ electron- hole pairs. Device scaling strongly increases the soft error sensitivity. The use of high energy implant can alleviate soft error immunity, as said before, for the built-in electric field that repels minority carriers. In addition the use of a buried p^+ layer in the case shown in Fig. 7.27(b) reduces the extension of the depletion layer and then the

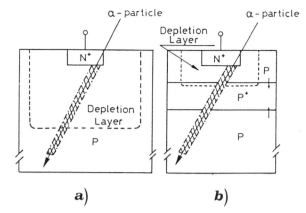

Figure 7.27 *Effect of a buried doped layer in reducing the depletion volume where minority carriers are generated by α· particles. (a) conventional diode, (b) buried p^+ layer below the n^+p junction.*

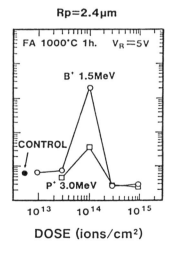

DOSE (ions/cm²)

Figure 7.28 *Leakage current density of a p^+n junction as a function of 1.5 MeV B or 3.0 MeV P implanted dose after 1000^0C · 1hr. The diode was reverse biased at 5V (from ref. 7.54).*

volume available for electron-hole generation.

The peculiar shape of the high energy implant has been applied to programme masked Read Only Memory (ROM) after the formation of interconnesions [7.53]. These memories are programmed according to the requirements of the customers and the use of high energy reduces the interval from receipt of order to

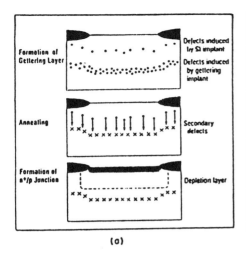
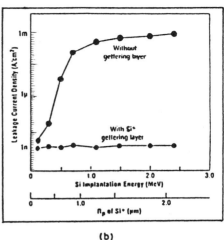

(a) (b)

Figure 7.29 *(a) Schematic of gettering of microdefects created by a low energy implant to secondary defects created by high energy - high dose ($> 3 \times 10^{14}/cm^2$) implants. (b) Leakage current density at 10V reverse bias of n^+p junction as a function of $1 \times 10^{14}/cm^2 Si$ implant at different energies without and with a gettering buried layer created by 2.8 MeV·$10^{15}/cm^2 Si$ ions implant (from ref. 7.56).*

shipment of devices. The programming is obtained by changing, raising or lowering, the MOS threshold voltage of a memory cells family by ion implantation. For istance B implant in a NMOS device increases V_T and if above the voltage power supply the memory is "off" programmed. P implant in NMOS induces instead a permanent short path. The use of high energy allows the implant and then the change to be made at the end of the fabrication process. The programming step can be performed either after gate fabrication and the post-implant annealing temperature can be higher than 550^0C or after metal-interconnections. If Al is adopted the post-implant annealing temperature is limited to 500^0C. Boron ions are normally used for the implant in view of their low mass and then high depth penetration.

A similar process is adopted for channel stop formation after field oxidation. The concentration below the oxide must be increased to avoid inversion and this can be realized by an implant of the appropriate dopant at a suitable high energy to locate the peak just below the oxide. This procedure simplifies the layout. The number of masks can be reduced with a gain in yield and a

Table 7.2 *Applications of high energy implants.*

Implant step	Tipical dose & energy (0.7 μm process)	Device performance benefits	Manufacturing benefits
Retrograde wells	$1 \times 10^{13}/cm^2$ @ 400 keV (B) $1 \times 10^{13}/cm^2$ @ 800 keV (P)	Reduced n^+/p^+ spacing Improved latch-up control Reduced sensitivity to soft failures Punch-through suppression	Reduced drive-in time Fewer mask levels Reduced thermal budget
Profiled wells	varies	Optimize many transistor characteristics independently	Fewer mask levels
Field isolation (channel stop)	$1 \times 10^{13}/cm^2$ @ 250 keV (B) $1 \times 10^{13}/cm^2$ @ 400 keV (P)	Improved isolation Reduced n^+/p^+ spacing Reduced dopant diffusion during oxidation	
Deep buried gettering layers	$1 \times 10^{15}/cm^2$ @ 1.5 MeV (B) $1 \times 10^{15}/cm^2$ @ 3.0 MeV (P)	Capture defects and contamination Junction leakage control in aggressively scaled CMOS devices	Eliminate need for epitaxial silicon
Buried collectors	$1 \times 10^{14}/cm^2$ @ 800 keV (B) $1 \times 10^{14}/cm^2$ @ 1.5 MeV (P)	Improved isolation	Reduced thermal budget
ROM programming	$1 \times 10^{13}/cm^2$ @ 400 keV (B) $1 \times 10^{13}/cm^2$ @ 800 keV (P)		Process flexibility Fast turnaround

Figure 7.30 *Several applications of high energy implants for different beam energies and doses.*

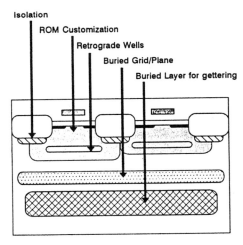

Isolation
ROM Customization
Retrograde Wells
Buried Grid/Plane
Buried Layer for gettering

reduction in cost and in contamination.

The previous applications require usually doses below $5 - 10^{13}/cm^2$ for the formation of the buried layers. With increasing dose the layer has a beneficial effect on the electrical characteristic of the built-in device because it getters not only metallic contamination but also point and micro defects. The gettering effect can be determined for istance by measuring the leakage current of a diode built near-by the buried layer. The leakage current density of a p^+n junction is shown in Fig. 7.28 as a function of $1 \cdot 5 MeV B^+$

or $3.0 MeV P^+$ implant dose [7.54].

Leakage current gradually increases with dose, up to 3 $10^{14}/cm^2$ where a drastic decrease occurs. The depletion layer at the adopted reverse bias voltage is substantially lower than the B^+ or P^+ projected range. A transmission electron microscopy analysis of the cross sections has indicated the absence of defects in the depletion region and the presence of extended defects at the projected range of the high energy ions. The effect is another example of defect engineering. The damage created by the high energy implant causes the precipitation of oxygen above the solid solubility as we have shown in sect. 6.6. These precipitates are known to getter metallic impurities such as gold, copper, nickel and iron. In addition secondary defects if above a certain density appear to getter the residual microdefects in the upper region.

These defects act as a recombination-generation centers and if located in the depletion region increase junction leakage current. This phenomenon is named "self-gettering" because the gettering sink consists of the secondary defects formed by the implant of the dopant itself. The relative position of the secondary defects with respect to the depletion region of the *pn* junction is very important in the reduction of leakage current, i.e. the secondary defects should be outside the depletion layer [7.55].

The damage can be created by high energy silicon implant and no dopant effect is introduced. The gettering mechanism of buried defects is illustrated schematically in Fig. 7.29(a), and the influence on the leakage current of a diode in Fig. 7.29(b) [7.56]. The applications so far discussed are summarized schematically in Fig. 7.30 and in table 7.2.

7.8 High - Speed Bipolar Transistors

The bipolar transistor was briefly presented in sect. 1.4 and illustrated in Fig. 1.11. The sequence of the adopted steps for the fabrication of bipolar transistors was shown in Fig. 1.12. In the present section we will describe in slightly more details the use of ion implantation in the fabrication of high-speed bipolar transistor [7.57] paying some attention to the formation of polysilicon emitter [7.58]. For clarity a schematic view of an n-p-n bipolar transistor is reported again in Fig. 7.31 and the several resistances and capacitances are also indicated together with the three main regions, emitter base and collectors.

The two contiguous junctions (n-p emitter base, p-n base collector) are separated by a distance W_B, the base width. As mentioned already (see sect. 1.4) the transistor action is due to

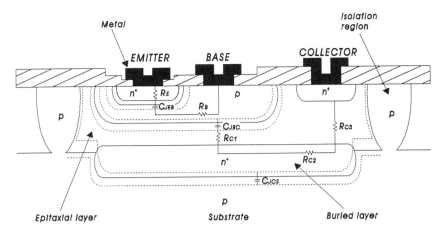

Figure 7.31 *Schematic cross section of a bipolar transistor.*

the modulation of the current flow in one p-n junction by means of a charge in the bias of the nearby junction. Bipolar transistors are used as discrete or as integrated components. The planar process has allowed bipolar transistors and other components such as resistors to be fabricated simultaneously on a single silicon chip. They can be used as analogue devices to amplify faithfully small-voltage-time amplitude time-variable signals or as digital devices switching from an ON (low voltage drop and high current) to an OFF (high voltage and very low current) state by a small current.

The different modes of working are due to the different biasing configurations. The forward-active mode is characterized by the emitter-base junction forward biased (~ 0.7 V) and by the base-collector junction reverse biased ($V_{CB} > 1V$). In the cut-off mode both junctions are reverse biased and the current flow between emitter and collector is due to leakage current, and in general must be kept as low as possible by the different technological processes. In the saturation mode both junctions are forward biased, as a result the emitter-collector voltage approaches zero (ON) and the base current no longer controls the collector current that varies with the power-supply voltage according to the Ohm's law. A simple circuit to illustrate the different operation modes of a bipolar transistor is shown in Fig. 7.32a and the transfer characteristics between the output voltage at the collector and the input voltage at the base for a common emitter circuit are reported in Fig. 7.32b.

Transistors can be used in digital logic gates and memory arrays, as transistor-transistor logic (TTL) and emitter-coupled logic (ECL). A bipolar transistor in the common emitter configu-

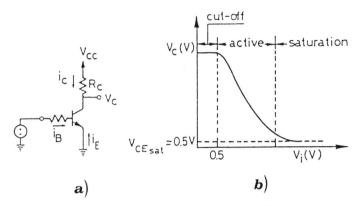

a)   b)

Figure 7.32 *(a) A simple circuit used to illustrate the different modes of operation of a bipolar transistor by varying the input voltage v_i. If v_i is smaller than about 0.5 V, the device is in the cut-off mode. If v_i has a such value that the E-B is forward biased at $v_{BE} \simeq 0.7$ volt the transistor is in the active mode. A further increase in v_i causes v_c to be less than 0.7 V, the device has left the active region and entered the saturation region (ON). (b) Transfer characteristics of the bipolar transistor.*

ration and two external resistors can be used as an inverter (see Fig. 7.32). Transistors are also used for power devices to handle thousand of voltage and thousand of amperes.

The desired device characteristics of bipolar transistors include high current gain, high frequency, alternating current operation, fast switching speed, high device breakdown voltages, minimum device size and high reliability of device operation. A trade-off among all these different characteristics should be found and the device design and fabrication must be optimized according the specified application. In any case the parasitic resistance (R_E, R_B, R_C) and the parasitic junction capacitances (C_{EB}, C_{CB}, C_{CS}) shown in Fig. 7.31 must be minimized.

Among all the several useful electrical parameters adopted for the description of a bipolar transistor we will consider only the current gain and the cut-off frequency being limited our analysis to high speed bipolars. In the common base configuration the current gain is defined by $\alpha \equiv I_C/I_E$ and in the common-emitter configuration by $\beta \equiv I_C/I_B$, with $\beta = \alpha/(1 - \alpha)$ and I_C, I_B and I_E the collector, base and emitter current. A transistor is characterized by $\alpha \simeq 1$ and small change in I_B will bring about $\beta(\sim 10^2)$ times change in I_C. I_B acts as the input signal and I_C

as the output signal. In an n-p-n γ is given by the ratio of current
in the base due to the injection of electrons from the emitter and
the total emitter current (this last includes the injection of holes
into the emitter from the base), α_T is defined as the ratio of the
electron current reading the collector and the electron current
injected into the base. For a transistor with large β on can write

$$\frac{1}{\beta} \simeq 1 - \gamma\alpha_T, \simeq \frac{1}{2}\left(\frac{W_B}{L_{nB}}\right)^2 + \frac{D_{pE}GN_B}{D_{nB}GN_E} \qquad 7.7$$

The first tern in the last member represents the recombination of
e-h pairs into the base and the second the injection into the emitter. L_{nB} is the electron diffusion length in the base, D_{pE} and D_{nB}
the diffusion coefficient of holes and of electrons in the emitter and
base region respectively, GN_B and GN_E the base and the emitter
Gummel numbers i.e. the total number of active impurity atoms
in the base and in the emitter regions respectively. GN_B ranges
between 10^{12} and $10^{13}/cm^2$, while GN_E ranges between 10^{15} and
$10^{16}/cm^2$. These values are substantially higher ($\sim$ two orders
of magnitude) than those used for the threshold voltage control
of MOSFET and for this reason the use of ion implantation in
the fabrication of bipolar transistors was retarded. High current
implanters are needed.

A high current gain requires then a heavy doped emitter and
a light doped base. There are limits to the increase in the emitter
doping and to a decrease in the base doping. Heavy doping affects
the transistor behaviour in three ways. The carrier mobility is
degraded for the large amount of scattering events on ionized impurities [7.59]. The band gap of silicon decreases with increasing
dopant concentration in excess of $10^{19}/cm^3$. The band narrowing
[7.60] has the effect of reducing the effective doping concentration
and hence also the gain of the bipolar transistor. The recombination rate of minority carriers increases in the heavy doped
material due to Auger recombination [7.61], i.e. a three-particle
process in which the energy and momentum released by band-to-
band recombination of e-h pair is transferred to a majority carrier
(electron or hole). The base doping cannot be decreased because
of the presence of high injection effects. The collector-base junction is reverse biased in the active mode and a change in the
voltage varies the width of the depleted region that may decrease
the neutral base width (Early Effect) [7.62]. For thin base width
and low dopant concentration all the base may be depleted at a
suitable collector- base bias. This condition is referred as punch-
through and the emitter and collector are then connected by a
single depletion region and a large current flows.

The other figure of merit of relevance in high-speed bipolar transistors is the cut-off frequency f_T. It is defined as the frequency at which the extrapolated common emitter, short circuit load, small signal current gain drops to unity. In the advanced bipolar transistor it is above 10GHz. Several factors influence f_T, such as the emitter delay, τ_E, the emitter/base depletion region transit time, τ_{EBD} the base transit time τ_B and the collector/base depletion region transit time, τ_{CBD}. In addition the charge and discharge of the several capacitances through the corresponding resistances must be taken into account, e.g. the junction capacitance C_{jBC} of the collector-base through the collector serie resistance R_C. The three resistances R_C, R_E and R_B are the semiconductor resistances between the active transistor area and the collector, emitter and base resistance respectively. The capacitances should include also those due to the store of excess minority carriers in the quasi-neutral regions (diffusion capacitances). Without going into a detailed discussion of all the physical phenomena that contribute to the frequency response of a bipolar transistor it must be remarked that τ_E and τ_{EBD} are generally small compared with τ_B and τ_{EBD}. τ_B is given for a uniformly doped base by

$$\tau_B = \frac{W_B^2}{2Dn_B} \qquad\qquad 7.7a$$

and τ_{CBD} by

$$\tau_{BCD} = \frac{W_{CBD}}{2v_{sat}} \qquad\qquad 7.7b$$

where W_{CBD} is the width of the collector/base depletion region and v_{sat} is the saturation velocity, $\sim 2 \times 10^7 cm/s$ [7.57]. It can be seen that the transit time in the base decreases with the base width W_B, and the collector/base depletion layer delay with increasing the doping concentration in the collector.

To summarize the equation of the cut-off frequency becomes

$$2\pi f_T = [\tau_F + R_c C_{jBC} + r_d(C_{jEB} + C_{jBC})]^{-1} \qquad\qquad 7.8$$

with r_d the differential resistance of the E-B forward biased I-V diode characteristic, and τ_F the forward transit time $= \tau_E + \tau_{EBD} + \tau_B + \tau_{BCD}$. The series resistance of the extrinsic base and of the emitter have been neglected in the previous equation. Extrinsic base is the p layer from the contact to the intrinsic base just below the emitter. The base resistance limits the rate at which the input capacitance can be charged, and the emitter resistance becomes relevant in the submicron devices and for polysilicon emitters.

The design and the performance of high speed bipolar transistor have been substantially implemented by the use of ion implantation, polysilicon emitters and self-aligned processing techniques. The consumer market of these devices is growing year by year at an elevated rate in particular in the field of cellular phones where working frequencies in the few GHz range are required. Silicon VLSI technology allows cost fabrication and high through-put and has several advantages compared with heterostructure transistors based on compound semiconductors. Bipolar transistors with Si_xGe_{1-x} base have also a quite high-frequency cut-off for the narrowing of the band gap and the decrease of the base resistance but their technological development is still in R&D stage [7.63].

The TEM cross section of an advanced high speed bipolar transistor is shown in Fig. 7.33a. Contact to the emitter is made through a polysilicon n^+ doped layer. A polysilicon layer heavily doped by boron implant, contacts the intrinsic base and forms after diffusion the extrinsic base region. The intrinsic base is fabricated by self-aligned procedure through oxide- nitride sidewall spacer. Implants of low energy B^+ or BF_2^+ ions are used together with a relatively low-temperature annealing to restore the lattice structure. Care must be paid to contact the intrinsic with the extrinsic base. The base resistance decreases because p^+ extrinsic base is self-aligned to the polysilicon emitter.

The emitter is obtained by a deposition of a polysilicon layer 200nm thick, implanted with high dose $As^+ \sim 10^{16}/cm^2$ and then followed by a thermal anneal to diffuse the dopant into the underlying silicon single crystal. Isolation of contiguous devices is performed by a trench. This is a multi-step process involving trench etching, refilling with thermal oxide and nitride, deposition of polycristalline silicon and planarizations. Trench isolation has the advantage of giving a reduced collector/isolation capacitance and an increased collector/isolation breakdown voltage. Metallization with TiW diffusion barriers completes the process.

Polysilicon emitter overcomes all problems of implantation-induced damage, and the other major advantage is its suitability to produce very shallow emitter/base junctions and the compatibility with self-aligned fabrication techniques. Effective scaling to smaller lateral geometries can be achieved only if the emitter/base depth is also scaled to reduce the contribution of the peripheral junction sidewall capacitance to the total capacitance.

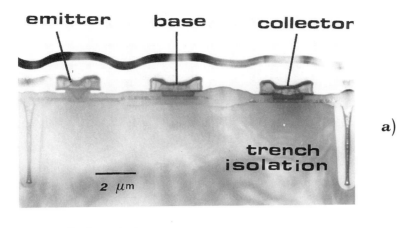

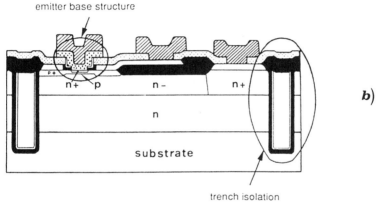

Figure 7.33 *(a) Transmission electron micrograph cross section of an advanced high speed bipolar transistor (Courtesy of C.Spinella and P.Ward), (b) schematic illustration of the bipolar cross section.*

As said before high cut-off frequencies requires low parasitic capacitances and resistances. Assuming an emitter/base junction of $0.2\mu m$ the peripheral capacitance contributes to 50% of the total for a $1.5 \times 1.5\mu m$ emitter size and to 70% for a $0.5 \times 0.5\mu m$. The polysilicon emitter structure was also shown in Figs. 6.33 and 6.34 and the interface with the single crystal was detailed by high resolution TEM cross section. For clarity the cross-sectional TEM image of the emitter structure is reported in Fig. 7.34 [7.64]. The poly-Si1 is the p^+ heavily doped extrinsic base, the nitride spacer allows the self-aligned process and the poly-Si2 for the emitter is separated by an oxide from the poly-Si1.

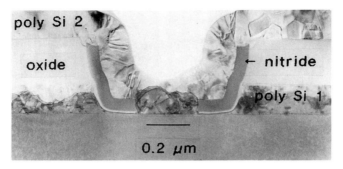

Figure 7.34 *Cross-sectional TEM image of the strip structure for the emitter (from ref. 7.64).*

A critical step in the formation of the emitter is the realignment of the polylayer during the thermal process, several phenomena occur: rupture and balling-up of the native oxide layer at the polycrystalline-single crystal substrate, grain growth and enhanced grain growth by the presence of dopant, realignment and columnar growth of epitaxial islands. The process is schematically illustrated in Fig. 7.35 for As doped and undoped samples.

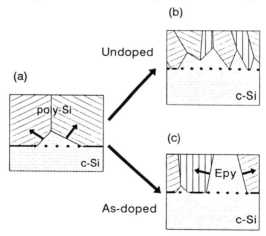

Figure 7.35 *Schematic drawing of an epitaxial protuberance attached at a grain boundary termination at the poly-Si/c − Si interface (a), of the planar realignment of undoped films (b) and of the columnar realignment mode of As doped films (c) (from ref. 7.65).*

In the case of undoped films the polycrystalline grains do not grow during the rapid heat treatment at $\sim 1000^0C$ for several seconds, leading thus to a large density of grain boundary terminations where the realignment transformation can take place (fig. 7.35b). The closely distributed growing epitaxial protuberances appear as a rough interface that moves in a planar manner toward the film's surface. In the presence of As doping, the growth kinetics is enhanced so that during the early stages of the heat treatment the polycrystalline grains grow to dimensions comparable to the film thickness before the realignment process begins. The large dimensions of the polycrystalline grains result in a smooth interface and a reduced density of grain boundary terminations at which the epitaxial realignment can take place. The reduced density of realignment sites is offset by the enhanced grain growth ($\sim 10-100$) caused by the As doping. A rapid extension of source large epitaxial columns occurs. The lateral walls of the realigned columns (see Fig. 7.35a) also form preferred realignment sites at the interface, allowing their further lateral growth [7.65].

The electrical characteristics of the polysilicon emitter depends on the structure of the interface. A thin oxide layer is present at the interface between the polysilicon and the single crystal structure, if after the thermal process it maintains the integrity the current is transported mainly by tunneling and an improved current gain of one or two order of magnitude is obtained [7.66]. The rupture of the oxide and the formation of balls reduces the current gain to a few times that of a crystal emitter, but the resistance of the emitter is improved and this is of relevance in the sub-micron geometry as in the case shown in Fig. 7.34.

Another effect so far neglected is the influence of grain size dimension compared to the lateral extent of the polysilicon film. The plan view TEM micrographs of Fig. 7.36 show the structure of $1\mu m$ (7.36a) and $0.25\mu m$ (7.36b) wide emitter strips, from a sample annealed at 975^0C for 85s. The bright regions represent in the dark field conditions portion of the poly-Si 2 (Fig. 7.34) film realigned to the substrate. In the large strip the film is completely realigned while in the narrow strip (7.36b) is still partially polycrystalline [7.67]. The different behaviour for the realignment process must be associated to a geometrical effect induced by the reduction of the lateral extent of the strips. Fig. 7.36c reports the variations of the realigned areal fraction f versus the annealing time for $1\mu m$ (squares) and for the $0.25\mu m$ (circles) wide strip. The continuous lines are best fits to the experimental data using the classical model of nucleation

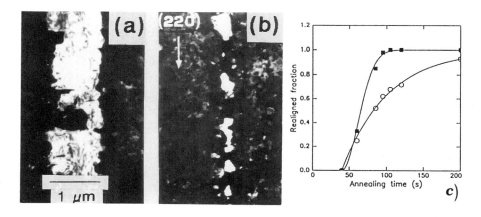

Figure 7.36 *Plan view TEM images of As doped poly-Si films deposited onto 1μm (a) and 0.25μm (b) wide strips after annealing at 975°C for 85 s.(C)Areal realigned fraction vs annealing time at 975°C for 1μm (■) and 0.25μm (0) wide strips. The continuous lines are least squares best fits of the experimental data (from ref. 7.67).*

given by

$$y(t) = 1 - exp\left[-\left(\frac{t-\tau_0}{\tau_c}\right)^n\right] \qquad 7.9$$

where τ_0 is the delay time for the interfacial oxide rupture and τ_c is a characteristic time for realignment related to the density of epitaxial columns N_c, to their growth rate v. If the nucleation rate is negligible the value of n is equal to 3, 2 or 1 for three-two and one-dimensional growth respectively. The fits of Fig. 7.36c were obtained using n=2 for the 1μm wide strip and 1 for the 0.25μm wide strip. Thus the in-plane isotropy loss induced by the reduction of the lateral extent of the poly-Si, reduces the dimensionality of the growth of the epitaxial columns. The thermal budget needed for the realignment is then dependent on the wide strip and indicates the influence of mesoscopic dimension on the metallurgical behaviour.

The base is usually formed by low energy B^+ or BF_2^+ implant followed by a low thermal process to reduce the diffusion. It is the same problem encountered for the formation of shallow junctions in MOSFET devices. Here the shallow junction reduces the base width and then the transit time increasing the cut-off frequency. Figure 7.37a shows the vertical doping profile along the emitter-base-collector of the self-aligned high speed bipolar

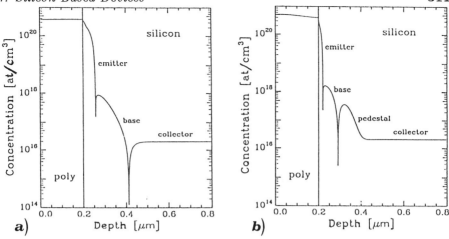

Figure 7.37 *Vertical doping profiles for a 16 GHz (a) and for 25 GHz (b) silicon bipolar transistor. The main difference between the two profiles is the pedestal 400 keV P^+ implant added in the fabrication of the second transistor (b) (Courtesy of V.Raineri).*

reported in Fig. 7.32. The emitter/base junction depth is $0.06\mu m$ and the metallurgical basewidth $0.26\mu m$. The base was obtained by 15 keV - $4 \times 10^{13}/cm^2 BF_2^+$ implant annealed at $950^0 C$ - 30s. The emitter was realized by $1 \times 10^{16}/cm^2 80 keV As^+$ implant in a polysilicon 200 nm thick layer and processed at $1000^0 C$ for 30 s. The f_T is 16GHz. A better optimized doping profile is reported in Fig. 7.37b and the f_T is 25 GHz. The emitter/base junction depth is reduced now to $0.02\mu m$ and the metallurgical basewidth to $0.10\mu m$. A high energy implant of P at 400 keV was adopted as pedestal collector [7.69] at the base collector region. This implant has the advantage to reduce the base/collector junction depth decreasing the transit time in the base and in the base-collector depletion region (i.e. τ_B and τ_{BCD}).

The high energy implant has infact two effects. It compensates the boron in the tail profile with a narrowing of the base region with a little loss of the integrated base electrical dose. The collector doping just under the base region becomes high enough to prevent base-widening effects at high currents. As a result the cut-off frequency is enhanced.

High energy implants can be applied to the formation of buried collector and the usual implanted collector plus the epitaxial silicon overgrowth is replaced by a direct P implant into the silicon substrate. For istance according to the geometry and the maximum allowable buried n^+ sheet resistance the following

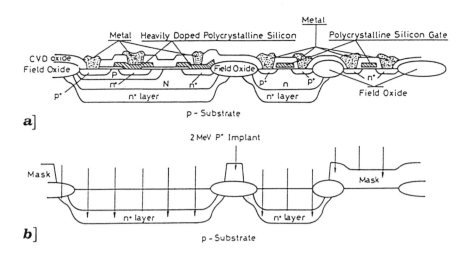

Figure 7.38 *Cross-section of a bipolar-CMOS structure (a), mask configuration for the use of 2.0 MeV P^+ - implant to form the n^+ subcollector and the retrograde n-well. (from ref. 7.71)*

P^+ implants are used: $1\mu m - 130\Omega/\square - 800 keV \cdot 1 \times 10^{15} P^+/cm^2$; $0.5\mu m - 300\Omega/\square - 700 keV - 3 \times 10^{14}/cm^2$. These two cases refer to bipolar transistor for high speed emitter-coupled logic. In the case of low power current-mode logic at $0.5\mu m$ geometry and $1700\Omega/\square$ sheet resistance on implant of 600 keV - $7 \times 10^{13}/cm^2 P^+$ is adopted [7.6].

CMOS and bipolar transistor emerge together to form BIC-MOS devices in the same chip. CMOS has emerged as the dominant technology for VLSI circuits, with a low power consumption and a large noise margin but with a limited ability to drive large capacitive loads. Bipolar transistors have instead a high transconductance and then a greater driver current per unit silicon area. If CMOS and bipolar transistors are combined on a single chip, high density MOS circuits are combined with high current-drive bipolar transistor. The most important advantage of BICMOS processes is that they offer a simple means of integrating a wide variety of different analogue and digital building blocks onto a single chip. Microprocessors and memories can be freely integrated with analog/digital converters, amplifiers, filters etc. [7.70].

A cross section of a simple BICMOS device is shown in Fig. 7.38 [7.71]. The main features of this process are the use of an n^+ buried layer to reduce the collector resistance of the bipolar transistor and a polysilicon gate emitter. Three additional masking steps are required, one for the buried layer, one for the base and one for the polysilicon emitter. The n-channel drain/source implant is used to introduce the n^+ collector region and the p-channel source/drain implant to introduce a p^+ extrinsic base [7.72]. The process is essentially a CMOS process modified to include a bipolar transistor. It is possible to modify the bipolar process to include MOS transistors. The use of high energy P implant avoids a masking step. The implant of $2.0 MeV - 1 \times 10^{14}/cm^2 P^+$ forms simultaneously the n^+ bipolar subcollector and the retrograde n-well. The figure 7.38b illustrates the use of masking to define the n^+ pattern.

CHAPTER 8

ION IMPLANTATION IN COMPOUND SEMICONDUCTOR AND BURIED LAYER SYNTHESIS

8.1 Introduction

In the previous chapters silicon was considered as the only substrate in which to perform the implants. More than 99% of the total production of solid state devices is based on silicon wafers and this might justify the choice. Regardless of the small quantity in which are produced devices based on compound semiconductors, e.g.. $GaAs$ and InP, are of extremely relevance in optoelectronics, in high frequency applications and in the photon-electrical carrier conversion [8.1].

Main advantages of $GaAs$ over Si are the higher electron mobility, five times, and the direct gap so that photons can induce electron transitions without the need of phonons creation and/or annihilation for momentum conservation. III-V compound semiconductors stacked in sequence present a lot of interesting properties [8.2]. Potential wells for electrons and holes, allowing one to confine electron motion to two dimensions and giving still higher electron mobility, are obtained by single crystal heterojunctions between $GaAs$ and $GaAlAs$. The band gap can be changed by varying the composition of the ternary or quaternary compound semiconductors, new materials can be made by the use of molecular beam epitaxy or chemical vapor deposition procedures [8.3]. Photo detectors, lasers, optical radar, light emitting diodes are used in many fields.

In spite of their better electrical and optical characteristics compound semiconductors and in particular $GaAs$ are limited to a niche market. The higher electron mobility reduces the power dissipation relative to silicon for the same level of performance. The higher energy gap of $GaAs$ (1.4 eV) allows intrinsic performance by the substrate, which reduces parasitic capacitive coupling among circuit elements. The market niche comes from economical reasons, a technology or device wins when it makes up for worse intrinsic properties by means of greater circuit complexity at lower cost. This is the main reason why volume applications of $GaAs$ are practically delayed to the far future as material to replace silicon.

Ion implantation is also in these compound materials a relevant technology. It allows the selective doping of certain regions and provides a quite convenient method to isolate, by creation of trap centers, electrically one device from the nearest one [8.4]. There are several problems and the annealing procedure is more complex for the large number of different atoms involved and for the lack of a reliable oxide layer such as SiO_2. Some of these aspects peculiar to a compound material will be considered in view of applications to device.

Ion implantation offers also the possibility to form buried dielectric and metal layers overlayed by a thin single crystal. The formation of SiO_2, N_3Si_4 and $CoSi_2$ by implants of oxygen, nitrogen and cobalt in silicon will be described. As an example the fabrication of MOSFET in the ultrathin, $\sim 50nm$ thick, surface Si layers isolated by the dielectric SiO_2 is considered. The structure present a better performance with respect to latch-up, short channel effects, hot electrons, and radiation hardness [8.5]. In the case of metal buried layer the fabrication of permeable base transistor is possible, although their use is still limited [8.6].

8.2 Ion implantation in GaAs

The mobility of electrons in $GaAs$ is ten times that of holes, n-type $GaAs$ layers are then required for the fabrication of faster devices. Unlike to silicon no diffusion technology exists for the introduction of n-type dopants in $GaAs$, so n-type epitaxial layers or implanted undoped material are used. High purity $GaAs$ substrates became available in the late seventies and ion implantation became the standard method for doping. In this wide-gap material ion implantation is also used to form high-resistivity regions on a wafer that contains doped layers. The damage created by the implant reduces the carrier mobility in the material and forms deep level centres which trap free carriers. The material after bombardment, but before annealing, has a high resistivity. This forms the basis for rather simple device isolation scheme that prevails in III-V technology. Simple masking with photoresist or other materials in order to define the areas to be bombarded is sufficient to achieve device isolation, provided that the ion energies and the fluences are chosen correctly. In contrast to Si, higher beam energies are required being layer the stopping powers in $GaAs$ than in Si. Implanters of 200-400 KV accelerating voltage are needed.

The distribution of the implanted ions in the compound semiconductors can be described following the same treatment developped in chapter 3 for silicon. The same considerations are true for the damage distribution. Again the critical dose for amorphization depends very strongly on the temperature of the sample during implantation. With an increase in the dose rate, the critical dose usually decreases because the damage is able to accumulate before it self-anneals. Dynamic annealing at high dose rate can occur also in these substrates. The critical dose for the creaction of an amorphous layer in $GaAs$ for RT implants is $\sim 10^{15}/cm^2$ for C, $2 \times 10^{14}/cm^2$ for Si and $3 \times 10^{13}/cm^2$ for Zn and Cd. The previous estimates refer to low energy implants [8.7].

As for Si the most important step in the ion implantation process for semiconductor compound doping is the annealing of damage. The main differences, with respect Si, are the following ones: the need to avoid amorphization of III-V during the implant, the formation of local non stoichiometric regions resulting from the different displacement properties of Ga and As (or In and P etc), the need to protect the surface of wafer from dissociation during annealing at temperature above 500^0C, and the fact that the implanted species should occupy only one lattice site after annealing.

The two atomic species forming the compound have different masses and after collision recoil with different energy. The lighter elements will recoil further leaving an excess of the heavier element near the surface at distances shallower than R_p and an excess of the lighter element at greater depths (between R_p and $R_p + \Delta R_p$). Examples of calculations for Si implants in $GaAs$ and for Se in InP are shown in Fig. 8.1 [8.8]. The calculations yield directly four concentration profiles: namely, gallium (indium) and arsenic (phosphorus) vacancy distributions (from the origin points of high energy recoils) and gallium (indium) and arsenic (phosphorus) recoil distributions (from the final stopping points of the same recoils). The net displaced atom distribution is obtained by subtracting the vacancy distribution from the recoil distribution for each atom type. The net excess atom distribution requires a further step, namely the subtraction of the net displaced arsenic (phosphorus) distribution from the net displaced gallium (indium) distribution.

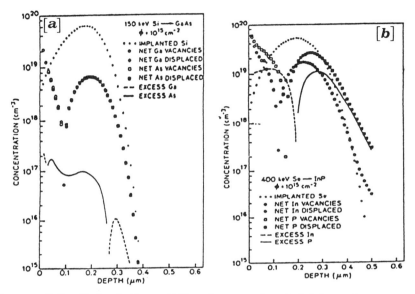

Figure 8.1 *Calculation of the changes in stoichiometry caused by* $10^{15}/cm^2 - 150keV\,Si$ *in* $GaAs(a)$ *and by* $10^{15}/cm^2 - 400keV\,Se$ *in* $InP(b)$ *(from ref. 8.8).*

The lost of stoichiometry impaires the solid phase regrowth of amorphous regions. In fact the reordering that occurs at relatively low temperature cannot be accomplished because the diffusion lengths of the atomic species are not long enough to reach the correct lattice position. As shown for Si substrate, in chapter 5, the reordering takes place at the interface α-single crystal and does not involve long range migration of defects. The epitaxial regrowth in $GaAs$ is then limited to a few tens of nanometers and then stops with the formation of twins, stacking faults, polycrystals and a lot of dislocations [8.9]. Amorphization must be avoided then and the high dose implants should be done maintaining the substrate at a temperature in the $200 - 400^0C$ range. The defects created during the implant are mobile and can recombine or annihilate each other. A partial activation of the implanted species occurs already just after the implant but in any case an annealing at higher temperature is necessary to recover the crystal structure in the substrate and to diffuse the dopant to a lattice position.

The annealing treatment must be performed, to be effective, at temperature where the group V element is preferentially lost. The vapor pressure of As is higher than that of Ga and increases exponentially with temperature. At 600^0C it is 5×10^{-11} atm, at 1000^0C 10^{-3} atm. The flux of emitted molecules is given by [8.10]

$$J = 3.5 \times 10^{22} p \frac{1}{\sqrt{MT}} \qquad 8.1$$

with p, vapor pressure in torr, M molecular weight in g and T temperature. One second at 840^0C decomposes a layer 60 nm thick! To preserve the surface stechiometry the annealing is performed by providing an over pressure of As in the ambient, or by encapsulating the surface with a thin dielectric layer of Si_3N_4 or SiO_2. In some cases the wafer is placed face-to-face with another uncapped $GaAs$ wafer. At elevated temperatures a small amount of As will be lost from both wafers, creating an overpressure between them and suppressing further loss. The three most used annealing methods are shown schematically in Fig. 8.2 [8.11].

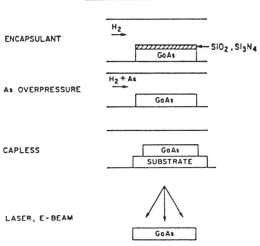

Figure 8.2 *Schematic representation of some annealing methods for GaAs (from ref. 8.11).*

The thin capping layer should in principle satisfy the following requirements: prevent surface degradation and loss of stoichiometry in the near surface region, act as a barrier to the out diffusion of impurities from the substrate or to the substrate from the ambient, must be deposited at low temperatures, does not react with the surface, does not induce strain in the near surface region of the semiconductor as those caused by different thermal expansion coefficient, has a good thickness uniformity ($\sim 1-2nm$) and can be easily removed after annealing. The most commonly used encapsulant is Si_3N_4 plasma deposited. Two other encapsulants of interest are the reactively sputtered AlN and the spin-on PSG (phosphosilicate glass), both of which have similar expansion coefficients to $GaAs$.

The various stages of damage removal and dopant activation in implanted $GaAs$ are summarized in Fig. 8.3. As in Si the primary damage can consist of amorphous continuous layers, or amorphous regions embedded in a single crystal environment or extended defects according to the implant conditions [8.12].

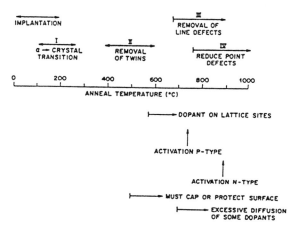

Figure 8.3 *Schematic of damage annealing and dopant activation steps in implanted GaAs (from ref. 8.12).*

Amorphous layers recrystallize at $150^0C - 200^0C$, temperature range but the layer is quite defective with twins, stacking faults. In the $400^0C - 600^0C$ - range twins are removed and dislocation loops are left. Annealing at higher temperature, $\sim 700^0C$, is necessary to remove the line defects and to reduce the point defect clusters, involving vacancies and anti-site defects i.e. a Ga atom that occupies an As site, and viceversa. Extended defects as seen by TEM have only a secondary effect on the electrical characteristics of the implanted layer, point defects and small complexes play instead a dominant role. Infact these small clusters will compensate the electrical activity of dopants. Carrier activation needs then most of the implant damage, even point defects, be removal by the annealing at high temperature, as shown in Fig. 8.4 [8.13].

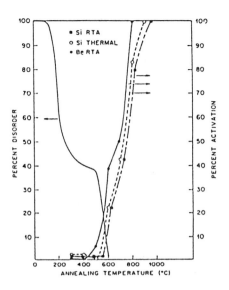

Figure 8.4   *Temperature dependence of carrier activation and damage annealing in GaAs implanted with Si or with Be at room temperature (from ref. 8.13).*

n-type layers are obtained by implantation of one of the following species: Si, Se, Te and S. Si is an amphoteric impurity, i.e. if replaces Ga behaves as donor, if replaces As behaves as acceptor. Si at not too high concentration is mainly donor and is widely used for the formation of n or n^+ regions. Of the other group VI elements, Se has a good activation for low dose, RT implant, Te has instead a low activation if implanted at room temperature. S diffuses during implant.

The sheet carrier concentration is shown in Fig. 8.5 for room temperature Si implants as a function of the implant dose. The annealing was performed at 900^0C for 10 sec by rapid thermal process [8.14].

At high doses the silicon ions occupy both Ga and As lattice sites and the compensation between donor and acceptor takes place with a reduction in the efficiency. If 60% of Si occupy Ga sites and 40% As sites the net donor concentration amounts only to 20%. The maximum achievable concentration for Si is $2 \times 10^{18}/cm^3$. Sligthly higher values $\sim 10^{19}/cm^3$ can be reached with Se at high doses.

For this non amphoteric species the limitation in concentration is probably associated to the formation of donor - arsenic vacancy complexes that are acceptors. The case of the amphoteric species Si is quite complex. The silicon atom in a Ga lattice site, Si_{Ga}, if bound with an arsenic vacancy V_{As}, can make a site

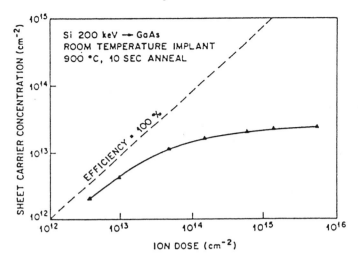

Figure 8.5 *Sheet carrier density as a function of Si implanted dose into $GaAs$ and subsequently annealed at 900^0C for 10 sec. (from ref. 8.14).*

interchange according to

$$Si_{Ga} + V_{As} \rightarrow Si_{As} + V_{Ga} \qquad\qquad 8.2$$

leading to the formation of the $(Si_{As} - V_{Ga})$ complex. In addition there will be also the formation of the $Si_{Ga} - Si_{As}$ neutral pairs. The compensation mechanism would be more complex in this case with the presence of Si_{Ga} donors, Si_{As} acceptors, $(Si_{As} - V_{Ga})$ and $(Si_{As} - Si_{Ga})$ neutral complexes, and other higher order defect complexes. It seems V_{As} to be the predominant stable defect in implanted $GaAs$, gallium interstitial, Ga_i, is mobile at very low temperatures and probably recombine with V_{Ga} and $(V_{Ga} - Ga_{As})$ complexes. The formation of EL2 centers, related to As antisite (As_{Ga}) complexes, may also take place according to the reaction and V_{As} compensates donors in n-type material.

$$V_{Ga} + As_{As} \rightarrow As_{Ga} + V_{As} \qquad\qquad 8.3$$

This should explain the saturation doping for n-type impurities [8.15].

The formation of p-type layers is quite straightforward in $GaAs$, but n-type material is preferred for the high mobility of electrons. All the acceptors species, such as Be, Mg, Zn and Cd, show high activation to a much higher dose level than do the

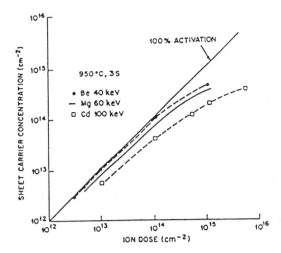

Figure 8.6 *Sheet carrier densities as a function of ion dose for Be, Mg or Cd implanted into GaAs and subsequently annealed at 950⁰C for 3 sec. (from ref. 8.16).*

donors. The sheet carrier concentration as a function of ion dose for Be, Mg, or Cd implanted into $GaAs$ is reported in Fig. 8.6 [8.16].

It is higher for lighter species and is much higher than for the donors. Peak concentration values above $10^{19}/cm^3$ are easily obtained and a maximum concentration of $\sim 10^{20}/cm^3$ has been reported.

The relationship between the solubility of the implanted dopants and the electrical activity is not necessarily 1:1 as in Si. The presence of native defects like antisite complexes, vacancies, etc. can compensate or trap charge carriers. The substitutional lattice site occupation of an impurity is not a sufficient condition for the electrical activity. An example is given by Te implants, the substitutional fraction as measured by channeling effect technique can be 90% after annealing, but the electrical efficiency is of about 5%. This is probably due to the formation of $Te - V_{As}$ complexes at high Te concentration [8.17].

In these last years quite relevant improvements have been reached in the quality of the substrates. The liquid encapsulant crystallization is now the common method to produce high resistivity $GaAs$ wafers of 4" or 5". Of course the control reached for the Si crystal growth is not obtained for the $GaAs$ growth and then according to the characteristic of the substrate the implant parameters and the annealing procedures should be checked often.

This is particular important at low donor concentration. The undoped wafers are made highly resistive by compensation with deep level stoichiometry-related defects created during crystal growth the shallow acceptors (principally C) and donors (principally Si and S) unintentionally introduced during growth. All of these impurities or defects are present at the $10^{15} cm^{-3}$ range, but the compensation can result in materials with net carrier densities in the $10^7 - 10^8 cm^{-3}$ range. After annealing the sheet resistance is not uniform over the wafer. The non uniformity is due to a direct correlation or anti-correlation with the dislocation distribution in the wafer and it results from the nature of the impurity cloud gettered to the dislocations. The quality of the substrate is improved by the addition of In to harden the crystal and by the reduction of the thermal stress during growth. In these wafers the corresponding improvement of the resistivity uniformity is quite marked [8.18].

The typical dose of implanted Si ranges between 2 and 5 $\times$ $10^{12}/cm^2$ at 60-100 keV for the channel doping of the FET. As in Si channeling tails are also present and techniques such as preamorphization cannot be used for $GaAs$. In some cases, if necessary, the electrical profile is modified by co-implantation of an acceptor species at low dose to compensate the deep tail or of a neutral species to introduce deep level, making high resistivity material.

Fast heating at high temperatures for short times is in principle an interesting annealing procedure to remove the damage and to activate electrically the dopant implanted into $GaAs$ or compound semiconductors. The requirements on the capping layers are less stringent, although thermal stress can induce slip lines in the processed wafers. The optimum annealing conditions for $GaAs$ is an high temperature ($900^0C - 950^0C$) and short duration (5-10 sec) cycle. The standard surface proximity technique in which the wafer is placed face-to-face with another uncapped wafer can be used. However as device density increases, the presence of microscratches and baked-on contamination on the wafer surface becomes a problem and encapsulating layers are preferred. Recently the use of a capped layer at the start of the fabrication procedure and the accomplishement of the implant through it, is gaining interest.

To reduce temperature gradient and then thermal stress between the edge and the centre of the wafer SiC-coated graphite susceptors have been used with success. Slip-line formation during implant activation treatments is eliminated together with the surface degradation. The wafers (one or more, according to the design) are placed in the depressions milled into the bottom plate and covered with a graphite lid. The As overpressure is provided by coating the inside surface of the susceptor with As deposited from a previous heating-up of sacrificial $GaAs$ wafers or providing the susceptor with small holes that can be filled with material that provides As partial pressure. The heating and cooling rates of the wafer in the susceptor are much slower than those obtained in the proximity geometry. The reduced rates are advantageous in reducing slip generation.

Rapid annealing gives electrical properties in the implanted layer similar or better to the furnace treated wafers. The dopant redistribution during annealing and out-diffusion are also considerably reduced. This is quite important in the high temperature processing of multilayer structures. The extremely high electron mobilities in selectively doped heterostructures are degraded by diffusion of Si in the undoped $GaAs$ layers during conventional furnace annealing. This undoped layer is the region where the two dimensional electron gas at high mobility resides and where impurity scattering is absent. The intermediate layer is typically 2-10 nm thick and diffusion must be prevented. RTA is able to activate the implanted dopant without a considerable degradation of the electron mobility [8.19].

8.3 Ion Implantation in InP

Indium phosphide is the third semiconducting material for importance in device fabrication. It is now being used for active layers and substrates in optoelectronic, microwave and millimeter wave devices. Compared to $GaAs$ it has higher peak and saturation electron drift velocities, a larger energy gap, higher thermal conductivity, greater avalanche breakdown resistance. The dependence of the electron drift velocities on the electric field are shown in Fig. 8.7 for $Si, GaAs$ and InP.

The implantation process in InP is still more complex than in $GaAs$ because of its relatively low surface decomposition temperature, the larger mass ratio of In to P, the redistribution during annealing of Fe impurities present in the substrate the formation of several deep levels by annealing and encapsulation stress.

Different devices are realized with InP. The low Schottky barrier height ($\phi_\beta < 0.5eV$) for metals on n-type InP prevents the formation of MESFET device and alternative technologies based on junction field effect transistor (JFET) and on metal-insulator field effect transistor (MISFET) are tried [8.20]. Ion implantation is widely used to form n^- and n^+ type layers and p-n junction in

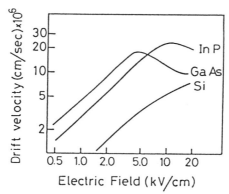

Figure 8.7 *Dependence of the electron drift velocity in $Si, GaAs$ and InP on the applied electric field. The dopant concentration is of $10^{17}/cm^3$.*

semi-insulating InP substrates.

As for the other semiconductors the characteristics of any device rely on the quality of the substrate. The only semiinsulating substrate material commercially available is LEC-grown Fe-doped InP. These substrates have a high dislocation density $\sim 10^4 - 10^5 cm^{-2}$ and a non uniform radial distribution of Fe atoms. Fe impurities in substitutional sites introduce midgap level at Ec-0.65 eV that acts as acceptor to compensate the n-type background impurities in undoped InP. Fe redistributes during typical annealing conditions with a change of the electrical characteristics of the substrate. The requirements on the InP wafers to be used for IC technology are not only a drastic reduction in dislocation density but also the availability of high-resistivity undoped material, i.e. without compensating dopants as with $GaAs$.

Amorphization should be avoided during implantation in InP. The epitaxial recrystallization occurs at $150^0C - 200^0C$ but the layer is highly defective with stacking faults, twins, and other defects. The problem is again the local loss of stoichiometry due to the unequal recoil of In and P. The amorphization dose is considerably lower for InP. For 300 keV Se^+ implant at room temperature it amounts to $2 \times 10^{13}/cm^2$. With increasing temperature, at $400 - 500^0C$ a high density of dislocation loops is left in the sample, and the optimum electrical activity is reached at $700^0C - 750^0C$. In this temperature range the InP surface must be protected with encapsulating layer against dissociation by P loss. Common, although not of ideal properties, layers are Si_3N_4, SiO_2, AlN and PSG. Implantation at elevated temperatures, $\sim 150^0C$ of heavier ions (S, Se, Zn) prevents the formation of an amorphous layer, with a better activation than in the presence of amorphous layer. Implantation of nearly all group II, IV and VI ion species into InP has been reported, [8.21] in the following only the most common dopants will be considered. Si and Se implantation are used to produce n-type layers.

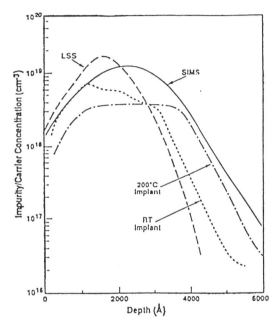

Figure 8.8 *SIMS and carrier concentration profiles of 3 × $10^{14}/cm^2 - 180 keV Si$ implanted into InP at room temperature or at 200^0C after a subsequent annealing at 750^0C - 15 min. (from ref. 8.22).*

Figure 8.8 shows the atomic and carrier concentration of 180 keV $Si - 3 \times 10^{14}/cm^2$ implanted into InP [8.22]. The carrier concentration of the 200^0C implant is flatter and wider than that of RT implant and is close to the observed SIMS profile. The asymmetric RT carrier profile is reasonably close to the theoretical LSS profile on the deep side but not to the SIMS profile. These differences are attributed to the amphoteric nature of Si, to the redistribution of Fe and to damage. RTA has also been used for implanted InP wafers. Relatively long (> 5 sec) times at high temperature $(800^0C - 900^0C)$ are required to obtain high quality

Si implanted n-type layers. With *Se* implants carrier concentration in the $10^{19}/cm^3$ range is obtained, but in many cases the dopant is substitutionally located but not electrically active. A variety of complexes with native defects or with point defects can be formed by the dopants, and these complexes as for *GaAs* are inactive.

p-type layers in *InP* are formed by *Be* and *Zn* implants, unlike n-type implants, electrical activation of p-type is relatively poor (< 50%) and broadening and in-diffusion occurs after annealing. p-type native donor defects act to compensate the acceptors. The rapid diffusion of the p-type dopants in *InP* at high annealing temperatures can be reduced by a co-implantation of the p-species with *P*. The acceptor is forced to occupy an In site and enough *In* vacancies are created to ensure a high degree of substitutional over interstitial dopants. Carbon in contrast to the *GaAs* case acts as a donors in all implants and annealing conditions in *InP*. p-n junctions of quite good quality are formed by selective sequential implantation of *Si* and *Be* into *Fe*-doped semiinsulating *InP*.

8.4 Isolation of III-V Semiconductors

High resistivity layers are used to isolate electrically one device from the other. In compound semiconductors a useful approach is afforded by ion implantation. One bombarding ion through the formation of deep level damage can remove tens or even hundreds of mobile carriers, whereas one implanted dopant ion is required to create one carrier. $1MeV B^+$ ion in *GaAs* removes 200 carriers, and $100keV B^+$ or O^+ 50 carriers respectively. The implant dose for damage isolation is then 10-100 times lower than the dose of dopant ions needed to dope the layer. In addition to the damage isolation, ion implantation can be formed with selected species that create a chemically active deep-level state, associated to the particular impurity, such as *Cr* in *GaAs* and *Fe* in *InP*.

The temperature behaviour of these two implant approaches to the formation of high resistivity layers is quite different, as shown schematically in Fig. 8.9 [8.11]. Bombardment-induced isolation is effective up to temperatures at which the damage anneals out, typically 500^0C in *InP* and 600^0C in *GaAs*, whereas chemically-induced isolation requires high annealing temperatures because the species must occupy a substitutional lattice site. The stability with temperature depends on the kind of carrier removal if by damage or by chemical effect. Ion implantation offers also a in-depth selectivity by changing the

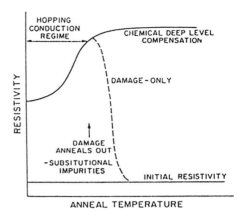

Figure 8.9 *Annealing-temrature dependence of sheet resistance in a III-V compound semiconductor implanted with a species that creates a deep level (from ref. 8.11).*

beam energy. Isolation can be achieved at a certain depth or by multi-energy implants for a certain thickness.

Let us consider at first the behavior of damage isolation in $GaAs$.

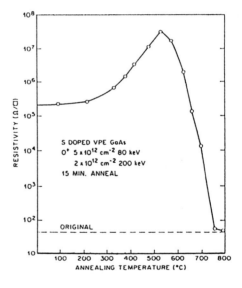

Figure 8.10 *Sheet resistance of oxygen - implanted GaAs as a function of post-implant annealing temperature (from ref. 8.11).*

Fig. 8.10 reports the sheet resistance trend of $GaAs$ wafers implanted with oxygen ions at two energies as a function of the post-implant annealing temperature. The as-implanted resistivity is four orders of magnitude higher than the initial value. With increasing temperature the resistivity increases and reaches a maximum value at 500^0C. Further annealing at higher temperatures restores the original value. The implant produces both extended and point defects. At 500^0C the electrical characteristics of the implanted wafer are dominated by the residual complexes that act as carrier traps. At still higher temperatures their concentration decreases and if below the carrier concentration, electrons in n-type or holes in p-type material populate again the conduction or the valence band respectively. Bombardment with heavier ions such as O, B and F is more effective than with protons and lower doses ($\sim 10^{12} - 10^{13}/cm^2$) are required [8.11].

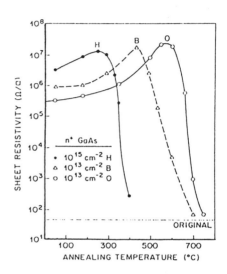

Figure 8.11 *Thermal stability of high- resistance regions formed in GaAs by proton, boron or oxygen implants (from ref. 8.14).*

The temperature stability of the damage-deeps level increases with the ion mass as shown in Fig. 8.11 for H, B and O implanted in n-type $GaAs$. Note the changes in dose from $10^{15}/cm^2$ for H, to $10^{13}/cm^2$ for B and O and those in temperature from 300^0C for H to 650^0C for O. The same trend is also found in n-type $GaAs$ because implant damage in this material creates electron and hole traps so that both n- and p-type material can be made semiinsulating [8.11].

The data of Fig. 8.10 and 8.11 show the as-implanted values of the resistivity to be lower than those achievable by a subsequent anneal. If the implants are performed in a semi-insulating material the as-implanted damage induces then conduction.

A detailed investigation on the dose and temperature dependence of the sheet resistance in bombarded semi-insulating $GaAs$ has shown that the conduction is due to the hopping of the trapped electrons from one damage site to another closely damage site. This conduction mechanism is similar to that found in amorphous materials [8.23] and makes the dose dependence of the effectivenes of the isolation a critical parameter. The dose must be enough to completely compensate the doped layers but not so much to induce hopping conduction effects. In any case there is now a clear evidence that point defects are responsible for the compensation effect in implanted III-V compounds.

The damage levels introduced by bombardment in $GaAs$ are localized mainly at E_c-0.55 eV, E_c-0.71 eV and $E_V + 0.70 eV$ The first two are the two most common levels for electron traps and are stable up to 850^0C, while the third one, hole trap, anneals out around 500^0C. Similar behaviour has been also found in bombarded $AlGaAs$, ternary As based material, although the presence of Al can be responsible of several other types of complexes [8.24].

Implant isolation is not so effective in InP and in In-based compounds like $InGaAs$. The difference is due to the location in the band gap of the damage level. The schematic behaviour of the sheet resistance vs post-bombardment temperature is represented in Fig. 8.12 for Ga and for In based compound semiconductors.

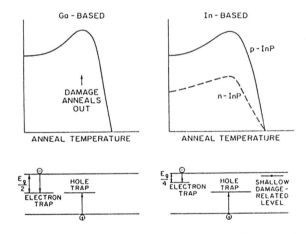

Figure 8.12 *Annealing - temperature dependence of sheet resistance of $Ga - In$ based compound semiconductors irradiated with a non-dopant species, together with the relative energy positions in the forbidden gap of the damage - induced levels (from ref. 8.11).*

The relative positions of the damage levels in the band gap are reported in the lower part of the figure. The Fermi level in the Ga based materials is pinned in the middle of the bandgap, because midgap electron and hole traps are created by implantation. The material becomes then of high resistivity. In In-based compound the Fermi-level is pinned in the upper half bandgap for n-type material and the resistivity can be increased only to $10^3 - 10^4 \Omega \cdot cm$ range. Damage-related hole traps are in the middle of the band gap and so p-type material can be made highly resistive.

The chemically-induced isolation requires the implantation of species that create stable deep levels in the considered material. The resistivity remains high at high temperatures annealings in the chemically-active implanted material. In $GaAs$ chromum is a deep acceptor with an energy level of E_c-0.75 eV. It occupies a Ga site and has a maximum solubility around $5 \times 10^{16}/cm^3$. This impurity is not common implanted because of its small projected range and of its solubility too low for compensating n^+ material. Moreover Cr redistributes at 650^0C. In n-type InP the implantation of Fe led to the formation of highly resistive material after a suitable annealing to locate the Fe in substitutional site. Semi insulating regions have been produced by Fe implants in light wave devices [8.25].

8.5 Isolation of Superlattice and Quantum Well Structures

Band gap engineering relies on the possibility to fabricate in a controlled and designed way multilayer structures of compound semiconductors. The most common method is the Molecular Beam Epitaxy (MBE) under ultra high vacuum conditions. In addition to the heterostructures, multiquantum wells (MQW) and superlattices (SC) are opening a new field of research and applications on photonic devices. Using quantistic structure and MQW it is possible not only to improve the performance of already experimental devices, such as laser and photodetectors but also to realize new devices that require non-linear optical properties such as optical modulators and switches or bistable memory devices [8.26].

In many of these devices it is necessary to change selectively the properties of the material as in guided wave devices and in optical confinement structures. Ion implantation offers also in this case several advantages compared to the thermal diffusion. With respect the previous investigation on monoelemental or compound semiconductors, a multilayer, such as $GaAs/AlAs$ is a metastable system. The stable state would be a completely intermixed and uniform alloy. normally, the large energy barriers for atomic migration and the small equilibrium concentration of intrinsic point defects under these systems quite stable at moderate temperatures. Disordering in superlattice structure has been also observed by thermal impurity diffusion. The impurity enhances locally the intermixing and the composition of selected area can be modified [8.27].

The situation is different under ion implantation where impurities and defects are present at high concentration. Ion implantation causes an intermixing of the structures at considerably lower temperature and shorter times than those required by thermal disordering. Either in a multi period $GaAs/AlAs$ superlattices or for a QW structure, a thin $GaAs$ barrier layers, the compositional disordering induced by the implant leads to a single layer of bulk $AlGaAs$ alloy whose composition depends on the initial compositions and thicknesses of the original layers, and on the experimental conditions producing the intermixing. Selective implantation gives rise to areas of different bandgap integrated on the same substrate. This method has been already applied to the fabrication of heterostructure lasers [8.28].

The formation of selected areas with a specified composition requires the knowledge of the interaction between the different defects with the host atoms. Defects in semiconductor are electrically charged because of their effect on the carrier concentration in the conduction and valence bands of these materials. Defect migration and agglomeration process can be strongly influenced by internal electric fields and by the band structure of the material being implanted. In the case of $AlAs/GaAs$ multilayer heterostructure implanted at 77 K with 320 Ga^+ and annealed at 150^0C it has been found a substantial variations in the interfacial composition of the top $AlAs/GaAs$ interface but not of the low $GaAs/AlAs$ interface. The broadening can be reversed by the top and the bottom interface by reversing the sign of an external applied electric field. The strong difference in the behaviour of adiacent interfaces is due to the drift of the implantation damage through the sample in internal electric fields. The drift can be prevented by a combination of bandgap discontinuity as the defect attempts to cross from one layer to the next [8.29].

8.6 Synthesis of Buried Dielectric SiO_2 and Si_3N_4 Layers

The main application of ion implantation is the doping of semiconductor materials. To this aim the concentration of the dopant rarely exceeds the 1 atomic % and then the dose is in the $10^{15}/cm^2$ range. Increasing the dose of several orders of magnitude it is possible to change the composition of the material substantially and to form alloys or compounds. Being the atomic density of typical solids of $5 \times 10^{22}/cm^3$ the formation of a 100 nm thick compound AB, with A the substrate material, layer requires a dose of $2.5 \times 10^{17}/cm^2 B$ atoms. The possibility of buried compound is then strictly tied to the achievement of high and very high current implanters. Some machines have been designed to fulfill the requirements of buried layer formation and oxygen currents in the 100 m A range are available in dedicated implanters [8.30].

The first buried compound formed by ion implantation was silicon oxide, SiO_2, and the process was termed SIMOX as the acronym of $\underline{S}$eparation by $\underline{I}$mplanted $\underline{O}$xygen [8.31]. This is today one of the best method to form single crystal silicon layer on insulators. Such structures provide complete dielectric isolation of individual devices and often some advantages over other isolations in conventional integrated circuits. These advantages include radiation hardness, high voltage capabilities, increased packing densities, higher operating speed, elimination of latchup and better tolerance of elevated temperatures as detailed in the next section. The radiation hardness is related to the small volume from which electron-hole pairs generated by an energetic particle can diffuse to the device. In SOI structures the device volume is 10^3 times smaller than in conventional substrates. Circuits have then much greater resistance to photocurrents caused by ionizing radiation. Soft errors in memory chip associated for istance to the charge created by alpha particles from residual radio activity in the package material are reduced in SOI devices. Soft errors were discussed in more details in the section 7.6.

The SiO_2 formation was reported schematically in Fig. 1.11. A high oxygen dose above $10^{18}/cm^2$ is implanted at energies exceeding 150 keV into (100) Si substrates at substrate temperature of $500^0 C$. During implantation and the subsequent anneal at high temperature oxygen atoms reacts with silicon to form a continuous stoichiometric oxide layer buried under 0.1 - 0.3 μm of Si. A 100 nm thick layer of SiO_2 contains 4.4×10^{17} oxygen atom/cm^2, but the formation by implantation due to the straggling of the ions requires at least 1.5×10^{18} at cm^2. The process of oxide growth starts during implantation with the formation of small oxide precipitates. Heat of formation of SiO_2 is about 210 Kcal/mole. Soon as the oxide precipitates form a continuous layer the subsequent growth occurs by migration of oxygen toward the near surface Si/SiO_2 interface [8.32]. After implantation the top silicon layer is highly defective and another annealing at temperature above 1300^0C is necessary to obtain a layer nearly free of defects and suitable for the fabrication of devices [8.33].

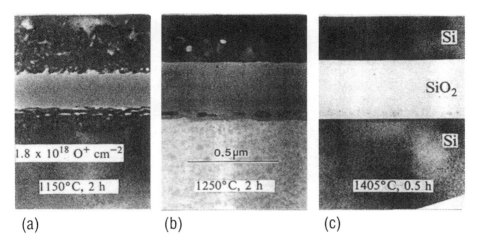

Figure 8.13 *TEM cross section of* $1.8 \times 10^{18}/cm^2 - 200keV0^+$ *implanted samples annealed at* 1150^0C-*2h(a),* 1250^0C - *2h(b) and* 1405^0C, *0.5h(c) (from ref. 8.34).*

The TEM cross section micrographs of Fig. 8.13 show the evolution of the structure with temperature [8.34]. Up to 1250^0C the quality of the very top Si layer is acceptable only as a substrate for the epitaxial growth of Si by chemical vapor deposition. Oxide precipitates are dispersed in the matrix and the interfaces Si/SiO_2 are rough. The quality improvement at 1400^0C - 0.5 h (8.13·c) is due to the dissolution of the small oxide precipitates that are unstable at this temperatures. The released oxygen migrates toward the interfaces of the buried layer that can be considered of infinite radius. The critical radius of the precipitates increases infact with the temperature. As a result, $Si/SiO_2/Si$ multilayer structures with atomically-abrupt interfaces can be produced by SIMOX.

The remaining defects in the silicon single crystal layer are threading dislocations. During these last years the dislocation density has been considerably reduced from $10^9/cm^2$ to $10^3/cm^3$ by a suitable choice of the implantation temperature and dose. Best results have been obtained by multiple implants followed by annealings or by implantation in channeling directions at low dose rates. Only a fraction of the total oxygen dose is implanted each time with a reduction of the stresses in Si below the threshold for dislocation formation and with the elimination of Si inclusion that otherwise remain trapped in the Si the lower Si/SiO_2 interface [8.35].

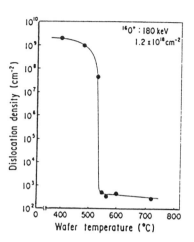

Figure 8.14 *Relationship between dislocation density and wafer temperature during oxygen implantation* $(1.2 \times 10^{18}/cm^2 - 200keV)$ *for the formation of SOI layers. Post-implant annealing at* 1300^0C *- 4h in* $Ar - O_2$ *ambient (from ref. 8.36).*

Fig. 8.14 shows the relationship between the dislocation density and the wafer temperature during oxygen implantation. The experiments were performed with $1.2 \times 10^{18}/cm^2 - 200keV$ O^+ and the post-implant annealing was performed at 1300^0C, for $4h$ in $Ar - O_2$ gas mixture. For the wafers implanted at temperatures lower than 500^0C the dislocation density is approximately $10^9/cm^2$. It decreases drastically to an order of $10^2/cm^2$ at wafer temperatures higher than 550^0C. This trend was observed only in a region of doses lower than $1.5 \times 10^{18}/cm^2$. At high dose $\sim 2 \times 10^{18}/cm^2$ the dislocation density remains quite high on the order of $10^9/cm^2$ independent of the wafer temperature [8.36].

Formation of dielectric layers is not limited to SiO_2, Si_3N_4 layers have been obtained by nitrogen implantation. The process is quite similar to that of SIMOX. A dose of $\sim 10^{18}/cm^2$ nitrogen at 200 keV energy is implanted into Si at temperatures around 500^0C. In contrast to SiO_2 the layer process does not synthesise stoichiometric Si_3N_4 even if the dose exceeds the critical value. This effect is due to the low diffusivity of nitrogen in nitrogen-rich silicon and the excess nitrogen at the peak of the distribution is trapped-out as nitrogen bubbles [8.37].

Annealing at 1200^0C causes a redistribution of nitrogen and the profiles become more rectangular with a flat top as shown in the channeling spectras of Fig. 8.15. The interfaces between the

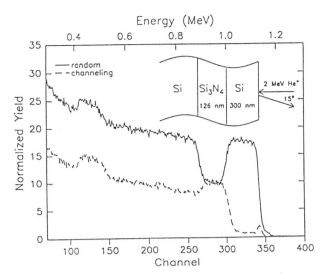

Figure 8.15 *Energy spectra of $2.0 MeV {}^4He^+$ backscattered from a Si sample implanted at room temperature with 180 keV - $2 \times 10^{17}/cm^2 N^+$ ions and annealed at $1200^0 C$ for 2h. The channeling spectrum indicates the good crystallinity of the top silicon layer 300 nm thick. The configuration shown in the insert has been obtained by simulation (Courtesy of G.Galvagno).*

buried layer and the silicon overlayer and the substrate are quite abrupt. The minimum yield is that of a perfect single crystal.

The value of the dose is critical in contrast to the SiO_2 formation. At low doses a single layer of polycrystalline $\alpha Si_3 N_4$ is obtained with huge grains as shown by the TEM cross section of Fig. 8.16a for $7.5 \times 10^{15}/cm^2 - 180 keV N^+$ implant in Si at $530^0 C$. In the case of high dose implant above $1 \times 10^{18}/cm^2$ a quite complex structure has been detected in the cross section TEM analysis (see Fig. 8.16b). It includes a top good Si single crystal, a polycrystalline $\alpha Si_3 N_4$, a porous $Si_3 N_4$, a nitrogen bubbles and amorphous $\alpha Si_3 N_4$, and a defective silicon transition layer. The sample analyzed in Fig. 8.16b was implanted at $530^0 C$ with $1.25 \times 10^{18}/cm^2 - 180 keV N^+$. The post-implant annealing temperatures was $1200^0 C$ - 2h in N_2 ambient.

The synthesis of silicon nitride by implantation occurs through the nucleation and growth of silicon nitride precipitates in regions where the volume concentration of nitrogen equals or exceeds that of the stoichiometric compound. The growth of the precipitates occurs by gettering of nitrogen from the tails of the

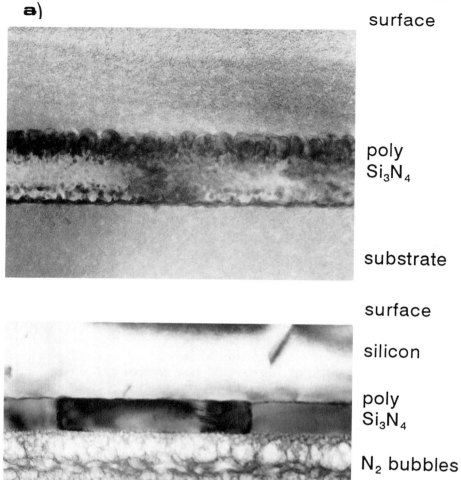

a)

surface

poly
Si$_3$N$_4$

substrate

surface

silicon

poly
Si$_3$N$_4$

N$_2$ bubbles

amorphous
Si$_3$N$_4$

substrate

0.5 um

Figure 8.16 *TEM cross section of a Si sample implanted with* $1.25 \times 10^{18}/cm^2$ - *180 keV* N^+ *(a) and with* $7.5 \times 10^{17}/cm^2$ − $180keV N^+$ *(b) at* 530^0C. *Both samples were annealed at* 1200^0C *for 0.5h. The structure of the different layers is shown in the right-hand side (Courtesy of G.Galvagno and C.Spinella).*

distribution. The excess nitrogen in the peak of the concentration is unable to migrate due to its low diffusivity in nitrogen rich silicon and in silicon nitride. These nitrogen atoms are trapped as bubbles between the two surrounding layers of polycrystalline $\alpha Si_3 N_4$. The diffusivity of nitrogen in silicon is instead high and as the implanted doses are below the critical value all the implanted atoms are consumed by the growing nitride layer gain. The driving force for both SiO_2 and SiN_4 layer formation is the large gain in free energy due to the large heat of formation of the compounds.

8.7 Devices in SOI Substrates

The fabrication of devices, mainly FET, using silicon layer on insulators was driven initially by space and military applications. Recently a wide range of potential new high performance integrated circuits is designed in SOI materials. For this reason the present section will illustrate briefly this topic, being again the base material related to ion implantation. Among the several methods adopted to fabricate SOI we have considered in the previous section the SIMOX procedure. There is another method that deserves at least a brief presentation, it is the bonded process. An oxide layer of the desired thickness $(0.25 - 1.0 \mu m)$ is grown on a standard silicon wafer. The wafer is then bonded at high temperature to another wafer, with the oxide sandwiched between. One of the wafers is then ground to a thickness of a few microns by a mechanical tool such as grinding and polishing and then by more complex procedure as chemical and plasma etch [8.38] to maintain silicon film thickness and uniformity standards. Film thickness of about $1 \mu m$ is suitable for bipolar transistor, but thinner for CMOS.

Thin silicon layers $(< 50 nm)$ allow the fabrication of fully depleted FET (see Fig. 8.17), i.e. of devices whose region under the gate is depleted of carriers in the depletion mode. These devices as we will illustrate present some advantages, but they need thin layers. Implants at high dose $(\sim 2 \times 10^{18}/cm^2)$ and energy (150-200 keV) of oxygen form ~ 200 nm thick Si layer that must be thinned to fabricate fully depleted FET. Recently [8.39] thin oxide buried layers and surface silicon layers have been obtained by the use of industrial available implanters on 6" or 8" wafers. Typical process is done at 30 keV - $2.2 \times 10^{17}/cm^2$ oxygen dose. The wafer temperature during the implant is $530^0 C$ and the subsequent annealing is performed at $1300^0 C$ - 6hr in nitrogen

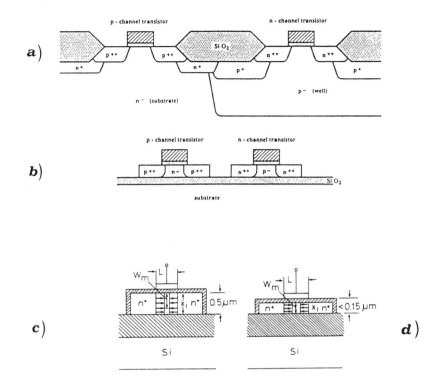

Figure 8.17 *(a) Schematic representation of a conventional CMOS structure built on a Si substrate, (b) on SOI substrate, (c) non-fully depleted mode, (d) fully-depleted mode of operation.*

containing a small amount ($\sim$ 0.25%) of oxygen. The sample is heated by high intensity heater lamp during the implant. The buried SiO_2 layer results 44 nm thick and the Si surface layer $\sim$ 50 nm thick.

As in all the other procedures a compromise must be found between dose, implant temperature and annealing temperature to optimize the quality of the oxide layer, the thickness and the amount of residual defects [8.40]. Threading dislocations with a density of $10^3/cm^2$ are still the major source of defects. Pinholes in the buried SiO_2 layer are present at a density as low as $0.1\mu m/cm^2$. The SIMOX process requires a quite specified and dedicated control of impurities due to the very high dose of implant. Heavy metals such as Fe, Cu, Ni and light elements such as Al and C can be introduced during implant and anneal procedures should be kept at low temperatures. The placement of Si shields inside the implanter reduces the Al contamination in the $10^{15} - 10^{14}/cm^3$ range. Carbon is reduced through a specialized pump - down and cleaning procedures in the implanter. Being a long time process of the order of several hours per batch of wafers particulate contamination must be carefully controlled.

The SOI oxide layer prevents metallic contaminations from being removed by the usual back-side gettering process. The lack of internal "gettering" regions in the material structure requires the reduction of impurities by 1 to 2 orders of magnitude compared to bulk wafers. The up-to date technology allows the fabrication of SIMOX substrates for VLSI production at industrial level. The SIMOX process is compatible with the standard fabrication line equipment and procedures. The high cost of SIMOX wafers is partially offset by the ability to eliminate several masking and implant steps during device fabrication.

The early use of SIMOX was for the fabrication of radiation hard devices for military applications mainly. Now it is also of commercial relevance for large scale production of MOSFET. As discussed in sect. 7.7 the soft errors caused by the ionizing radiation emitted from the packaging material are a serious problem for the 64 and more for the 256 Mbit DRAM and for any other high density memory based on MOS devices. SIMOX offers a natural immunity to this single event up set. The volume available to the ionization process is considerable reduced (see Fig. 8.17b). Only the depletion in volume bounded by the gate, buried oxide and junction edges is susceptible to charge collection. The MOSFET are separated by the insulating dielectric layer which serves also as protection against the vertical contact "spiking" through very shallow junctions which would otherwise occur as CMOS is scaled in the submicrometer regime.

The drastic reduction of p-n junction depletion volume reduces substantially the leakage current limited only at the much smaller vertical drain/channel junction. As matter of fact an overall reduction of the junction area of 3-4 orders [8.41] of magnitude is obtained and the devices can operate at high temperature ($\sim 300^0 C$). This characteristic is of relevance for automotive and aerospace applications [8.42]. In addition the stand-by current is reduced as the power consumption in both active and static mode.

Figure 8.17 compares a conventional CMOS structure (a) built on a Si substrate with a MESA isolated CMOS structure

(b) built on SOI substrate. The latter reduces inter-well (n^+ to p^+) design distance to less than $1\mu m$. The LOCOS (local oxidation of Si) step is eliminated and the packing density is increased. The active area with SOI construction is only the small cross section lying directly below the gate of the transistor. The device is characterized by a natural immunity to latch-up and by a less relevance of the drain-induced barrier lowering. These advantages would allow scaling of MOSFET to continue towards the 1 Gbit generation with design rule of $0.1\mu m$ [8.43].

Because the structure of Fig. 8.17b does not have a substrate contact, there is no direct access to the majority carriers in the body of the device. As a result when the external contacts deplete the substrate, the majority carriers cannot leave the device except by recombination or diffusion into the junctions. When the drain bias is large enough to cause e-h pair creation, the generated majority carriers accumulate until a balance is reached between generation and recombination, and the device body can become charged due to excess carriers. The substrate "floats" to a potential dependent upon the number of stored majority carriers; this potential is determined by internal charge storage and current and is not controlled directly by external bias. In this case the MOSFET built on SIMOX operates in the non-fully depleted mode (Fig. 8.17c) [8.44]. The depletion width W_M is higher than the thickness d of the silicon thin surface layer.

In the fully depleted mode (Fig. 8.17d), $W_M < d$, no charge neutral region exists and the device behaviour depends on both the top and bottom interface. This two-sided behaviour lowers the electric fields inside the device, tending to reduce hot-electron effects, drain-induced barrier lowering (see sect. 7.4), short-channel and to increase the driving ability of the device [8.45].

The surface Si layer must be thin enough for the entire region under the gate to be depleted in the fully-depletion mode. The depletion width under these conditions is determined by the thickness of the Si layer. The threshold voltage becomes dependent on the thickness of the surface Si layer and strict control thickness is needed for tight device parameter control.

Several deleterious effects that degrade the performance of MOSFETS in the submicron region are in part or totally eliminated using SIMOX structures. The shrink of MOS in bulk Si is done usually not at scaled supply voltage for constant electric field and higher doping concentrations are needed to avoid punchthrough with an increase of the internal electric fields. At high fields the carrier velocity saturates and the mobility decreases. As a consequence the transconductance, i.e. the change of drain current to gate voltage change, degrades together with the switching speed. Both quantities are proportional to the mobility. These effects are reduced in both non-fully and fully depleted mode MOSFETS fabricated on SIMOX structures.

Channel doping can be lowered (up to $10^{14} - 10^{15}/cm^3$) independently of the channel length and the lateral electric field drain to source is also reduced. The mobility increases and a much higher drive current flows in these fully depleted devices. The combination of increased transconductance and reduced junction capacitance leads to a much higher switching speed than in comparable bulk Si devices. Under full depletion through the lightly doped Si surface layer the transistors become very sensitive to changes at the "back interface" and oxide. The presence of two Si/SiO_2 interfaces in the active region increases the process sensitivity of the film because two interfaces need to be passivated and controlled. Additional thermal processing steps are needed to anneal the damage at the interface. The buried oxide must be of very high quality and thickness uniformity. Again the improvment of implant and annealing procedures will determine the future of large scale industrial applications of SIMOX structures for MOSFET fabrication. We have just illustrated the fabrication of MOSFET, of course bipolar transistors can be also built on SOI substrates, but they will not considered here.

8.8 Buried Metal Layer Formation

Not only dielectric but also metallic layers have been synthesized by ion beam. The most investigated material has been so far $CoSi_2$ although several other compounds such as $NiSi_2$ and $FeSi_2$ have been formed [8.46]. Silicides are good conductors with high mechanical and thermal stability, are stable in oxidizing atmosphere and have several applications in microelectronics (see chapter 6). Some silicides can be formed epitaxially by the metal deposition on a carefully cleaned Si surface and by a subsequent thermal annealing. The epitaxial growth requires a small lattice mismatch of the compound with that of the silicon substrate $\cdot CoSi_2$ is one of the best silicides that grows epitaxially on both (110) and (100) silicon wafers. It has the lattice structure of CaF_2 and a lattice constant of 0.5364 nm, 1.2% lower than that of silicon.

The first epitaxial silicides to be synthesized with ion implantation was infact $CoSi_2$. The buried $CoSi_2$ layer was fabricated by implantation of 200 keV Co into (100) Si with dose of $3 \times 10^{17}/cm^2$ at a nominal wafer temperature of 350^0C and subsequent anneal at $600^0 C$ for 1h and $1000^0 C$ for 1/2h. The RBS analysis of the as-implanted sample is shown in Fig. 8.18a [8.46].

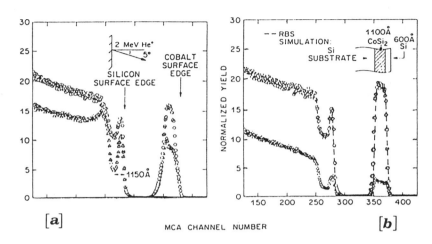

Figure 8.18 *RBS spectra for a 200 keV, $3 \times 10^{17}/cm^2 Co^+$ implant before (a) and after (b) annealing at $600^0 C$ - 1h and $1000^0 C$ - 0.5h. The dotted line is the result of a RBS simulation for the configuration shown in the insert (from ref. 8.46).*

The random spectrum exibits a nearly Gaussian Co distribution and a chip in the Si signal at the depth where most of the Co is located. The channeling spectrum indicates some degree of crystalline order. The two-step annealing changes dramatically the distribution. The Co signal in the random spectrum of Fig. 8.17b has become rectangular in shape with a height corresponding to the stoichiometric composition of $CoSi_2$. The thickness of the top Si layer was 60nm and that of the buried $CoSi_2$ layer 110nm. The low channeling yield indicates a good crystallinity of both the top Si layer and the buried silicide layer. This last is aligned with the substrate, with abrupt and smooth interface. This process was named mesotaxy, to distinguish it from epitaxy, which means ordered arrangement on the surface.

The synthesis of buried silicides by heavy metal-high dose implantation is limited, in principle, by the sputtering (see chapter 3). If each incoming ion sputters away more than one of the substrate atoms (the sputtering yield is 1:5-2 for Co in Si), eventually the implanted ions themselves will be eroded, limiting the retained dose. Initial experiments on high dose implants of transition metal at RT indicated the formation of a broad impurity profile, limited by the sputter. If the implant is performed at elevated temperatures amorphization of the target is prevented, and the metallic impurities become mobile in the Si crystal. The energy loss of the incoming ion is increased due to the presence of substantial Co in the Si and it acts to reduce the range of Co ions and keep the profile narrow.

It has been found that for the synthesis of buried $CoSi_2$, as for the other compounds, the Co peak concentration should be higher than 18%, named "the threshold peak concentration". At energy of 200 keV ion doses exceeding $10^{17}/cm^2$ are necessary. At the critical dose for the formation of a silicide layer of sto-ichiometric composition is $1.8 \times 10^{17}/cm^2$. At the critical dose the Co concentration is 19%. The critical value depends on the implantation and annealing conditions [8.47].

If the implanted dose exceeds the threshold value ϕ_{crit}, nearly all the implanted atoms will be found in the synthesized layer after high-temperature annealing. If the sputtering of the implanted atoms is negligible, as it is except for very high doses where surface layers are produced, the thickness d of the buried layer is given by

$$d = \frac{\phi}{N_M}, for \ \phi > \phi_{crit} \qquad 8.4a$$

where N_M denotes the atomic density of the metal atoms in the silicide. Using a density of 4.95 gcm^{-3} for $CoSi_2$, d becomes

$$d = \left(\frac{\phi}{2.59}\right) \times 10^{-15} nm \qquad 8.4b$$

The calculation of the implanted profile is not straighforward because at high dose implantation the following effects have to be taken into account:

    i) continuous change of the chemical composition of the target
    ii) surface erosion by sputtering
    iii) redistribution of implantation profile due to the compound formation for doses in excess of the threshold value.

Several attempts have been made to take into account these effects in the most common simulation code, such as TRIM. The formation is a quite complex phenomenon because it depends on the implantation temperature T_i, annealing temperature T_a and the implantation dose ϕ [8.48].

The evolution of the microstructure in (100) and (111)Si is illustrated schematically in Fig. 8.19 [8.48]. At $T_i < 350^0C$ the sample becomes amorphous and the annealing produces non-uniform, non-single crystal layers. In (100)Si rombohedral and

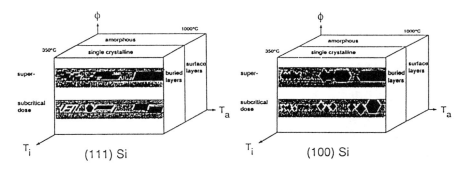

Figure 8.19 *Schematic illustration of precipitate growth and mesotaxial layer formation for (100) and (111) Si as a function of implantation temperature Ti, implantation dose ϕ and annealing temperature T_a. The formation of a buried of continuous layer epitaxial layer requires a supercritical dose, a temperature above the amorphization threshold, and an annealing temperature around $1000^0 C$ (from ref. 8.48).*

dodecahedra-shaped precipitates are formed while in $(111)Si$ plate-like precipitates are observed. The growth and coalescence of precipitates during annealing depends on the volume fraction of the precipitates at the maximum of the implantation profile. Only for peak concentrations above 18% the $CoSi_2$ precipitates form a connected array before the high temperature anneal. For subcritical doses coalescence of precipitates does not occur and thus layer and isolated precipitates growth. Planar layer formation is observed if precipitate coalescence occurs at an early stage before the final annealing. In summary the three requirements are, T_i above the amorphization threshold, $\phi > \phi_{crit}$ and $T_a \sim 1000^0 C$ to complete the layer formation and to eliminate the lattice defects.

The process of silicide formation is quite different in mesotaxy as compared to solid phase epitaxy. During mesotaxy the nuclei are formed under extreme non-equilibrium condition from a supersaturated solid solution rich in silicon. The implanted atoms must diffuse over short distance to form a supersaturated solid solution rich in silicon. The implanted atoms must diffuse over short distance to form nuclei and the diffusion is enhanced by the damage defects. As a consequence the barrier kinetic is lowered and precipitates of the silicon rich phase ($CoSi_2$) form already during implantation. The subsequent evolution implies the coarsening of the precipitates and a spatial arrangment in a nearly two-dimensional percolation network at the depth where the layer is evolving.

The buried $CoSi_2$ silicide develops rough interfaces at 1100^0C, the complete disintegration of the layer and the formation of islands is observed at 1200^0C.

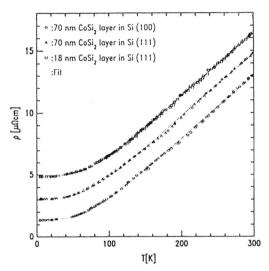

Figure 8.20 *Specific electrical resistivity versus temperature for $CoSi_2$ layers with different orientations and thickness. The solid lines are fits of the Bloch-Grüneisen formula (from ref. 8.49).*

The electrical characteristics of the buried $CoSi_2$ layers are even superior to those of conventionally grown films [8.49]. The electrical resistance of buried $CoSi_2$ layers in (111) and (100) silicon substrates is shown in Fig. 8.20 as a function of temperature. (100) layers have usually lower values than comparable (111) layers. The resistivity depends on the thickness of the layer, the curves are parallel, indicating that the difference lies mainly in the residual resistivity. The trends of the curves vs temperature is typical of a metal and can be described by the Matthiesen's rule

$$\rho(T) = \rho_0 + \rho_s(T) \qquad 8.5$$

where ρ_0 is the temperature-independent contribution due to scattering processes by defects and impurities and $\rho_s(T)$ the phonon contribution. The elastic scattering lenght e_a, given by

$$e_a = mv_F/n_n e^2 \rho_0 \qquad 8.6$$

where m is the effective mass of the carriers, v_F the Fermi velocity, and n_n the electrical carrier density. Using $n_n = 2.5 \times 10^{22}/cm^3$, and $\rho_0(4.2K) = 1.0\mu\Omega$, e_a becomes $\simeq$ 145 nm and 14 nm at 4.2 K and 290 K respectively. In $CoSi_2$ layers prepared by ultra high vacuum deposition and annealing the e_a values are lower at 4.2 K but comparable at 290 K. The best quality of the synthesized film may be due to the inherent cleanliness of the ion implantation process. Another possibility is that the composition of the buried layer is closer to true stechiometry, because formation occurs in a Si-rich environment without much mass transport. The high temperature anneal ($\sim 1000^0C$) may improve further the crystallinity of the buried layer, but this high temperature annealing cannot be done in the deposited layer because they tend to break apart. Hall effect measurements have been performed in those buried $CoSi_2$ layers. The most important result of these measurements is the positive Hall coefficient, indicating that holes are the dominant carrier species. The Hall coefficient allows also in combination with the resistivity measurement the evaluation of the hole density that results $2.5 \times 10^{22}/cm^3$. This value was used in eq. 8.6 to determine the elastic mean free path l_e of the electric carrier. By comparing this value with the unit-all density of $CoSi_2(2.59 \times 10^{22}/cm^3)$ it results that one hole per unit cell contributes to the carrier density [8.50].

8.9 Compound Semiconductor Based Devices

$GaAs$ is an ideal material to build discrete device and integrated circuits that operate at high speed. The effective mass of the electron is 4 times lighter that in silicon. The electron mobility in a $GaAs$ JFET n-channel is one order of magnitude higher than that of electrons in n-channel Si MOSFET. The saturation velocity of electrons in $GaAs$ is $5 \times 10^7 cm/sec$ to compare with $10^7 cm/sec$ for Si. The hole mobilities in $GaAs$ and in Si are instead comparable (400 $cm^2/V \cdot s$ for $GaAs$ and 480 $cm^2/V \cdot s$ in Si). For these reasons in all the FET devices fabricated with $GaAs$ the current is transported by electrons. The large band gap energy (1.4 eV at RT) allows the $GaAs$ devices to work in wide temperature range from 200^0C to 200^0C. The main limitations are the absence of a stable oxide, such as SiO_2 in Si and the lack of impurity diffusion process to form bipolar transistor. For these reasons the basic structure of $GaAs$ device is the MESFET, so called Metal - Semiconductor - Field - Effect - Transistor. The acronym indicates that a metal - semiconductor junction is used,

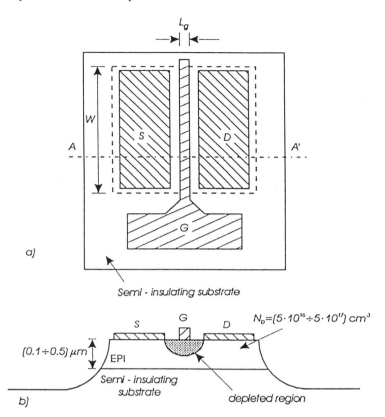

Figure 8.21 *Schematic of a GaAs MESFET isolated by a mesa structure (a), cross section of the device along the line AA' (b).*

i.e. a Schottky barrier.

An old and simple MESFET in *GaAs* is shown schematically in Fig. 8.21 with a MESA structure. The substrate is semi-insulating *GaAs* and on it a thin layer of lightly-doped n-type *GaAs* is grown epitaxially, to form the channel region of the FET. The electron concentration ranges between $5 \times 10^{16}/cm^3$ and $5 \times 10^{17}/cm^3$. The flux of the channel current between source and drain is limited to the width of the region between the depleted zone of the reverse-biased Schottky barrier and the semi-insulating substrate. The device works in the depletion mode and is called D-MESFET [8.51].

The photolitographic processing consists of defining patterns in the metal layers for source and drain ohmic contacts (e.g. $Au - Ge$) and for the Schottky barrier gate (e.g. Al or $Ti/W/Au$).

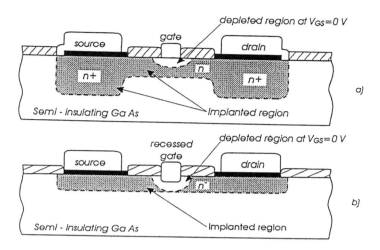

Figure 8.22 *Schematic of depletion-mode or D-MESFET (a) and of enhancement mode or E-MESFET. The semiinsulating layer is doped by ion implantation.*

By reverse biasing the Schottky gate, the channel can be depleted down to the semi-insulating substrate and the resulting I-V characteristics are similar to that of the MOSFET device. Since no diffusion is involved in the fabrication process close geometrical tolerances can be achieved and the MESFET can be made very small. Gate lengths, lower than one micrometer are common in these devices. The devices are isolated each other by etching part of the epitaxial layer up to the semiinsulating substrate [8.52].

    This structure presents several disadvantages. The uniformity and the reproducibility of the epitaxial layer thickness is not of easy control, the threshold voltage changes then with the process and the yield is reduced. The speed of the device is limited by the parasitic capacitance of the walls in the MESA structure. These and other disadvantages were eliminated by the use of ion implantation.

    The simplest structures of ion implanted MESFET are shown in Fig. 8.22a and 8.22b for depletion mode device and for an enhancement mode device respectively. In the enhancement mode the MESFET is normally "off" when gate is biased at 0 volt, and "on" when the gate is positive biased. The thickness of the channel layer and the doping should be chosen in such a way that the built-in voltage of the Schottky barrier is such to deplete completely the channel. The n^+ and n^- regions are usually obtained by Si or Se

implants, using the procedures described in sect. 8.2.

For istance low noise MESFET are fabricated in semiinsulating material, resistivity higher than $10^{17}\Omega \cdot$ cm by two implants of $^{28}Si - 40keV - 1 \times 10^{13}/cm^2$ and $100keV \cdot 5 \times 10^{12}/cm^2$. Both conventional furnace ($820^0C\cdot20$ min) or RTA ($850^0C - 30s; 900^0C$ - 10s) process is used to anneal the damage and to activate the dopant. The channel region is very thin ($< 0.1\mu m$) but heavily doped ($n > 10^{17}/cm^3$). This optimize the transconductance and decreases the gate voltage for pinch-off. In power - MESFET with a breakdown voltage between gate and drain higher than 20 V the $5 \times 10^{12}/cm^2$ - 100 keV implant is replaced by a $5 \times 10^{12}/cm^2$ - 300 keV implant. The maximum current flow between source and drain cannot be reduced and it is necessary below the active region to dope at a concentration of about $2 \times 10^{17}/cm^3$ a relatively thick layer ($d_p \sim 0.3\mu m$). This is realized by the high energy implant. Fig. 8.23 [8.53] illustrates schematically the process (SAINT) of self-aligned implanted n-type transistor fabrication.

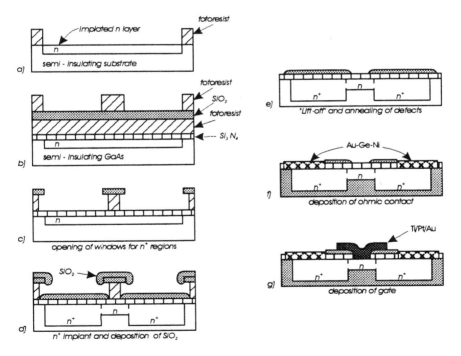

Figure 8.23 *Sequence of processing-steps for the fabrication of self-aligned implanted n-type enhancement mode - MESFET (from ref. 8.53 -* [(c)]*1982 IEEE).*

It makes use of selective etching, lift-off on a multilayer structure resist $/SiO_2/$ resist to fabricate the T shaped mask for the implant. The development of $GaAsIC$ has also required some improvements in the ion implantation technique. For istance planar channeling should be avoided otherwise the threshold voltage cannot be controlled accurately over the devices built on different regions of the wafer. This can be achieved by carring out the implant with a parallel-scanned system on wafer which is tilted 7^0 - 10^0 from the normal direction to prevent the axial-channeling effect and also rotated $10^0 - 30^0$ from the fundamental axis to prevent planar-channeling effect. The FET channel layer is made by $^{28}Si - 6 \times 10^{12}/cm^2$ - 60 keV implant in In-doped LEC $GaAs$ wafer. The samples are covered with 150 nm Si_3N_4 plasma enhanced CVD and then annealed at 800^0C for 20 min in flowing N_2 gas. A second step to improve the uniformity of $V_{..}$ is the formation of a buried p-layer by Be implant at an energy of 70 keV and a dose of $7 \times 10^{11}/cm^2$. The n^+ were realized by 120 keV $Si - 4 \times 10^{13}/cm^2$ implant through the 150 nm Si_3N_4 mask. Faster $GaAsIC$ are fabricated by a thinner channel layer with high carrier concentrations. Several processes can be used: low energy implantation and RTA, SiF_3 molecules ion implantation at low energy, same as BF_3 for shallow junction in Si and RTA.

Isolation between the devices is now performed with oxygen or proton implantation, as shown in Fig. 8.24.

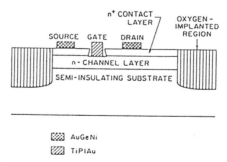

Figure 8.24 *Schematic of implant - isolated GaAs MESFET.*

This process increases the backgate sensitivity of the device. A negative charge is sometime present at the interface channel layer/semi-insulating substrate, a space-charge region results in the channel with a decrease of its width and with a modulation and a reduction of the source-drain current. The effect is called "backgating" because it is analogus to what happens in the presence of a reverse-bias or negative potential on a nearby electrode [8.54].

Implantation of protons or oxygen increases the number of traps to be filled by the current and extends the backgating threshold to voltages above those normally used in MESFET. The leakage current between the gate and an ohmic contact through an isolated layer is several orders of magnitude lower for an ion-implanted-insulated material than for a semi-insulating $GaAs$ material.

Band gap of a III-V compound, such as $Al_x Ga_{1-x} As$ changes with the composition of the alloy. Molecular beam epitaxy and metal organic chemical vapor deposition allows the fabrication of heterostructure, i.e. thin layers of different composition. Due to the change of the band gap, heterojunction is formed with two such layers as $AlGaAs/GaAs$. Let us consider the behaviour of an heterojunction FET, [8.55] made by a heavily doped, wide-band gap $AlGaAs$ grown on an epitaxial layer of high-impurity, high mobility $GaAs$. The electrons associated with the donors in the $AlGaAs$ transfer in the $GaAs$ and become trapped there by the band bending (see Fig. 8.25). The outer heavily doped $AlGaAs$ layer is separated from the $GaAs$ by an undoped $AlGaAs$ spacer layer 10 nm thick.

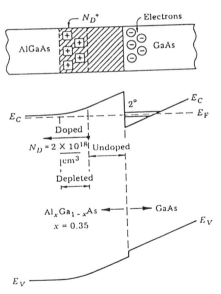

Figure 8.25 *Energy-band diagram on AlGaAs-GaAs hetero-junction, showing the formation of an electron gas at the interface (from ref. 8.55 $^{(c)}$ 1984 IEEE).*

Since the donors are in the $AlGaAs$ rather than in the $GaAs$ layer, there is no impurity scattering of electrons in the well. The electrons in the channel form a high-electron mobility transistor (HEMT). This configuration is called "modulation doped". The energy band discontinuity is the key to the spatial separation of the electrons and donors as clearly indicated in Fig. 8.24. The transport of electrons in the channel is now dominated by lattice vibrations. At room temperature the mobility is a factor two higher than in impurity-doped $GaAs$. At 77K (LN_2T) the electron mobility can exceed $10^5 cm^2/v.s.$ The cross section of a conventional HEMT is shown in Fig. 8.25. The top $n^+Al_{0.3}Ga_{0.7}As$ layer is 45 nm thick, the following intrinsic $Al_{0.3}Ga_{0.7}As$ layer 3nm [8.56].

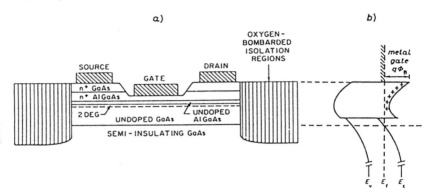

Figure 8.26 *Schematic of typical HEMT structure (a), and of band structure (b) (from ref. 8.56).*

The depletion region under the gate extends through the $AlGaAs$ layer. Planar fabrication of HEMTs for circuit is possible using implant isolation rather than mesa isolation. Implant of high dose oxygen in the $n^+AlGaAs$ layer creates a thermally stable high-resistivity region. Other multienergy oxygen implants form a uniform damaged material all the way to the substrate to isolate neighboring devices from each other.

Heterojunction bipolar transistors (HBT) have an higher transconduttance and excellent threshold-voltage control. Because the flow of the current is vertical the speed of the device depends on the thickness of the base region instead of the gate lenght as in FET. A typical HBT structure is shown in Fig. 8.27. The emitter is $n-AlGaAs$, the base $p-GaAs$ and the collector $n-GaAs$. The heavily doped n^+ layer in the emitter is used for ohmic contact to the $AuGeNi$ metallization skeme, while the n^+GaAs layer reduces the collector resistance. The isolation is provided by oxygen and proton implantation. The advantage with the use of oxygen is that the collector is usually doped in the $(3\text{-}5)\times10^{16}/cm^3$ range, and so thermally stable compensation can be achieved.

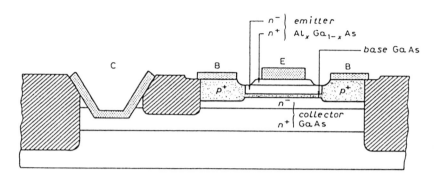

Figure 8.27 *Schematic of typical heterostructure bipolar transistor. The emitter is formed by AlGaAs, the base by $p - GaAs$ and the collector by $n - GaAs$.*

This allows the selective Be implants for base contacts and the oxygen implants for the reduction of the extrinsic base-collector capacitance to be performed sequentially followed by an anneal to activate the Be and to promote the oxygen onto a substitutional site. The doping in the p^+ layer is not effected by the isolation implant.

HBT operates at $25GH_Z$ and using a process incorporating $AlInAs$ and $GaInAs$ on an InP substrate it reaches $48GH_Z$. The speeds are possible primarily because of the high current gain inherent to the bipolar transistor structures. High current gain allows faster charging and discharging of the base, interconnect and other stray capacitances in the chip.

Many other devices based on $GaAs$ and on InP have been beneficied by ion implantation, e.g. impatt diodes, heterostructure lasers and optical waveguides, either as dopants or isolation.

$CoSi_2$ buried layer could be applied for metal-base and permeable-base transistors, [8.58] as a buried collector contacts for high speed bipolar transistor or as buried interconnects for ultrahigh-speed integrated circuits. The small thickness of the top silicon layer is not a limitation for device fabrication because it can serve as a seed layer for epitaxial Si overgrowth. Depending on the implantation energy, the thickness of the top layer ranges from 10 to 90nm for 20 to 200 keV implantations. The surface is cleaned chemically after implantation and the thermal stability of the buried layer allows heating above 800^0C to remove under UHV conditions the thin native SiO_2 layer. The Si overlayer is deposited by low pressure vapor phase epitaxy (LPVPE) at 800^0C

or by MBE.

Metal-base or metal gate transistor is made by a semiconductor-metal- semiconductor sequence. The thickness of the continuous metal layer should be thinner than the electron mean free path ($\sim$ 7 nm at room temperature) to enable ballistic carrier transport. The transistor works with hot-electrons, i.e. with electrons at energy greater than the Fermi level in the metal and at energy KT/q or more above the conduction-band edge in semiconductor. In the hot-electron transistors, an excess of energetic electrons is produced by acceleration across the semiconductor-emitter-metal-base barrier. For a silicon emitter with a $CoSi_2$ base, the electrons have approximately 0.7 eV of excess energy above the metal Fermi level. If the hot electrons impinge on a collector barrier of 0.7 eV or less before having an electron collision, they may be collected.

These transistors have the potential to give better microwave performance than the bipolar, but are limited by the low current gain. Quantum-mechanical reflection of electrons at the barrier and phonon scattering reduce the gain. It must be pointed out that replacing the thin metal base by a heavily doped semiconductor layer the δ-doped heterojunction bipolar transistor (HBT) is formed. By increasing the base layer thickness and using identical semiconductor for the base and the emitter, the classic bipolar transistor is obtained.

The buried metal layer in the MBT should be not only thin by also free of pinholes. So far the successful fabrication of a metal-base transistor even with $CoSi_2$ material has not been achieved. Different is the case for the fabrication of permeable-base transistor (PBT) with buried $CoSi_2$ gate. In the PBT the metal base is not uniform, it consists of a grid of submicron stripes and the current flows preferentially through a large number of thin portions. PBT is in principle a vertical MESFET [8.58] with the unique feature of the submicrometer periodicity of Schottky-barrier greating that modulates the vertical flow of electrons from the source to drain. PBT is a useful device for high speed digital logic circuits and for high frequency power amplifiers or oscillators.

The fabrication step of permeable-base transistor using ion beam synthesis are shown schematically in Fig. 8.28. The 200 KeV Co^+ implantation at $2 \times 10^{17}/cm^2$ dose was performed through layered metal/SiO_2 mask into a 420nm thick epitaxial layer (n-doped $4 \times 10^{16}/cm^3$) on n-type $4m\Omega \cdot cmSi(100)$. The metal mask was removed before the annealing at $750^0C - 30s$ and $1150^0C - 10s$. The overgrowth of $Si(\sim 820nm$ thick) was made by LPVPE, the metallization of the top contact was

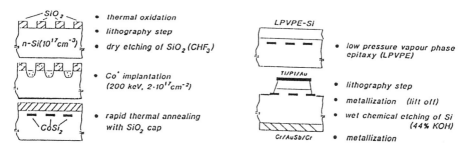

Figure 8.28 *Processing steps for a permeable - base transistor (from ref. 8.59).*

$Ti/Pt/Au$ alloy. The gate contact was fabricated by mesa-etched by reactive ion etching. The bottom contacts ($AuSb/Cr$) were evaporated [8.59]. The gate spacing was $1.0\mu m$.

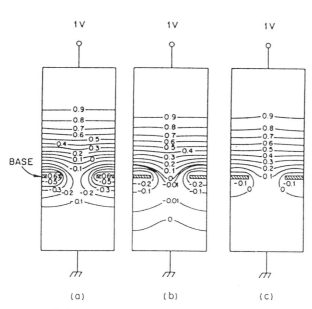

Figure 8.29 *Calculated equipotential lines in the unit cell of a permeable-base transistor shown in cross section for three different base-bias conditions. (a) $V_{BE} = 0V$, (b) $V_{BE} = 0.3V$, (c) $V_{BE} = 0.5V$, (from ref. 8.60 -* [(c)]*1979 IEEE).*

Figure 8.29 shows the calculated equipotential lines in a unit cell shown in cross section for three different bias conditions [8.60].

If the collector is positive biased and the emitter and the base are at zero bias, the electrons, travelling from the emitter to the collector, encounter a negative potential barrier (Schottky-barrier) and then have to pass over an energy barrier. The barrier is lower in the center of the opening between the $CoSi_2$ layer and highest at the $CoSi_2/Si$ interface (see Fig. 8.29a). The large barrier causes a small current at the collector. If the base is forward biased (Schottky barrier forward biased), the barrier height decreases and a large current flows to the drain (Fig. 8.29b). At larger bias, carrier accumulation develops and the current is now space-charge limited (Fig. 8.29c).

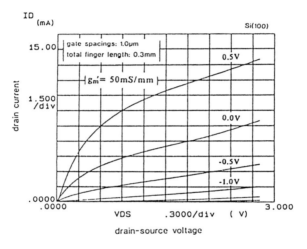

Figure 8.30 *Output - characteristics of a $Si/CoSi_2$ (100) permeable - base transistor (from ref. 8.61).*

The electrical characteristics of the permeable-base transistors fabricated with buried $CoSi_2$ are equivalent or even superior to the best PBT fabricated with the overgrowth of Si on superficial epitaxial $CoSi_2$ layer. The output characteristic of such a Si (100) PBT is shown in Fig. 8.30. The maximum transconductance exceeds 50ms/mm at a drain-source voltage, V_{DS}, of 2V and a gate voltage, V_G, of 0.55 V [8.61].

SELECTED REFERENCES

Ion Implantation

Carter, G., W.A. Grant *Ion Implantation of Semiconductors*, London 1976

Dearneley, G., J.H. Freeman, R.S. Nelson, J. Stephen *Ion Implantation Technology* North Holland, Amsterdam, 1973

Hirvonen, J.K. *Ion Implantation* Academic Press, New York 1981

Mayer, J.W., L. Eriksson, J.A. Davies *Ion Implantation in Semiconductors* Academic Press, New York, 1970

Ryssel H., I. Ruge *Ion Implantation* John Wiley & Sons, Chichester 1986

Townsend, P.D., J.C. Kelly, N.E.W. Hartley *Ion Implantation, Sputtering and their Applications*, London, 1976

Williams, J.S., J.M. Poate eds. *Ion Implantation and Beam Processing*, Academic Press, New York (1984)

Ziegler, J.F. editor *Handbook of Ion Implantation Technology* North Holland Amsterdam, 1992

Ziegler, J.F. editor *Ion Implantation: Science and Technology* Academic Press, Orlando Florida, 1984

Solid State Devices

Baliga, B.J., *Modern Power Devices* Wiley, New York, 1987

Butcher, P.N., N.H. March, M.P. Tosi *Crystalline Semiconductor Materials and Devices* Plenum Press, New York, 1986

Ghandhi, S.K. *VLSI Fabrication Principles*, Wiley, New York, 1983

Levy, R.A. editor *Microelectronic Materials and Processes*, Kluwer, Dordrecht, The Netherlands, 1989

Mayer, J.W., S.S. Lau *Electronic Materials Science for Integrated Circuits in Si and GaAs* Macmillan Publishing Company, New York, 1990

Muller, R.S., T.I. Kamins *Device Electronics for Integrated Circuits*, second edition, Wiley, New York, 1986

Murarka, S.P., M.C. Peckerar *Electronic Materials Science and Technology* Academic Press, London, 1989

Sah, C.T., *Fundamentals of Solid State Electronics* World Scientific, Singapore 1991

Streetman, B.G., *Solid State Electronic Devices* third edition, Prentice - Hall Englewood Cliffs, N.J., 1990

Sze, S.M. *Semiconductor Devices, Physics and Technology* Wiley, Singapore 1985

Sze, S.M. eds, *VLSI Technology* 2nd edition Mc Graw-Hill, New York, 1983

Wolf, S., R.N. Tauber *Silicon Processing for the VLSI Era* Lattice Press, Sunset Beach, California 1986

REFERENCES

CHAPTER 1

[1.1] Dataquest, December 1993

[1.2] Sah, C.T., *Fundamentals of Solid State Electronics* World Scientific, Singapore 1991

[1.3] Tiwari, S., *Compound Semiconductor Device Physics* Academic Press, San Diego, Ca. 1992

[1.4] Baliga,B.J., *Modern Power Devices* Wiley, New York, 1987

[1.5] Beadle, W.E., J.C.C. Tsai, R.D. Plummer eds. *Quick Reference Manual for Silicon Integrated Circuit Technology* Wiley, New York, 1985

[1.6] Shockley, W., *Proc. IRE* **40**, 1289 (1952)

[1.7] Streetman, B.G., *Solid State Electronic Devices* third edition, Prentice - Hall Englewood Cliffs, N.J. 1990 ch.3

[1.8] Sze, S.M. *Semiconductor Devices, Physics and Technology* Wiley, Singapore 1985 ch. 1 and 2

[1.9] Muller, R.S., T.I. Kamins *Device Electronics for Integrated Circuits*, second edition, Wiley, New York, 1986 ch. 1

[1.10] Baliga,B.J., ref. [4] ch. 3

[1.11] Milnes,A.G., *Deep impurities in semiconductors* Wiley, New York, 1973

[1.12] Streetman,B.G., ref. 7 ch. 8

[1.13] Sze,S.M., ref. 8 ch. 5

[1.14] Barrett,C.R., *Silicon Valley, what next? MRS Bulletin* Vol. XVIII, July 1993, p. 3

[1.15] Beadle, W.E., J.C.C. Tsai, R.D. Plummer eds ref [5] p. 9-15

[1.16] Mayer, J.W., S.S. Lau *Electronic Materials Science for Integrated Circuits in Si and GaAs* Macmillan Publishing Company, New York, 1990 ch. 4

[1.17] Agray - Gurena, J.P.T. Panousis, B.L. Morris *IEEE Trans Electron Devices*, ED-27, 1397(1980)

[1.18] Sze, S.M. eds, *VLSI Technology* 2nd edition Mc. Graw Hill, New York 1983, Chapter II.

[1.19] Shockley, W., *Forming Semiconductive Devices by Ionic Bombardment* U.S. Patent #2,787,564 (Filed Oct. 28, 1954; Granted April 2, 1957)

[1.20] Mayer, J.W., L. Eriksson, J.A. Davies *Ion Implantation in Semiconductors* Academic Press, New York, 1970

[1.21] Ziegler, J.F., editor *Ion Implantation Technology* North Holland Amsterdam, 1992

[1.22] MacPherson, M.R. *Appl.Phys.Lett.* **18**, 502 (1971)

[1.23] Frosch, C.J., L. Derick *J.Electrochem.Soc.* **104**, 547 (1957)

[1.24] Hill, C., P. Hunt *Nucl.Instr.Meth.Phys.Res.* **B55**, 1, (1991)

[1.25] Murphy, B.T., *Proc. IEEE* **52**, 1537(1964)

[1.26] Cowern, N.E.B., K.T.F. Janssen, H.F.F. Jos *J.Appl.Phys.* **68**, 6191(1990)

[1.27] Hill, C. *Nucl.Instr.Meth.* **B19/20**, 348(1987)

[1.28] Moore, G. *Technical Digest 1975* **2**EDM(1975), p.11

[1.29] Dennard, R.H., F.H. Gaensslen, H. Yu, V.L. Rideout, E. Bassons, A.R. Le Blance *IEEE J.Solid State Circuits*, **SC-9**, 256(1974)

[1.30] Brews, J.R., W. Fichtner, E.H. Nicollain, S.M. Sze, *IEEE Electron Devices Lett.* **EDL-1**, 2(1980)

[1.31] Galvagno, G., A. Cacciato, F. Benyaich, V. Raineri, F. Priolo, E. Rimini, S. Capizzi, P. Romano *Mat. Science and Engineering* **B10**, 67(1991)

[1.32] Tsai, M.YY., B.G. Streetman *J.Appl.Phys.* **50**, 183(1979)

[1.33] Celler, G.K., T.E. Seidel, *Applied Solid State Science*, Part C, Kahng ed. Academic Press, New York, 1982

[1.34] Poate, J.M., J.W. Mayer eds *Laser Annealing of Semiconductors* Academic Press, New York, 1982

[1.35] Keyes, R.W. *Contemporary Physics*, **32**, 403(1991); Physics Today, August 1992, p.42

[1.36] Celler, G.K., P.L.F. Hemment, K.W. West, J.M. Gibson, *Appl.Phys.Lett.* **48**, 532 (1986)

[1.37] Jaumi, K., M. Doken, H. Ariyoshi *Electron.Lett.* **14**, 593(1978)

[1.38] Reeson, K.J. *Nucl. Instr. Meth.* **B19/20**, 269(1987)

[1.39] White, A.E., K. T. Short, R.C. Dynes, J.P. Garno, J.M Gibson *Appl.Phys.Lett.* **50**, 95(1987)

[1.40] Lombardo, S., F. Priolo, S.U. Campisano, S. Lagomarsino *Appl. Phys. Lett.* **62**, 2335(1993)

[1.41] Patton, G.L., J.H. Comfort, B.S. Meyerson, D.F. Crabbe, G.J. Scilla, E. De Fresart, J.M.C. Stork, J.Y.-C. Sun, O.L. Harame, J.N. Burghartz *IEEE Electron Dev. Lett* **EDL-11**, 171(1993)

[1.42] Bean, J.C. *Phys. Today* October 1986, pp. 2-8

[1.43] Meyerson, B.S. *Proceedings of IEEE* **80**, n.10(1992)

[1.44] Michel, J., L. Benton, R.F. Ferrante, D.C. Jacobson, D.J. Eaglesham, E.A. Fitzgerald, Y.J. Xie, J.M. Poate, L.C. Kimerling *J. Appl. Phys.* **70**, 2672(1992)

[1.45] Priolo, F., S. Coffa, G. Franzó, C. Spinella, A. Carnera, V. Bellani *J. Appl. Phys.* **74**, 4936(1993)

[1.46] Lombardo, S., S.U. Campisano, G.N. van der Hoven, A. Cacciato, A. Polman, *Appl. Phys. Lett.* **63**, 1942(1993)

[1.47] Shioshansi , P. *Nucl. Instr. Meth.* **B24/25**, 767(1987)

[1.48] *Handbook of Ion Beam Processing Technology* edited by J.J. Cuomo, S.M. Rossnagel and H.R. Kaufman, Noyes Publications, Park Ridge 1989

CHAPTER 2

[2.1] Current, M.I., W.A. Keenan, *Solid State Technology* **28**(2) 139(1985)
[2.2] Benian, D.W., R.E. Kaim, J.W. Vanderpot, J.F.M. Westendorp *Nucl. Instr. Meth.* **B37/38**, 500(1989)
[2.3] Stephens, K.G., in *Handbook of Ion Implantation Technology* edited by J.F. Ziegler, p. 455 - Elsevier 1992
[2.4] Brown, I.G., editor *The Physics and Technology of Ion Sources*, John Wiley & Sons, New York 1989
[2.5] Freeman, J.H. *Nucl. Instr. Meth.* **22**, 306(1963)
[2.6] Simonton, R.B. in *Handbook of Ion Implantation Technology* J.F. Ziegler ed. p.501, Elsevier 1992
[2.7] Aitken, D., in *Ion Implantation Techniques* H. Ryssel, H. Glawischnig eds. p.23, Springer-Verlag (1982)
[2.8] Von Ardenne, M., *Atomkernenergie* **1**, 121(1956)
[2.9] adapted from Genus brochure, see also O' Connor, J.P., N. Tokoro, J. Adamik, *Nucl. Instr. Meth.* **B74**, 18(1993)
[2.10] Moak, C.D., H.E. Banta, J.N. Thurston, J.W. Johnson and R.F. King, *Rev. Sci. Instrum.* **30**, 694(1959)
[2.11] Bunge, A.V., C.F. Bunge *Phys. Rev.* **A19**, 452(1979)
[2.12] Middleton, R., C.T. Adams *Nucl. Instr. Meth.* **118**, 329(1974)
[2.13] O'Connor, J.P., L.F. Joyce *Nucl. Instr. Meth.* **B21**, 334(1987)
[2.14] Mc Intyre, E., D. Balek, P. Boisseau, A. Dart, A.S. Denholm, H. Glavish, C. Hayden, L. Kaminski, B. Libby, N. Meyyappan, J. O' Brian, F. Sinclair, K. Whaley *Nucl. Instr. Meth.* **B55**, 473(1991)
[2.15] Simonton, R.B., in *HandBook of Ion Implantation Technology*, ed. by J.F. Ziegler, North Holland, Amsterdam, 1992, p. 532.
[2.16] Feynman, R.P., R.B. Leighton and M.Sands *The Feynman Lectures on Physics*, Addison-Wesley 1964, Vol. II 29-3
[2.17] Purser, K.H., N.R. White, in *Ion Implantation Science and Technology* 2nd edition, J.F. Ziegler ed., Academic Press, San Diego 1988 p.291
[2.18] Kaim, R.E., J.F.M. Westendorp, *Energy Contamination in Ion Implantation Solid State Technology*, April 1988 p.65
[2.19] van der Meulen, P.F.H.M., S. Mehta, R.E. Kaim, *Nucl. Instr. Meth.* **B55**, 45(1991)
[2.20] Massey, H.S.W., H.B. Gilbody, *Electronic and Ionic Impact Phenomena IV*, University Press, Oxford(1974)
[2.21] Downey, D.F., R.B. Liebert, *Nucl. Instr. Meth.* **B55**, 49(1991)
[2.22] Mack, M.E. in *Handbook of Ion Implantation Technology* J.F. Ziegler editor, Elsevier p.599(1992)
[2.23] Ryding, G., M. Farley, *Nucl. Instr. Meth.* **189**, 295 (1981)
[2.24] Tami, T., M. Diamond, B. Doherty, P. Splinter *Nucl.Instr. Meth.* **B55**, 408(1991)
[2.25] Beanland, D.G. in *Ion Implantation and Beam Processing* J.S. Williams, J.M. Poate eds., Academic Press, New York (1984) p.261
[2.26] King, M., P.H. Rose *Nucl.Instr. Meth.* **189**, 169(1981)
[2.27] Benveniste, V., in *Ion Implantation Technology* M.I. Current, N.W. Cheung, W. Weisenberg, B. Kirby eds. North-Holland, Amsterdam 1990 p.366
[2.28] Sinclair, F. *Nucl. Instr. Meth.* **B55**, 115(1991)

[2.29] Holmes, A.J.T. *Phys.Rev.* **A19**, 389(1979)

[2.30] Ryding, G., M. Farley, M. Mack, K. Steeples, V. Gillis *10th International Conference of Electron and Ion Beam Science and Technology*, Montreal (1982)

[2.31] Nasser-Ghodshi, M., M. Farley, J. Grant, D. Bernhardt, M. Foley, S. Holden, T. Bowe, C. Singer, K. Dixit, G. Angel *Nucl. Instr. Meth.* **B55**, 398(1991)

[2.32] Felch, S.B., L.A. Larson, M.I. Current, D.W. Lindsey *Nucl. Instr. Meth.* **B55**, 563(1981)

[2.33] Smith, A.K., W.H. Johnson, W.A. Keernan, M. Rigik, R. Kleppinger, *Nucl. Instr. Meth.* **B21**, 529(1987)

[2.34] Rosencwaig, A. in *Photoacustic and Thermal Wave Phenomena in Semiconductors*, A. Mondelis ed., Elsevier New York 1987, p.78

[2.35] Opsal, J., A. Rosencwaig *MRS Bulletin*, Vol. XIII, April 1988, p.28

[2.36] Smith, W.L., A. Rosencwaig, D.C. Willenborg, J. Opsal, M.W. Taylor *Solid State Technology* **29**, 85 Jan. 1986

[2.37] Martini, R., C. Wichard, W.L. Smith, M.W. Taylor *Solid State Technology* **30**, 89 May 1987

[2.38] Rosencwaig, A., in *VLSI Electronics: Microstructure Science*, N.G. Einspruch ed. Academic Press, New York 1985 p. 227

[2.39] Taylor, M., K. Hurley, K. Lee, M. Le Mere, J. Opsal, T. O'Brien *Nucl. Instr. Meth.* **B55**, 525(1991)

[2.40] Larson, L.A., W.A. Keenan, W.H. Johnson *Solid State Technology* **36**, 67 Oct. 1993

[2.41] Smith, T.C. *Nucl. Instr. Meth.* **B37/38**, 486(1989)

[2.42] Current, M.I., L.A. Larson in *Mat. Res. Soc. Proc.* **147** p.365(1985)

[2.43] Bergholz, B., G. Zoth, F. Gelsdorf, B.O. Kolbesen in *Defects in Silicon* 2nd ed. W.M. Bullis, F. Shimura and U.Goséle, The Electrochemical Soc. Pennington, N.J., 1991, p.21

[2.44] Conrad, J.R., J.L. Radtke, R.A. Dodd, F.J. Worzala, N.C. Tran *J .Appl. Phys.* **62**(1987) p. 459

[2.45] Tendys, J., J.J. Donnelly, M.J. Kenny, J.T.A. Pollock *Appl.Phys. Lett.* **53** 2143(1988)

[2.46] Cheung, N.W. *Nucl. Instr. Meth.* **B55** 811(1991)

[2.47] Qian, X.Y., D. Carl, J. Benasso, N.W. Cheung, M.A. Lieberman, I.G. Brown, J.E. Galvin, R.A. MacGill, M.I. Current *Nucl. Instr. Meth.* **B55**, 884(1991)

[2.48] Qian, X.Y., N.W. Cheung, M.A. Lieberman, M.I. Current *Nucl. Instr. Meth.* **B55** 821(1991)

[2.49] Kaim, I. *Solid State Technol.* **33** 103(1990)

[2.50] Mizuno,B., I.Nakayama, N.Aloi, M. Kubota, T.Komeda *Appl.Phys. Lett.* **53**(1988)2059

[2.51] En, B., N.W. Cheung *Nucl. Instr. Meth.* **B74**, 311(1993)

CHAPTER 3

[3.1] Schiott, H.E. *Mat. Fys. Medd. Dan. Vidensk. Selsk.* **35** (9), 1 (1966)

[3.2] Lindhard, J., M. Scharff, M.E. Schiott *Mat. Fys. Medd. Dan. Vidensk. Selsk.* **33**(14),1(1963)

[3.3] Goldstein, H., *Classical Mechanics*, Addison-Wesley, Reading MA, 1956, ch. 3

[3.4] Nelson, R.S. *The observation of atomic collisions in crystalline solids* North Holland, Amsterdam 1968

[3.5] Nielsen, K.O., in *Electromagnetically Enriched Isotopes and Mass Spectroscopy*, M.L. Smith ed., Academic Press, N.Y. (1956), p.68

[3.6] Winterbon, K.B., P. Sigmund, J.B. Sanders *Mat. Fys. Medd. Dan. Vidensk. Selsk* **37**(14),1(1970)

[3.7] Gombas, P. *Die Statistische Theorie des Atoms und ihre Anwendungen* Springer-Verlag, Austria 1949

[3.8] Winterbon, K.B. *Ion Implantation Range and Energy Deposition Distribution* Vol. 2, Plenum, New York (1975)

[3.9] Ziegler, J.F., J.P. Biersack, U. Littmark *The stopping and range of ions in solids*, Pergamon Press, New York 1985

[3.10] Lindhard, J., A. Winter *Mat. Fys. Medd. Dan. Vidensk. Selsk.* **34**(4),1(1964)

[3.11a] Xia, Y., C. Tan *Nucl. Instr. Meth.* **B13**, 100(1986)

[3.11b] Cheschire, I., G. Dearneley, J.M.Poate *Phys. Lett.* **12**, 129(1964)

[3.12] Bethe, H., *Ann. Phys.* **5** 325(1930)

[3.13] Davies, J.A., *Surface Modification and Alloying*, J.M. Poate, G. Foti, D.C. Jacobson eds., Plenum New York(1983)p.189

[3.14] Seidel, T.E., in *Ion Implantation VLSI Technology* S.M. Sze ed., Mc.GrawHill, New York, 1983 ch. 6

[3.15] Ghandhi, S.K. *VLSI Fabrication Principles*, Wiley, New York 1983

[3.16] Hofker, W.K. *Philips Res.Rep.Suppl.* **8**,(1975)

[3.17] Gibbons, J.F. *Handbook on Semiconductors* Vol. 3, T.S. Moss ed., North-Holland, Amsterdam(1980) ch. 10

[3.18] Peterson, W.P., W.Fichtner, E.H. Grosse *IEEE Trans. Elec. Dev.* **30**,1011(1983)

[3.19] Tsukamoto, K., Y.Akasaka, K.Kijima *Jpn. J. Appl. Phys.* **19**, 87(1980)

[3.20] Giles, M.D. *Ion Implantation in VLSI Technology*, 2nd ed. S.M.Sze ed., McGraw-Hill, New York 1988 ch. 8

[3.21] Cho. K., W.R. Allen, T.G. Finstand, W.K. Chu, J. Liu, J.J. Wortman *Nucl. Instr. Meth.* **B7/8**, 265(1985)

[3.22] Ishiwara, H., S.Furukawa, J.Yamada, M.Kawamura *Ion Implantation in Semiconductors*, S. Namba ed., Plenum, New York (1975) p.423

[3.23] Kennedy, E.F., C. Czepregi, J.W. Mayer, T.W. Sigmon *Ion Implantation in Semiconductors and Others Materials*, F.Chernow ed., Plenum, New York (1977) p.511

[3.24] Steinbauer, E., P. Bauer, J.P. Biersack *Nucl. Instr. Meth.* **B45**, 171(1990)

[3.25] Bottiger, J., J.A. Davies, P. Sigmund, K.B. Winterbon *Radiat. Eff.* **11**, 69(1971)

[3.27] Sigmund, P., *Phys. Rev.* **184**,383(1969)

[3.28] Andersen, H.H., H.L. Bay *J. Appl. Phys.* **46**, 1919(1975)

[3.28] Ryssel, H., and K. Hoffman *Process and Device Simulation for MOS-VLSI Circuits*, P.P. Antognetti, D. Antoniadis, R. Dutton and W. Oldham eds., NATO ASI Series Martinus Nijihoff, The Hague, Netherlands (1983)p.125

[3.29] Piercy, G.R., F. Brown, J.A. Davies, M.Mc Cargo *Phys. Rev. Lett.* **10**, 399(1963)

[3.30] Robinson, M.T., O.S.Oen *Phys. Rev.* **132**, 2385(1963)

[3.31] Lindhard, J. *Kgl. Danske Videnskab. Selsk. Mat. Fys. Medd.* **34**, No 14(1965)

[3.32] Appleton, B.R., C. Erginsoy, W.M.Gibson *Phys.Rev.* **131**, 330(1967)

[3.33] Raineri, V., G. Galvagno, E. Rimini, S. Capizzi, A. La Ferla, A. Carnera, G.
        Ferla *Semicond. Sci. Technol.* **5**, 1007(1990)
[3.34] (a) Kaim, R.E., P.F.H.M. van der Meulen *Nucl. Instr. Meth.* **B55**, 453(1991)
        (b) Ray, A.M., J.P. Dykstra, R.B. Simonton in *Ion Implantation Technology -
        92* edited by D.F. Downey, M. Farley, K.S. Jones, G. Ryding, Elsevier Science
        Publishers B.V., Amsterdam, 1993, p. 401
[3.35] Turner, N.L., M.I. Current, T.C. Smith, D.Crane *Solid State Technol.* **28**(2),
        163(1985)
[3.36] La Ferla A., G. Galvagno, V. Raineri, R. Setola, E. Rimini, A. Carnera, A.
        Gasparotto *Nucl. Instr. Meth.* **B66** 339(1992)
[3.37] Raineri V., R. Setola, F. Priolo, E. Rimini, G. Galvagno *Phys. Rev.* **B44**,
        10568(1991)
[3.38] Oen, O.S., M.T. Robinson *Nucl. Instr. Meth.* **132**, 647(1976)
[3.39] Schreutelkamp, R.J., V. Raineri, F.W. Saris, R.E. Kaim, J.F.M. Westendorp,
        P.H.F.M. van der Meulen, K.T.F. Janssen *Nucl Instr. Meth.* **B55**, 615(1991)
[3.40] Simonton, R., Al F.Tasch *Channeling Effects in Ion Implantation* in *Handbook
        of Ion Implantation Technology* J.F. Ziegler editor, Elsevier Science Publ. 1992
        p. 119
[3.41] Klein, K., C. Park, Al F. Tasch, R. Simonton, S. Novak, *J.Electrochem. Soc.*
        138, 2102(1991)
[3.42] Furukawa, S., H. Matsumura, H. Ishiwara *Jpn. J. Appl. Phys.* **11**, 134(1972)
[3.43] Tamura, M. *Damage Formation and Annealing of Ion Implantation in Si* in
        Mat. Sci. Rep. **6**, 141(1991)
[3.44] Giles, M.D. *IEEE Trans. CAD* **5**, 679(1986)
[3.45] Runge, H. *Phys. Status Solidi* **A39**,595(1977)
[3.46] Gong, L., L. Frey, S. Bogen, H. Ryssel *Nucl.Instr. Meth.* **B74**, 186(1993)
[3.47] La Via, F., C. Spinella, E. Rimini in *Proceeding of ESSDERC '93* edited by J.
        Borel, P. Gentil, J.P. Noblanc, A. Nouailhat, M. Verdone (Edition Frontieres,
        Gif-sur-Yvette, 1993) p. 489
[3.48] Privitera, V., W. Vandervorst, T. Clarysse *J. Electrochem. Soc.* **140**, 262(1993)
[3.49] Privitera V., V. Raineri, E.Rimini *J. Appl. Phys.* **74**, 2370(1993)
[3.50] Eckstein, W., *Computer Simulation of Ion-Solid Interactions*, Springer-Verlag
        Berlin, 1991
[3.51] Mazzone, A.M., in *Process and Device Modeling for Microelectronics* G. Bac-
        carani ed., (Elsevier, Amsterdam 1993) p.31
[3.52] Christel, L.A., J.F. Gibbons, S. Mylroie *J. Appl. Phys.* **51**, 6176(1980)
[3.53] Christel, L.A., J.F. Gibbons, S.Mylroie *Nucl. Instr. Meth.* **182**, 187(1981)
[3.54] Diaz de la Rubia, T., R.S. Averback, H. Horngming, R.Benedek *J. Mater. Res.*
        4, 579(1989)
[3.55] Ziegler, J.F. in *Ion Implantation Science and Technology* 2nd edition, J.F.
        Ziegler editor, Academic Press, San Diego 1981, p. 3
[3.56] Biersack, J.P., L.G.Haggmark *Nucl. Instr. Meth.* **174**, 257(1980)
[3.57] Biersack, J.P., W. Eckstein *Appl. Phys.* **34**(1982)73
[3.58] Raush, M.L.,T.D. Andrealis, O.F. Goktepe *Radiat.Eff.* **55**(1981)119
[3.59] Moller, W., W. Eckstein *Nucl. Instr. Meth.* **B2** (1984)814
[3.60] Eckstein, W., J.P. Biersack *Comput. Phys. Commun.* **51**(1988)355
[3.61] Chau, P.S., N.M. Ghoniem *J. Nucl. Mater.* **141-143**(1986)216
[3.62] Schonborn, A., N. Hecking, E.H. de Kaat *Nucl.Instr. Meth.* **B43**(1988)170
[3.63] Robinson, M.T., I.M. Torrens *Phys.Rev.* **B9** (1974)5008

[3.64] Rosselt, C. *Radiation Effect and Defect in Solid* **129**(1993)
[3.65] Crandle, B., J. Mulvaney *IEEE Electron Dev. Lett.* **11**(1990)42
[3.66] Kong, H.J., R. Shimizu, T. Saito, H. Yamakawa *J. Appl. Phys.* **52**(1987)2733
[3.67] Klein, K.M., C. Park *Appl. Phys. Lett.* **57**(1990)2701
[3.68] Klein, K.M., C. Park, F. Al Tash, *Nucl. Instr. Meth. Phys.* **B59/60** 60(1991)
[3.69] Fichtner, W. *Physics of VLSI Processing and Process Simulation*, in *Silicon Integrated Circuits* Part C, Dawon Kahng ed. Academic Press, New York 1985
[3.70] Lau, M.E., C. Rafferty, and R.W. Dutton, *Suprem IV Users Manual* Technical Report, Integrated Circuits Laboratory, Stanford University, July 1986
[3.71] Antoniadis, O.A., R.W. Dutton *IEEE J. Solid State Circuits* SC **14**, 412(1979)

CHAPTER 4

[4.1] Gibbons,J.F., *Proc. IEEE,* **60**, 1062(1972)
[4.2] Brown, W.L., *Mat. Res. Soc. Symp. Proc.* **51**,53(1985)
[4.3] Davies, J.A., *MRS Bulletin*, Vol. **17**, No. 6, 26(1992) Mat. Res. Soc.
[4.4] Haines, E.L., and A.B.Whitehead, *Rev. Sci. Instrum.* **37**, 190(1966)
[4.5] Davies, J.A., *Surface Modification and Alloying*, J.M.Poate, G.Foti and D.C.Jacobson eds., Plenum, New York (1983) p.189
[4.6] Narayan, J., D. Fathy, O.S. Oen and O.W. Holland, *Mater. Letters* **2**,211(1984)
[4.7] Gibson, J.M., *MRS Bulletin*, Vol. XVI, No. 3, 27(1991)
[4.8] Winterbon, K.B., *Rad. Effects*, **13**, 215(1972)
[4.9] Brice, D.K., *Appl. Phys. Lett.* **16**,3(1970)
[4.10] Walker, R.S., and D.A. Thompson, *Rad. Effects* **37**, 113(1978)
[4.11] Kinchin, G.H., and R.S.Pease, *Rep. Prog. Phys.* **18**, 1(1955)
[4.12] Sigmund, P., *Appl. Phys. Lett.* **14**, 114(1969)
[4.13] Thompson, D.A., *Rad. Effects* **56**, 105(1981)
[4.14] Raineri, V., G. Galvagno, E. Rimini, S. Capizzi, A. La Ferla, A. Carnera and G. Ferla, *Semicond. Sci. Technol.* **5**, 1007(1990)
[4.15] Gösele,U., *Advances in Solid State Physics*, Vol. XXVI, Vieweg, Braunschweig (1986)
[4.16] Watts, R.K. *Point Defects in Crystals* John Wiley & Sons, New York 1977
[4.17] Benedek, G., A. Cavallini, W. Schröter eds. *Point end Extended Defects in semiconductors* NATO ASI Series B; Physics Vol. 202, Plenum Press, New York 1989
[4.18] Watkins, G.D., J.R. Troxell, and A.P. Chatterjee in *Defects and Radiation Effects in Semiconductor*, ed. by J.H. Albany, (Inst. of Physics, London) Conf. Series No. 46, 16(1978)
[4.19] Fahey, P.M., P.B. Griffin, and J.D. Plummer, *Rev.Mod.Phys.***61**, 289(1989)
[4.20] Maroudas Dimitris, and Robert A. Brown, *Appl. Phys. Lett.* **62**,172(1993)
[4.21] Tan, T.Y., and U.M.Gösele, *Appl.Phys.* **A37**, 1(1985)
[4.22] Askeland, D.R., *The Science and Engineering of Materials* 2nd ed., Chapman and Hall, London 1990
[4.23] Cerofolini, G.F., L. Meda *Physical Chemistry of, in and on Silicon*, Springer-Verlag M.S. 8, Berlin, 1989 chapt. 9
[4.24] Honda, K., A. Ohsawa, N. Toyokura *Appl. Phys. Lett.* **46**,582(1985)
[4.25] Priolo, F., A. Battaglia, R. Nicotra, and E.Rimini, *Appl. Phys. Lett.* **57**, 768(1990)

[4.26] Coffa, S., L. Calcagno, M. Catania and E. Rimini, *Appl. Phys. Lett.* **56**, 2405(1990)

[4.27] Donovan, E.P., F. Spaepen, D. Turnbull, J.M. Poate, D.C. Jacobson *Appl. Phys. Lett*, **42**, 698(1983)

[4.28] Vook, F.L., in *Radiation Damage and Defects in Semiconductors*, ed. by J.E. Whitehouse (The Institute of Physics, London and Bristol, 1973) p.60

[4.29] Morehead, F.F. Jr., and B.L. Crowder, *Rad.Effects* **6** 30(1970)

[4.30] Nakata, J., and K. Kajiyama, *Appl.Phys.Lett.* **40**, 686(1982)

[4.31] Cannavó, S., A. La Ferla, E. Rimini, G. Ferla and L. Gandolfi, *J .Appl. Phys.* **59**,4038(1986)

[4.32] Servidori, M., S. Cannavò, G. Ferla, A. La Ferla, E. Rimini, *Appl.Phys.* **A44**, 213(1987)

[4.33] Tan, T.Y., H. Föll and S.M. Hu, *Philos. Mag.* **A44**, 127(1981)

[4.34] Rimini, E. *Basic aspects of ion implantation* in *Crucial Issues in Semiconductor Materials and Processing Technologies* edited by S. Coffa, F. Priolo, E. Rimini, J.M. Poate, NATO ASI Series E, Vol. 222, Kluwer Academic Publshers, Dordrecht, 1992, p. 167

[4.35] Campisano, S.U., S. Coffa, V. Raineri, F. Priolo and E. Rimini, *Nucl. Instr. Meth.* **B80/81**, 514(1993)

[4.36] Avrami,M., *J. Chem. Phys.* **7**,1103(1939), ib **8**,212(1940), ib **9**,177(1941)

[4.37] Leiberich, A., D.M. Maher, R.V. Knoell, and W.L. Brown, *Nucl. Instr. Meth.* **B19/20**, 457(1987)

[4.38] Battaglia, A., F. Priolo, E. Rimini, G. Ferla, *Appl. Phys. Lett.* **56**, 2622(1990)

[4.39] Linnros, J., R.G. Elliman, and W.L. Brown, *J. Mat. Res.* **3**,1208(1987)

[4.40] Campisano, S.U., S. Coffa, V. Raineri, F. Priolo, E. Rimini, *Nucl. Instr. Meth.* **B80/81**, 514(1993)

[4.41] Servidori, M., S. Cannavò, G. Ferla, S.U. Campisano and E. Rimini, *Nucl. Instr. Meth.* **B19/20**, 317(1987)

[4.42] Cerofolini, G.F., L. Meda, G. Queirolo, A. Armigliato, S. Solmi, F. Nava and G. Ottaviani, *J .Appl. Phys.* **56**, 2981(1984)

[4.43] Williams, J.S., R.G. Elliman, W.L. Brown and T.E. Seidel, *Phys. Rev. Lett.*,**55**,1482(1985)

[4.44] Priolo, F., A. La Ferla, and E. Rimini, *J. Mater. Res.* **3**,1212(1988)

[4.45] Priolo. F., C. Spinella, A. La Ferla, E. Rimini and G. Ferla *Appl. Surf. Sci.* **43**, 178(1989)

[4.46] Priolo, F., and E. Rimini, *Mat. Sci. Rep.* **5**, 319(1990)

[4.47] Priolo, F., C. Spinella and E. Rimini, *Phys. Rev. B* **41**,5235(1990)

[4.48] Kennedy, E.F., L. Csepregi, J.W. Mayer, and T.W. Sigmon, *J .Appl. Phys.* **48**, 4241(1977)

[4.49] Priolo, F., A. La Ferla, C. Spinella, E. Rimini, G. Ferla, A. Baroetto and A. Licciardello, *Appl. Phys. Lett.* **53**, (1988)2605

[4.50] Rimini,E., F.Priolo and C.Spinella, Il Nuovo Cimento **15D**,399(1993)

[4.51] Yu, A.J., J.W. Mayer, and J.M. Poate, *Appl. Phys. Lett.* **54**,2342(1989)

[4.52] Elliman, R.G., M.C. Ridgway, J.S. Williams, and J.C. Bean, *Appl. Phys. Lett.* **55**,843(1989)

[4.53] Ridgway, M.C., R.G. Elliman, J.S.Williams, *Nucl.Instr.Meth.*, **B48**, 453(1990)

[4.54] Williams, J.S., M.C. Ridgway, R.G. Elliman, J.A. Davies, S.T. Johnson, and G.R. Palmer, *Nucl. Instr. Meth.* **B55**, 602(1991)

[4.55] Jackson, K.A., *J. Mater. Res.* **3**, 1218(1988)

[4.56] Priolo, F., C. Spinella, E. Rimini, G. Ferla, *Appl. Phys. Lett.* **56**,24(1990)
[4.57] Custer, J.S., A. Battaglia, M. Saggio, F. Priolo, *Phys. Rev. Lett.* **69**, 700(1992)
[4.58] Coffa, S., F. Priolo, A. Battaglia, *Phys. Rev. Lett.* **70**, 3756(1993)
[4.59] Tamura, M., *Mat. Sci. Rep.* **6**, 141(1991)

CHAPTER 5

[5.1] Csepregi, L., E.F. Kennedy, J.W. Mayer, T.W. Sigmon, *J .Appl. Phys.* **49**,3906(1978)
[5.2] Olson, G.L., and J.A. Roth, *Mater. Sci. Rep.* **3**(1988)p.1
[5.3] Donovan, E.D., F. Spaepen, D. Turnbull, J.M. Poate, and D.C. Jacobson *J. Appl. Phys.* **57**, 1795(1985)
[5.4] Roorda, S., S. Doorn, W.C. Sinke,P.M.L.O. Scholte, and E. van Loenen, *Phys. Rev. Lett.* **62**, 1880(1989)
[5.5] Csepregi, L., J.W.Mayer, and T.W.Sigmon, *Appl. Phys. Lett.* **29**,92(1976)
[5.6] Foti G., L. Csepregi, E.F. Kennedy, J.W. Mayer, P.P. Pronko, and M.D. Reichtin, *Philos.Mag.* **A37**, 4234(1977)
[5.7] Spaepen, F., and D. Turnbull, in *Laser Annealing of Semiconductors* J.M.Poate and J.W.Mayer eds. (Academic Press, New York), (1982) p.15
[5.8] Williams, J.S., and R.G. Elliman, *Phys. Rev. Lett.* **51**, 1069(1983)
[5.9] Csepregi, L., E.F. Kennedy, T.J. Gallagher, J.W. Mayer, and T.W. Sigmon, *J. Appl. Phys.* **48**, 4234(1977)
[5.10] Kennedy, E.F., L.Csepregi, J.W.Mayer, and T.W.Sigmon, *J. Appl. Phys.* **48**, 4241(1977)
[5.11] Sumi,I.S., G.Goltz, M.G.Grimaldi, M.A.Nicolet, and S.S.Lau, *Appl. Phys. Lett.* **40**, 269(1982)
[5.12] Williams, J.S., in *Surface Modification and Alloying*, eds. J.M. Poate, G. Foti, and D.C. Jacobson, Plenum New York (1983)p.138
[5.13] Mader, S., in *Ion Implantation: Science and Technology* ed. J.F. Ziegler, Academic Press, Orlando Flo., 109(1984)
[5.14] Foell, H., T.Y. Tan, and W. Kracow, in *Defects in Semiconductors* eds. J.Narayan and T.Y.Tan, North-Holland, N.Y. 1981, Vol. 2 p.13
[5.15] Wu, W.K., and J. Washburn, *J .Appl. Phys.* **48**, 3747(1977)
[5.16] Fairfield, J.M., and B.J. Masters, *J .Appl. Phys.* **38**, 3148(1967)
[5.17] Sands, T., J. Wasburn, R. Gronsky, W. Maszara, D.K. Sadana, G.A. Rozgonyi in 13th *Intl. Conf. on Defects in Semiconductors* ed. by L.C. Kimerling, J.M. Parsey, (The Metallurgical Society of AIME, Wanendale, PA, 1987)p.531
[5.18] Rozgonyi, G.A., E. Myers, D.K. Sadana *Semiconductor Silicon, Proc. of Electr. Soc.* (ECS Press, (Pennington N.J. 1986)
[5.19] Narayan, J., and O.W. Holland *Phys. State. Solidi* (a) **73**, 225(1982)
[5.20] Tamura, M., *Philos.Mag.* **35**, 663(1977)
[5.21] Prussin. S., K.S. Jones *Materials Issues in Silicon Integrated Circuit Processing* eds. by M.Strathman, J.Stimmel, M.Wittmer, *Proc. Mat. Res. Soc.* **71**(1986)
[5.22] Tamura, M., T. Ando, and K. Ohyer, *Nucl. Instr. Meth.* **B59/60**, 572(1991)
[5.23] Tamura, M., and T. Suzuki*Nucl. Instr. Meth.* **B39**, 318(1989)
[5.24] Sadana, D.K., J. Washburn, G.R. Booker *Phil. Mag.* **B46**,611(1982)
[5.25] Priolo, F., A. Battaglia, R. Nicotra, and E. Rimini *Appl. Phys. Lett.* **57**, 768(1990)

[5.26] Cerva, H., K.H. Kusters, *J .Appl. Phys.* **66**, 4723(1989)

[5.27] Tamura, M., *Mat.Sci.Rep.* **6**,141-214(1991)

[5.28] Cerva, H., and H. Wendt, *Mat. Res. Symp. Proc.* Vol. **138**,533(1989)

[5.29] Kolbesen, B.O., W. Bergholz, H. Cerva, B. Fiegl, F. Gelsdorf and G. Zoth, *Nucl. Instr. Meth.* **B55**, 124(1991)

[5.30] Yoshihiro, N., T. Ikeda, M. Tamura, T. Tokuyama, T. Tsuchimoto in *Ion Implantation in Semiconductors*, S. Namba ed., Japanese Society for Promotion of Science, Kyoto, 1972, p.33

[5.31] Seidel, T.E., and A.V. McRae, in *Proc. 1st Intl. Conf. Ion Implantation* L.Chadderton and F.H.Eisen eds.(Gordon and Breach, New York 1971) p. 149

[5.32] North, J.C., and W.M. Gibson in *Proc. 1st Intl.Conf. Ion Implantation* L.Chadderton and F.H.Eisen eds. (Gordon and Breach, New York) (1971), p.143

[5.33] Wu, K.H., and J. Washburn, *J .Appl. Phys.* **48**,3742(1977)

[5.34] Hu, S.M., J.Vac.Sci.Technol.*J. Vac. Sci. Technol.* **14**,17(1977)

[5.35] Hu, S.M., in *Defects in Semiconductors*, ed. by J. Narayan, T.Y. Tan (North-Holland, New York 1981)p. 133

[5.36] Schimmel, D.J., *J. Electrochem. Soc.* **126**, 479(1979)

[5.37] Frank, W., U. Gösele, H. Mehrer, and A. Seeger, *Diffusion in Crystalline Solids*, G.E. Murch and A.S. Nowick, eds., Academic Press, New York (1984)p.64

[5.38] Antoniadis, D.A., and I. Maskowitz, *J .Appl. Phys.* **53**, 6988(1992)

[5.39] Wen, D.S., P.L. Smith, C.M. Osburn, and G.A. Rozgoni, *J. Electrochem. Soc.* **136**,466(1989)

[5.40] Calcagno, L., S. Coffa, C. Spinella, and E. Rimini *Nucl. Instr. Meth.* **B45**,442(1990)

[5.41] Seidel, T.E., D.J. Lischner, C.S. Pai, R.V. Knoell, D.H. Maher, and D.C.Jacobson, *Nucl.Instr. Meth.* **B7/8**,251 (1985)

[5.42] Jones, K.S., S. Prussin, R.W. Weber *Appl. Phys.* **A45**,1,(1988)

[5.43] Seidel, T.E.,C.S. Pai, D.J. Lischner, and S.S.Lau, *Extended Abst. Electrochem. Soc.*, **84-7**, 184(1984)

[5.44] Sato, T., *Jpn. J.Appl. Phys.* **6**, 339(1967)

[5.45] Vandenable, P., and K.Maex, *Microelectronic Engineering* **10**, 207(1991)

[5.46] Hodul, D., and S.Mehta *Solid State Technology*, May 1988 p. 209

[5.47] Hill, C., J. Jones, and D. Boys in *Reduced Thermal Processing for ULSI* eds. R.A. Levy NATO-ASI Proc. 207, p.143

[5.48] Scandurra, A., G. Galvagno, V. Raineri, F. Frisina, A. Torrisi *J. Electrochem. Soc.* **140**, 2057(1993)

[5.49] Galvagno, G., F. La Via, F. Priolo, and E. Rimini *Semicond. Sci. Technol.* **8**, 488(1993)

[5.50] Schröter, W., M. Scibt, O. Gilles in *Materials Sience and Technology*, Vol. 4, *Electronic Structure and Properties of Semiconductors*, edited by W. Schröter, VCH Publishers, New York 1991, pp. 539-589

[5.51] Galvagno, G., A. LaFerla, C. Spinella, F. Priolo, V. Raineri, L. Torrisi, E. Rimini, A. Carnera, G. Gasparotto *J. Appl. Phys.* **76**, 15 August(1994)

[5.52] Galvagno, G., A. Scandurra, V.Raineri, C.Spinella, A.Torrisi, A. La Ferla, V.Sciascia and E.Rimini, J.Electrochem. Soc. **140**, 2313(1993)

[5.53] La Ferla, A., L. Torrisi, G. Galvagno, E. Rimini, G. Ciavola, A. Carnera, G. Gasparotto, *Appl. Phys. Lett.* **62**,125(1993)

[5.54] Schreutelkamp, R.J., V. Raineri, F.W. Saris, R.E. Kaim, J.F.M. Westendorp, P.F.H.M. van der Meulen and K.T.F. Janssen, *Nucl.Instr. Meth.* **B55**,615(1991)

[5.55] Schreutelkamp, R.J., J.S. Custer,J.R. Liefting, W.X. Lu, and F.W. Saris, *Mat. Sci. Rep.* **6**, 1(1991)

[5.56] Liefting, J.R., V. Raineri, R.J. Schreutelkamp, J.S. Custer, *Mat. Res. Soc. Symp. Proc.* **235**, 173(1992)

[5.57] Schreutelkamp, R.J., J.S. Custer, J.R. Liefting, and F.W. Saris, *Appl.Phys. Lett.* **58**,2827(1991)

[5.58] Liefting, J.R., V. Raineri, R.J. Schreutelkamp, J.S. Custer, and F.W. Saris, *Mat. Res. Soc. Symp. Proc.* Vol. **235**, 173(1992)

[5.59] Wong, H., N.W. Cheung, P.K. Chu, J. Liu, and J.W. Mayer, *Appl. Phys. Lett.* **52**, 1013(1988)

[5.60] Baker, J.A., T.M. Tucker, N.E. Moyer, and R.C. Busher, *J .Appl. Phys.* **39**, 4365(1968)

[5.61] Liefting, J.R., J.S.Custer and F.W.Saris, *Mat. Res. Soc. Symp. Proc.* **235**,179(1992)

CHAPTER 6

[6.1] *Characterization of Semiconductor Materials*, Vol.1, ed. Gary E. Mc Guire (Noyes Publications, (Park Ridge, New Jersey) 1989

[6.2] Kartz, W., and J.G. Newman, *MRS Bulletin*, Vol. XII, n. 6 (1987) p.41

[6.3] *Secondary Ion Mass Spectrometry - SIMS II-V*; ed. by A. Benninghoven, (Springer-Verlag, New York 1979-1985)

[6.4] Benninghoven, A., F.G. Rudenauer, H.W. Werner *Secondary Ion Mass Spectrometry* (New York, Wiley 1987)

[6.5] Morgan, A.E. *Secondary Ion Mass Spectrometry* in ref. 6.1 p.51

[6.6] Williams, P., *Appl. Surf. Sci.* **1**,3(1982)

[6.7] Winters, H.F., *Rad. Effects* **64**,79(1982)

[6.8] Williams, P., and C.A. Evans Jr., *Surf. Sci.* **78**, 324(1978)

[6.9] Ion probe A-DIDA 300 - Atomika brochure

[6.10] Scandurra, A., G. Galvagno, V. Raineri, F. Frisina, A. Torrisi *J. Electrochem. Soc.* **140**, 2057(1993)

[6.11] La Ferla, A., E. Rimini, A. Carnera, A. Gasparotto, G. Ciavola, G. Ferla, *Nucl. Instr. Meth.* **B55**,561(1991)

[6.12] Kaiser, U., J.C. Huneke *MRS Bulletin*, Vol. XII, no.6 1987, p.48

[6.13] Mazur, R.G., D.H. Dickey *J. Electrochem. Soc.* **3**, 255(1966)

[6.14] Schumann, P.A., E.E. Gardner, *J. Electrochem. Soc.* **116**, 87(1969)

[6.15] Holm, R., *Electrical contact Theory and Application* Springer- Verlag, Berlin 1967

[6.16] Ehrstein, J.R., *Nondestructive evaluation of semiconductor material and devices*, ed. J.N. Zemal, Plenum Press New York 1979

[6.17] Berkowitz H.L., R.A. Lux *J.Electrochem. Soc.* **137**, 679(1981)

[6.18] Vandervorst, W., T. Clarysse, W.J. Herzog, H. Jorke, M. Pawlik, P. Eichinger, E. Frenzel, M. Baur *Esprit Project 519 - Dopant Profiling for Submicron structures* Final Report

[6.19] Hu, S.H., *J .Appl. Phys.* **53**,1499(1982)

[6.20] Vandervorst, W., and T. Clarysse, *J. Electrochem. Soc.* **137**,679(1990)

[6.21] Vandervorst, W., T. Clarysse, J. Vanhellemont and A. Romano Rodriguez, *J. Vac. Sci. Technol.* **10**,449(1992)

[6.22] Vandervorst, W., V. Privitera, V. Raineri, T. Clarysse and M. Pawlik *J. Vac. Sci. Technol.* **B12**, 276(1994)

[6.23] Privitera, V., W. Vandervorst, T. Clarysse *J. Electrochem. Soc.* **140**, 262(1993)

[6.24] Privitera, V., V. Raineri, W. Vandervorst, T. Clarysse, L. Hellemans and J. Snauwaert, in *Proceedings of ESSDERC '93* edited by J. Borel, P. Gentil, J.P. Noblanc, A. Nouailhat, M. Verdone (Editions Frontieres, Gif-sur-Yvette 1993) p.539

[6.25] Coffa, S., V. Privitera, F. Frisina, F. Priolo, *J. Appl. Phys.* **74**, 195(1993)

[6.26] Van der Pauw, L.J., *Philips Res. Rep.* **13**,1(1950)

[6.27] Fistul, V.J., *Heavy Doped Semiconductors* (Plenum, New York) (1968) p.77

[6.28] Felch, S.B., R. Brennan, S.F. Corcoran, G. Webster, *Solid State Technology*, Jan. 1993 p.45

[6.29] Johansson, N.G.E., J.W. Mayer, and O.J. Marsh, *Solid State Electronics*, **13**, 317(1970)

[6.30] Binger, Donovan, eds. *Fundamentals of Silicon Integrated Devices Technology* Vol. 1 - Prentice-Hall, Englewood, 1967

[6.31] Galvagno, G., A. Cacciato, F. Benyaich, V. Raineri, F. Priolo, E. Rimini, S.Capizzi and P.Romano, *Mat. Sci. Engin.* **B10**,67(1991)

[6.32] Cacciato, A., F. Benyaich, C. Spinella, E. Rimini, P. Romano and P.Wards, *Semicond. Sci. Technol.* **8**,327(1993)

[6.33] Chu, W.K., J.W. Mayer, and A.Nicolet *Backscattering spectrometry* (Academic Press, New York 1978)

[6.34] Feldman, L.C., J.W. Mayer and S.T. Picraux *Materials Analysis by Ion Channeling* (Academic Press, New York 1982)

[6.35] Feldman, L.C., and J.W.Mayer, *Fundamentals of Surface and Thin Film Analysis* North Holland, Amsterdam 1986

[6.36] Williams, J.S., *Nucl. Instr. and Meth.* **126**,205(1975)

[6.37] Campisano, S.U., G. Ciavola, E. Costanzo, G. Foti and E. Rimini, *Nucl. Instr. Meth.* **149**, 229(1978)

[6.38] Davies, J.A., Physica Scripta **28**,294(1983)

[6.39] Baeri, P., S.U. Campisano, G. Ciavola, G. Foti, E. Rimini *Appl. Phis. Lett* **26**, 154(1975)

[6.40] Czepregi,L., J.W.Mayer, and T.W.Sigmon, *Appl. Phys. Lett.* **29**, 92(1976)

[6.41] Foti, G., P. Baeri, E. Rimini, S.U. Campisano *J. Appl. Phys.* **47**, 5206(1976)

[6.42] Rimini, E., *Analysis of defects by channeling* in *Material characterization using ion beams*, edited by J.P. Thomas, and A. Cachard (Plenum Publishing Company, New York 1978) p.455

[6.43] Campisano, S.U., G. Foti, E. Rimini, and S.T. Picraux,*Nucl. Instr. Meth.* **149**, 371(1978)

[6.44] Picraux, S.T., E. Rimini, G. Foti and S.U. Campisano, *Phys. Rev.* **B18**, 2078(1978)

[6.45] Raineri, V., G. Galvagno, E. Rimini, S. Capizzi, A. La Ferla, A. Carnera, G. Ferla *Semic. Sci. Technol.* **5**, 1007(1990)

[6.46] Hirsch, P.B.,G., A. Howie, R.B. Nicholson, D.W. Pashley, M.J. Whelan *Electron Microscopy of Thin Crystals*, Ed. Butterworths London 1965

[6.47] Benyaïch, F., F. Priolo, E. Rimini, C. Spinella, P. Ward, *Mat. Chem. Phys.* **32**, 99(1992)

[6.48] Gong, L., L. Frey, S. Bogen, H. Ryssel *Nucl. Instr. Meth* **B74**, 186(1993)

[6.49] La Via, F., C. Spinella, E. Rimini in *Proceedings of ESSDERC '93* edited by J. Borel, P. Gentil, J.P. Noblanc, A. Nouailhat, M. Verdone (Editions Frontieres, Gif-sur-Yvette, 1993)p.489

[6.50] Wu, C.P., E.C. Douglas, C.W. Muller, R. Williams *J. Electrochem. Soc* **126**, 1982(1979)

[6.51] Spinella, C., V. Raineri, S.U. Campisano *J. Electrochem. Soc*, to be published (1995)

CHAPTER 7

[7.1] Tsukamoto, K., S. Komori, T. Kuroi, and Y. Akasaka *Nucl. Instr. Meth.* **B59/60**, 584(1991)

[7.2] Borland, J.O., R. Koelsch, *Solid State Techn.* Dec. 1993, p.28

[7.3] *High Energy Ion Implantation Technology* Nikkei Microdevices Special Report, Dec. 1991, p.94-121

[7.4] Rideout, V.L., F.H. Gaensslen, A. Le Blanc *IBM J. Res. Dev* **19**, 50(1975)

[7.5] Hill, C., *J. Vac. Sci. Technol.* **B10**, 289,(1992)

[7.6] Hill, C., P. Hunt *Nucl. Instr. Meth.* **B55**, 1(1991)

[7.7] Sze, S.M. *Semiconductor Devices Physics and Technology* Wiley & Sons., New York, 1985, p.482

[7.8] Haznedar, H., *Digital Microelectronics* The Benjamin/Gumming Publishing Co., 1991

[7.9] Hunt, R.W., *Memory Design and Technology* in M.J. Howes, D.U. Morgan eds. *Large Scale Integration* Wiley, New York, 1981.

[7.10] Hillenius, S.J., *VLSI process integration* in *VLSI Technology* 2nd edition, S.M. Sze ed., Mc Graw-Hill, New York, 1988, chap. 11, p.466

[7.11] Kolbesen, B.O., W. Bergholz, H. Cerva, F. Gelsdorf, H. Wendt, G. Zoth *Inst. Phys. Conf. Ser.* **104**, 421(1983)

[7.12] Muller, R.S., T.I. Kamins *Device Electronics for Integrated Circuits* 2nd ed. Wiley, New York, 1986

[7.13] Pickar, K.A., *Ion Implantation in Silicon - Physics, Processing and Microelectronic Devices* Appl.Sol.St.Sci. Vol. **5**, p.151 (Academic Press, New York 1975)

[7.14] Fair, R.B., J.C.C. Tsai *J. Electrochem. Soc.: Solid-State Science and Technology* **125**, 2050(1978)

[7.15] Hsu, F.C., *IEEE Trans. Electr. Dev.* **ED-29**, 1735(1982)

[7.16] Ogura, S., *IEEE Trans. Electr. Dev.* **ED-27**, 1359(1980)

[7.17] Koyanagi,M., H. Kanako, S. Shimizu *IEEE Trans. Elec. Dev.*, **ED-32**, 562(1985)

[7.18] Tsang, P.J., S, Ogura, W.W. Walker, J.F. Shepard, D.L. Critchlow *IEEE Trans. Elec. Dev.* **ED-29**, 590(1982)

[7.19] Hori,T., K. Kurimoto *IEEE Elec. Dev. Lett.*, **9**, 300(1988)

[7.20] Hori,T., K. Kurimoto *IEEE Elec. Dev. Lett.*, **9**, 641(1988)

[7.21] Simonton, R.B., D.E. Kamenitsa, A.M. Ray, *Sol. St. Techn.*, Nov. 1992 p. 28

[7.22] Solmi, S., F. Baruffaldi, R. Cantieri *J .Appl. Phys.* **69**, 2135(1991)

[7.23] Raineri V., R.J. Schreutelkamp, F.W. Saris, K.T.F. Janssen, R.E. Kaim *Appl. Phys. Lett* **58**, 922(1991)

[7.24] Schreutelkamp, R.J., J.S. Custer, V. Raineri, W.X. Lu, R.J. Liefting, F.W. Saris, K.T.F. Janssen, P.F.H.M. van der Meulen, R.E. Kaim *Mat. Sci. Eng.* **B12**, 307(1992)

[7.25] Ozturk, M., J.J. Wortman, C.M. Osburn, A. Ajmera, A. Rozgonyig, E. Frey, W.K. Chu *IEEE Trans. Elec. Dev.* **35**, 659(1988)

[7.26] Pfiester, J., J.Alvis *IEEE Elec. Dev. Lett.* **9**, 391(1988)

[7.27] Osburn, C.M., Q.F. Yang, M. Kellam, C. Canovai, P.L. Smith, G.E. McGuire, Z.G. Xias and G.A. Rozgonyi, *Appl. Surf. Sci.* **53**, 291(1991)

[7.28] Osburn, C.M. *J. Electronic Materials* **19(1)**, 67 (1990)

[7.29] Rozgonyi, G.A., and J.W. Honeycutt, *Material Res. Soc. Symp. Proc.* **147**, 3(1989)

[7.30] B. Davari, Y. Tour, D. Moy, F.M. d'Heurle, C.Y. Ting, *Proceedings of First International Symposium on Ultra Large Integration Science and Technology, The Electrochemical Society*, 87-11, 368(1987)

[7.31] Probst, V., P. Lippens, L. Van der Hove, K. Maex, H. Schaber, R. Dekeersmaecker, *Solid State Devices, ESSDERC '87*, edited by G. Soncini, P.U. Calzolari, Elsevier Science Publishers B.V. (North Holland), Amsterdam 1988, p.437

[7.32] La Via, F., C. Spinella, E. Rimini, *Proceedings of the 23rd European Solid State Device Reserch Conference* edited by J. Borel, P. Gentil, J.P. Noblanc, A. Nouailhat, M. Verdone (Editions Frontieres, Gif-sur-Yvette 1993), p.489

[7.33] Probst, V., H. Schaber, H. Mitwalsky, A. Kabza, H. Hoffman, K. Maex, L. Van der Hove, *J. Appl. Phys.* **70**, 693 (1991)

[7.34] La Via, F., V. Privitera, C. Spinella, E. Rimini, *Semicond. Sci. Technol.* **8**, 1196(1993)

[7.35] Sinha, A.K., S.M. Sze, R.S. Wagner *Silicon Devices of Integrated Circuit Processing* in M. Bever ed., *Encyclopedia of Materials Science and Engineering*, Pergamon, Oxford 1985

[7.36] Hu, G.J. *IEEE Trans. Electr. Dev.* **ED-31**, 62(1984)

[7.37] Lewis, A.G., J.Y. Chen, R.A. Martin, T.Y. Huang, *IEEE Trans. Electr. Dev.* **ED-34**, 1337(1987)

[7.38] Seidel, T.E., *Nucl.Instr. Meth.* **B21**, 96(1986)

[7.39] Simonton, R., F. Sinclair in *Handbook of Ion Implantation Technology*, J.F. Ziegler ed., Elsevier Science Publishers, Amsterdam 1992, p. 277

[7.40] Baliga, B.J., *Modern Power Devices* Wiley, New York, 1987 Chap. 4

[7.41] Frank, W., *Def. Diff. Forum* **75**, 121(1991)

[7.42] Coffa, S., N. Tavolo, F. Frisina, G. Ferla, S.U. Campisano, *Nucl. Instr. Meth.* **B/74**, 47(1993)

[7.43] Coffa, S., G. Calleri, L. Calcagno, S.U. Campisano *J. Appl. Phys.* **54**, 291(1988)

[7.44] Catania, M.F., F. Frisina, N. Tavolo, G. Ferla, S. Coffa and S.U. Campisano, *IEEE Trans. Electr. Dev.* **ED-39**, 2745(1992)

[7.45] Coffa, S., L. Calcagno, G. Ferla, S.U. Campisano *J. Appl. Phys.* **68**, 1601(1990)

[7.46] Miller, M.D., *IEEE Trans. Electr. Dev.* **ED-23**, 1279(1976)

[7.47] Borland, J.O., R. Koelsch, *Solid State Technology*, Dec. 1993, p.28

[7.48] McIntyre, E., L. Kaminsky, K. Whaley *Nucl.Instr. Meth.* **B74**, 535(1993)

[7.49] Saxena, A.N., D. Pramanik, *Mat. Sci. Eng.* **B2**, 1(1989)

[7.50] Wong, H., N.W. Cheung, P.K. Chu, E.M. Strathman, M.D. Strathman *Nucl. Instr. Meth.* **B21**, 447(1987)

[7.51] Mutoh, N., K. Orihara, Y. Kawakami, T. Nakano, S. Kawai, I. Murakami, A. Tanabe, S. Suwazono, K. Arai, N. Teranishi, M. Furumiya, M. Murimoto, K. Hatano, K. Minami, Y. Hokari *IEDM Techn. Dig.*, 563(1993)

[7.52] Mc Sean, F.B., and T.R. Oldham, *IEEE Trans. Nucl. Sci.*, **29**, 2018(1982)

[7.53] Cottrel, P., Digest of Technical Papers, IEEE IEDM-88, Dec. 1988, p. 586

[7.54] Kuroi, T., S. Komori, H. Miyatake, K. Tsukamoto, Y. Akasaka *Extended Abstracts, 22nd Conf. on Solid State Devices and Materials*, Sendai (1990) p.441

[7.55] Böhm, H.J., L. Bernewitz, W.R. Böhm, R.Köpe *IEEE Trans. Electron. Dev.* **ED-35**, 1616(1988)

[7.56] Tsukamoto, K., T. Kuroi, S. Komori, Y. Akasaka *Solid State Technology*, June 1992, p.21

[7.57] Ashburn, P., *Design and realization of bipolar transistors*, John Wiley & Sons - Chichester 1988

[7.58] Cuthbertson, A., and P. Ashburn, *IEEE Trans.Electron. Devices* **ED-32**, 2399(1985)

[7.59] Sze, S.M., J.C. Irvin, *Solid State Electronics* **11**, 593(1968)

[7.60] Van Overstraeten, R.J., H.J. De Man, R.P.Mertens, *IEEE Trans. Electron. Devices* **ED-20**, 290(1973)

[7.61] Blakemore, J.S., *Semiconductor Statistics* Pergamon Press, Oxford (1962)

[7.62] Early, J.M. *Proc. IRE* **40**, 1401(1952)

[7.63] Burghartz, J.N., J.D. Cressler, K.A. Jenkins, J.Y.-C. Sun, J.M. Stork, J.H. Comfort, T.A. Brunner, C. L. Stanis *Micr. Eng.* **15**, 11(1991)

[7.64] Priolo, F., F. Benyaich, S.U. Campisano, E. Rimini, C. Spinella, A. Cacciato, P. Ward and G. Fallico, *Nucl. Instr. Meth.* **B85**, 159(1994)

[7.65] Spinella, C., F. Benyaich, A. Cacciato, E. Rimini, G. Fallico and P. Ward, *J. Mater. Res.* **8**, 2608(1993)

[7.66] Patton, G.L., J.C. Brauman and J.D. Plummer, *IEEE Trans. Electron. Devices* **ED-33**, 1754(1986)

[7.67] Benyaich, F., E. Rimini, C. Spinella, A. Cacciato, G. Fallico, G. Ferla, P. Ward, *Appl. Phys. Lett.* **62**, 1895(1993)

[7.68] Avrami, M., *J. Chem. Phys.* **7**, 1103(1939), Johnson, W.A., R.F. Mehl, *Trans. AIME* **135**,416(1939)

[7.69] Wilson, M.C., *Proc. ESSDERC '90*, eds. W.Eccleston and P.Rosser, Adam Hilger, Bristol 1990 p. 349

[7.70] Zimmer, G., B. Hoefflinger, J. Schneider *IEEE Sol. State Circuits*, **SC-14**, 312 (1979)

[7.71] Cheung, N.W., C.L. Liang, B.K. Liew, R.W. Mutikainen, H. Wong *Nucl. Instr. Meth.* **B37/38**, 941(1989)

[7.72] Momose, H., H. Shibata, S. Saitoh, J. Miyamoto, K. Kansaki, and S. Hoyama *IEEE Trans. Electron. Devices* **ED-32**, 217(1985)

CHAPTER 8

[8.1] Sze, S.M. ed. *"High Speed Semiconductor Devices"*, Wiley, New York 1990

[8.2] Blakemore, J.S., ed. *"Gallium Arsenide"* American Institute of Physics - New York 1987

[8.3] Stradling, R.A., and P.C. Klipstein eds. *Growth and Characterization of Semiconductors* Adam Hilger, Bristol 1990

[8.4] Pearton, S.J., *Nucl.Instr. Meth.* **B59/60**, 970(1991)

[8.5] Davis, J.R., K.J. Reeson, P.L.F. Hemment, C.D. Marsh *IEEE Electron. Dev. Letts.* **8**, 291(1987)

[8.6] Schüppen, A., S. Mantl, L. Vescan, S. Woiwod, R. Jebasinski and H. Lüth, *Mater. Sci. Eng.* **B12**, 157(1991)

[8.7] Morgan, D.V., and F.H. Eisen in *"Gallium Arsenide, Materials, Devices and Circuits"*, M.J. Howes and D.V. Morgan editors, Wiley, New York 1985, p.161

[8.8] Christel, L.A., and J.F. Gibbons, *J. Appl. Phys.* **52**, 5050(1981)

[8.9] Pearton, S.J., J.M. Gibson, D.C. Jacobson, J.M. Poate, J.S. Williams, and D.O. Boerma, *Mater. Res. Soc. Symp. Proc.* **51**, 351(1986)

[8.10] Roth, A., *Vacuum Technology* North Holland, New York 1979

[8.11] Pearton, S.J., *Mat. Sci. Rep.* Vol. 4, 313(1990)

[8.12] Pearton, S.J., *Solid State Phenomena* Vol. 1 & 2, 247-280(1988)

[8.13] Cummings, K.D., S.J. Pearton and G.P. Vella - Coleiro, *J. Appl. Phys.* **60**, 163(1986)

[8.14] Blunt, R.T., in *Solid State Devices* 1985, edited by P. Balk, and O.G. Folberth, Elsevier Science, The Netherlands 1986 p. 133

[8.15] Bourgouin, J.C., H.J. von Bardeleben, and D. Stievenard, *J. Appl. Phys.* **64**, R65(1988)

[8.16] Pearton, S.J., K.O. Cummings, and G.P. Vella - Coleiro, *J. Appl. Phys.* **58**, 3252(1985)

[8.17] Pearton, S.J., J.M. Poate, F. Sette, J.M. Gibson, D.C. Jacobson and J.S. Williams, *Nucl. Instr. Meth.* **B19/20**, 369(1987)

[8.18] Kirkpatrick, C.G., R.T. Chen, D.E. Holmes, and K.R. Elliott, in ref. 8.7, chapter 2, p. 39

[8.19] Williams, J.S., and S.J. Pearton, *Mat. Res. Soc. Symp.* **35**, 427(1985)

[8.20] Kruppa, W., J.B. Boos *IEEE Trans. Electron. Dev.* **35**, 2279(1988)

[8.21] Pearton, S.J., and U.K. Chakrabarti, in *Indium Phosphide and related materials: processing, technology and devices*, A.V. Shay Katz ed., Artech Hause, Boston 1992, Chapter 7, *Ion Beam Processing of InP and related Materials* p. 211

[8.22] Bahir, G., J.L. Merz, J.R. Abelson, T.W. Sigmon, *J. Appl. Phys.* **65**, 1009(1989)

[8.23] Mott, N.F., E.A. Davis *Electronic Process in Non - crystalline Materials*, 2nd ed., Clarendon Press, Oxford (1979)

[8.24] Martin, G.M., P. Secordel, and C. Venger, *J. Appl. Phys.* **53**, 8706 (1982)

[8.25] Kukimoto, H., S. Miyazawa, eds. *Semi-insulating III-V Materials* North-Holland Hakona, 1986

[8.26] Capasso, F., S. Sen and F. Beltram *Quantum - Effect Devices* in *High-Speed Semiconductor Devices* edited by S.M.Sze, Wiley, New York 1990, chapter 8, p. 465

[8.27] Harrison, I., *J. Mat. Sci. Electr.* **4**, 1(1993)

[8.28] Eagleshman, D.J., J.M. Poate, D.C. Jacobson, M. Cerullo, L.M. Pfeiffer and K. West, *Appl. Phys. Lett.* **58**, 523(1991)

[8.29] Bode, M., A. Ourmezd, J. Cunningham. and M. Hong, *Phys. Rev. Lett.* **62**, 933(1989)

[8.30] Guerra, M.A., *Mat. Sci. Eng.* **B12**, 145(1992)

[8.31] Izumi, K., M. Doken, and H. Ariyoshi, *Electr. Lett.* **14**, 593(1978)

[8.32] Chater, R.J., J.A. Kilner, E. Scheid, S. Cristoloveneau, P.L.F. Hemment, K.J. Reeson, *Appl. Surf. Sci.* **30**, 390(1987)

[8.33] Van Ommon, A.H., and M.P.A. Vigiers, *Appl.Surf.Sci.* **30**, 383(1987)

[8.34] Celler, G.K., P.L.F. Hemment, K.W. West, and J.M. Gibson, *Appl.Phys. Lett.* **48**, 532(1986), Celler,G.K., *Solid State Technology*, 30(3) 93(1987)

[8.35] Hill, D., P. Fraundorf, and G. Fraundorf, *J. Appl. Phys.* **63**, 4933(1988)

[8.36] Nakashima, S., and K. Izumi, *Nucl.Instr. Meth.* **B55**, 847(1991)

[8.37] Stephens, K.G., K.J. Reeson, B.J. Sealy, R.M. Gwilliam and P.L.F. Hemment, *Nucl.Instr. Meth.* **B50**, 368(1990)

[8.38] Frye, R.C., J.E. Griffith, Y.H. Wong, *J. Electrochem. Soc.* **133**, 1673(1986)

[8.39] Meyyappan, N., J. Blake, F. Sinclair, T. Nakato *Third IUMRS International Conference on Advanced Materials*, Sept. 1993 - Tokyo Japan - Symp. U to be published by Elsevier (Amsterdam) 1994

[8.40] Li, Y., *J. Electrochem. Soc.* **140**, 178(1993)

[8.41] Colinge, J., *Silicon-on-insulator technology: materials to VLSI* Kluwer Academic Publishers, Boston, p.191(1993)

[8.42] Krull, W.A., and J.C. Lee, *Proc. IEEE SOS/SOI - Techn. Workshop* p. 69 (1988)

[8.43] Kasudev, P.K., *Solid State Technology* - Nov. 1990 p. 61

[8.44] Mckitterick, J.B., A.L. Caviglia, *IEEE Trans.Electr.Dev.* **ED 36**, 1133(1989)

[8.45] Colinga, J.P., *IEEE Electron. Dev. Lett.* **EDL-9**, 97(1988)

[8.46] White, A.E., K.T. Short, R.C. Dynes, J.P. Garno, and J.M. Gibson, *Appl. Phys. Lett.* **50**, 95(1987)

[8.47] Mantl, S., R. Jebasinski and D. Hartmann, *Nucl.Instr. Meth.* **B59/60**, 666(1991)

[8.48] Mantl, S., *Mat. Sci. Rep.* Vol. 8, 1(1992)

[8.49] White, A.E., K.T. Short, R.S. Dynes, J.W. Gibson, and R. Hull, *Mat. Res. Soc. Symp. Proc.* **100**, 3(1988)

[8.50] van Ommen, A.H., J.J.M. Ottenheim, C.W.T. Bull - Lienwma and A.M.L. Theunissen, *J. Appl. Phys.* **67**, 1767(1990)

[8.51] Turner,B., in *Gallium Arsenide, Materials, Devices and Circuits* edited by M.J. Howes and D.V. Morgan, Wiley, New York, 1985, chapter 10, p.361

[8.52] Di Lorenzo, J.V., and D.O. Khandelwal eds. *GaAs FET principles and technology*, Dedham, M.A., Artech House (1982)

[8.53] Yamasaki, K., K. Kuramada *IEEE Trans. Electron. Dev.* **ED-29**, 1772(1982)

[8.54] Nelson, D.A. Jr., Y.D. Shen and B.M. Welch, *J. Electrochem. Soc.* **134**, 2549(1987)

[8.55] Solomon, P.M., and H. Morkoc *IEEE Trans. Electron. Dev.* **ED-31**, 1015(1984)

[8.56] Dingle, R., M.D. Feurer and C.W. Tu in *VLSI Electronics Microstructure Science*, Vol. II, eds. N.G. Einspruch and W.R. Wisseman, Academic Press, New York 1985

[8.57] Eden, R.C., A.R. Livingston, and B.M. Welch, *IEEE Spectrum* **20**, 30(1983)

[8.58] Bozler, C.O., G.D. Alley *Proc. IEEE* **70**, 46(1982)

[8.59] Schüppen, A., S. Mantl, L. Vescan, and H. Lüth, *ESSDERC - 90*, Nottingham, September 1990 (Institute of Physics, London 1990) p.45

[8.60] Bozler, C.O., G. D.Alley, R.A. Murphy, D.C. Flanders, and W.T. Lindley *IEEE Techn. Dig. Int .Electron Device Meet.*, 1979 p. 384

[8.61] Oshima, T., N. Nakamura, K. Nakagawa and M. Miyao, *Thin Solid Films* **184**, 275(1990)

INDEX